to Richard
Tasoff

my statistics
student

Arthur L. Comey

# Elementary Statistics
*A Problem Solving Approach*

**The Dorsey Series in Psychology**

Editor   HOWARD F. HUNT   *Columbia University*

# Elementary Statistics

## A Problem Solving Approach

**ANDREW L. COMREY,** Ph.D.

*Professor of Psychology*
*University of California, Los Angeles*

 1975

**The Dorsey Press** *Homewood, Illinois 60430*
*Irwin-Dorsey International London, England WC2H 9NJ*
*Irwin-Dorsey Limited Georgetown, Ontario L7G 4B3*

*First Printing, March 1975*

ISBN 0-256-01675-5
Library of Congress Catalog Card No. 74–25809
*Printed in the United States of America*

*To Barbara, Corinne, and Cynthia*

# Preface

This book is based on the author's class notes for teaching elementary statistics. These notes have evolved over many years of teaching this subject to behavioral science students. This teaching assignment is particularly challenging because the students are heterogeneous in aptitude, preparation, and motivation. At one extreme there are students with little or no background in mathematics who are taking statistics only because it is a required course and who have no postgraduate education plans. At the other extreme there are talented students who are well prepared in mathematics, who recognize the need for statistics, and who plan to enter scientific careers following graduate work. Providing a meaningful learning experience in the same classroom for such a diverse collection of students requires careful planning. This textbook was designed to fit the type of course which the author has found to be most effective in dealing with this problem.

A unique feature of this book is that the subject matter has been organized into 42 "Problems," each representing a major concept or commonly used statistical procedure. These Problems are carefully organized to make it easy for students with little mathematical background to follow them step-by-step. This approach simplifies the task of mastering the essential statistical ideas and techniques. Students can concentrate on learning how to do one problem at a time, achieving success through the attainment of intermediate goals instead of having to confront a monolithic and seemingly insuperable obstacle. For those students with greater motivation and superior preparation, supplementary Problems and mathematical proofs are provided to enrich their learning experiences and to present them with challenges commensurate with their capacities.

The author would like to express his appreciation to the many students, teachers, and colleagues who have helped shape his thinking and the character of this volume. The author's first statistics teacher, J. P. Guilford, especially should be singled out for mention. Wilfrid Dixon, Milton R. Blood, Douglas C. Chatfield, John R. Knapp, and Roy S. Lilly offered ideas, suggestions, and criticisms that had a rather direct effect upon the manuscript. Debbi Kanoff typed most of the manuscript. Thanks are due to the Literary Executor of the late Sir Ronald A. Fisher, to Dr. Frank Yates, and to Longman Group Limited, Essex, for permission to reprint Tables III and IV from their book *Statistical Tables for Biological, Agricultural and Medical Research,* and to George W. Snedecor and the Iowa State University Press for permission to reproduce Table E from the fifth edition of *Statistical Methods.*

*February 1975*                                    ANDREW L. COMREY

# Contents

# chapter 1

# Introduction

Statistics seems to hold a special fear for many behavioral science students that borders on the pathological. This fear might be described as "math shock." Although it could be argued that the subject of elementary statistics certainly is not as difficult as these students think it is, such arguments do not impress students with math shock. To them, these fears are very real, seem based on reality considerations, and they do operate to interfere with effective performance. Such students are convinced that they will have difficulty and their certainty of this often induces behavior on their part that makes the situation worse than it need be.

Math shock is a product of the typical methods used to teach mathematics in the primary and secondary schools. Since it is a subject that is organized vertically rather than horizontally, understanding of mathematics depends upon building up the structure block by block with each block firmly in place before the next row of blocks is laid on. If very many blocks in the substructure are missing, the whole edifice is apt to come tumbling down. A gap in the student's knowledge in verbal fields, such as history, English, languages, and so on, is less serious since the student can perform adequately on most tasks, albeit at less than 100 percent efficiency overall. Gaps in mathematics, however, quickly accumulate to put the student in a position where he cannot understand the next thing to be added to his store of knowledge because his background is deficient. When this happens, the student typically does not attribute his failure to lack of background, however. Instead, he is apt to think he is dumb and tends to develop a deeply entrenched feeling of inferiority where mathematics is concerned.

What the student in this predicament often fails to realize is that

1

if he had studied mathematics in an appropriate way, instead of the way it is usually taught, he could absorb a great deal more mathematics than he thinks he can. There is no doubt that some students have more aptitude for mathematics than others, just as some students have more verbal capacity than others. Even students with mediocre verbal talents, however, learn to speak, read, and write adequately. By the same token, students with only a mediocre talent for mathematics can learn to function at an adequate level in that field given the right set of learning conditions. To make this possible, however, the less talented math student must master each step as he goes and not progress to the next course of stones in constructing his mathematical edifice until the foundation is solidly established at each level. He probably will take longer to do this than most students who become college mathematics majors, but the point is that he *can* do it. Any student with enough general intelligence to do well in social science courses has the mental capacity to absorb a great deal of mathematics, given the right attitude, the correct approach, and sufficient time to master each step in the sequence before going on to the next one.

The student who brings math shock with him to the elementary statistics course must first of all try to eliminate from his own thinking the idea that he is incapable of absorbing mathematics because of his unfortunate experiences with this subject in the past. He should try to understand that his difficulty in absorbing mathematics in the past was due to many factors, including improper pacing and inadequate foundation at the time new material was presented. The teaching procedures presently used in the primary and secondary schools are designed to accommodate only the top five or ten percent of the students in mathematical aptitude. These are the only students who learn the material in course number one solidly enough to be ready to progress to course number two. A proper management of mathematical instruction would require each student to master course number one at the A level before he progresses to course number two. Course number two, likewise, would have to be mastered to the A level of proficiency before going on to course number three, and so on. As it is now, however, B, C, and even D, students are passed right on to the next course along with the A students. At the second course level, the less talented students in math are competing not only with less talent, but with an improper foundation as well. Thus, with each succeeding course in mathematics, the gap between the rich and the poor steadily widens until the less talented math students are hopelessly lost and forever convinced of their own inability to absorb mathematics. Even the sight of a symbol is often quite sufficient to induce an anxiety response in such students, a reminder of past failures. These past failures, of course, are not absolute failures, necessarily, but often merely failures to achieve up to their own and/or their parents' expectations or to the level of their achievements in other kinds of courses.

If the student who fears statistics can understand the dynamics of his own fear and can accept the foregoing explanation for it, he may be helped to believe that his is capable of absorbing a good bit of mathematics if it is properly presented and if he goes about it in the right way. After 25 years of teaching statistics to social science students, the author has become convinced that the vast majority of college students are capable of learning the material in elementary statistics if they do not permit fear of failure to interfere with the learning process. The student who does have this fear is urged to try to accept intellectually the idea that he has enough ability to learn this kind of material—it will just take more time. By repeatedly telling himself that this is so, the student can come gradually to accept it more and more on an emotional level as well as on an intellectual level, thereby decreasing the magnitude of his emotional block to learning. This process can be accelerated by achieving some real success in mastering the material itself.

This book was designed, quite frankly, with the math shock student in mind. It would be fatuous to claim that this book will enable the student with math shock to root out in a short time a conditioned fear response that has been developed over many years. This book, however, together with a knowledge of the dynamics of his fear and an intellectual acceptance of the idea that it is possible for him to master the material, can go a long way toward helping the student to cope with his anxiety and to work up to his real capacity. More often than not, this will be at a higher level that the student believed possible.

At times, the material in this book may appear to be too easy for students with better mathematical backgrounds and no emotional impediment to mathematical learning. On the other hand, this is not a fatal flaw, since these better equipped students will be able to skip quickly over some things that they find to be too elementary but which are very necessary for their more poorly prepared colleagues. Both types of students should arrive at pretty much the same end point as far as knowing the statistical material is concerned except that the better prepared students can do it in less time. The math shock student must remind himself that he can master the material if he will take the greater amount of time needed. Each step in building the structure of statistical knowledge is comparatively easy when taken by itself, *if* the previous steps have been mastered thoroughly. The material selected for this book and the manner of presenting it have been developed with the idea of making this process as easy as possible for every student, no matter how much difficulty he has had with mathematical subjects in the past.

## Major Features of This Book

No new book in statistics will be completely different from other published texts, but this book has several features which when taken together make it unique among presently available elementary statistics books for students in the behavioral sciences.

1.   Although it is expected that the student will know some arithmetic and a little elementary algebra, every attempt has been made in this book to minimize the presumption of previous mathematical knowledge on the part of the student. Furthermore, every attempt has been made to explain processes in English for the verbally oriented student so that he does not have to rely exclusively on facility with symbolic nonverbal manipulations. At the same time, the author has not shirked the responsibility for teaching the student how to use symbolic processes of a mathematical nature where he previously lacked these skills.

2.   An attempt has been made to give the student an understanding of the basic logic underlying statistics rather than merely a set of cookbook rules to be applied by rote without understanding. Since statistics is a rather deep subject requiring an extensive knowledge of advanced mathematics for a complete understanding, some compromises were of necessity required. First of all, mathematical proofs have been used only where the mathematics involved are at a rather elementary level. Secondly, considerable dependence on intuitive understanding of concepts has been substituted for actual rigorous mathematical proof. Language used in demonstrations and proofs has been chosen for the purpose of conveying an intuitive grasp of the underlying logic rather than to satisfy the demands of elegant mathematical exposition.

3.   In deference to students with math shock, many of the proofs and demonstrations have been separated out from the main text and placed in Mathematical Note sections. Students may omit these sections entirely if they want to learn how to use statistical methods without for the moment being concerned about where they come from. It is often helpful for some students to leave theoretical material of this sort until a later stage in their learning before tackling it. The proofs and demonstrations left in the main body of the text are regarded by the author as immediately essential for all students as they progress through the material. The mathematical notes sections also serve as extra material which can be used to enrich the beginning course for students with greater mathematical facility and background.

4.   Material has been chosen for this book which deals with the most commonly encountered and needed statistical ideas and techniques. The variety and complexity of statistical methods utilized in published research articles in social science scientific journals today is bewildering indeed. A mastery of all these methods goes well beyond the scope of this book. Fortunately, however, a substantial proportion of the commonly used methods can be accommodated in an elementary textbook. A competent research person could conceivably work for a life time in research getting by with the statistical methods presented in this intro-

ductory text. Most of the more complicated types of statistical treatment encountered in the literature are used only occasionally. In many cases, where these more complicated data treatment methods were used, simpler experiments could have been designed to accomplish the same end result. Thus, although statistics is a very complicated and extensive subject, a surprising amount of what is really essential to the social science research worker has been included here. By confining the material included in this book to the essentials, the math shock student is spared the demoralizing effect of being confronted with what appears to be an overwhelming mountain of figures, symbols, formulas, and special cases.

5. Perhaps the most unique feature of this book, however, is its manner of presenting the material. Forty two "Problems" have been identified as commonly used techniques or processes in statistical description or inference. The student is given these 42 statistical procedures to master. Or, putting it another way, elementary statistics has been broken down into 42 distinct, graded units of study which can be checked off by the student, one by one, as he masters them. The student gains a discrete increment in his power to understand and apply statistical methods to research data as he masters each successive problem. Instead of presenting a great deal of abstract theory all at once to the student, leaving him to struggle with how to apply it, this approach takes up actual practical applications of statistical methods one at a time. Theory is introduced in connection with a specific problem where it is needed to understand the problem. The connection between theory and practice is kept much more immediate in this manner, although perhaps at some cost as far as mathematical elegance is concerned.

This method of presenting the material has been evolved by the author gradually over the years as he has searched for ways of helping less mathematically inclined students to cope with statistics. Breaking statistics down into discrete types of problems has proved to be the most effective of the learning methods tried by the author. Some of the reasons for this effectiveness of the problem approach are as follows: (a) Instead of being faced with a giant monolithic obstacle to overcome, the student has only a relatively small number of well defined and circumscribed tasks to master. The total job is thus broken down into a fairly small number of manageable units. (b) Many of these units, particularly at the beginning, are very easy, so the student gains success, a definite feeling of progress, and picks up confidence that the material can be mastered after all. This success helps to extinguish to some extent the conditioned fear response established by previous failures. (c) With the more positive climate established by these conditions, the math shock student is less likely to engage in selfdefeating behavior, e.g., avoiding studying, missing classes, postponing assignments and examinations, and so on. Examination anxiety also tends to be reduced, since the student feels more confident, permitting a performance that is closer to real capacity.

## Organization of the Book into Problems

It was pointed out in the section above that the material in this book has been organized around 42 specific problems in statistics, statistical procedures that are commonly encountered in practice. These 42 problems have been broken down into 36 "number series" problems and six "letter series" problems. The 36 number series problems are to be regarded as the primary, most essential problems to be covered. The six letter-series problems are supplementary problems that are less essential either because they are somewhat less commonly used or because they represent *merely* labor saving alternatives to other procedures. Several of the letter series problems, for example, deal with "coded value" methods of calculating certain commonly used statistics. These methods are labor saving devices which were formerly widely used but which have now become less important with the ready availability of electronic computing aids. Letter series problems can be omitted entirely at the option of the instructor, assigned as special problems for extra credit, or, of course, can be required of all students. Letter series problems can be omitted entirely without disturbing continuity of study with the number series problems. The letter series problems are sequential, however, so that one of the later letter series problems cannot be assigned without assigning the earlier ones. The letter series problems past any point, however, can be dropped even if earlier ones have been assigned. The 42 problems are listed below in order of their appearance in the text.

1. Prepare a Frequency Distribution. (Ch. 2)
2. Prepare a Frequency Polygon. (Ch. 2)
A. Prepare a Histogram. (Ch. 2)
3. Determine the Mode of a Distribution. (Ch. 3)
4. Determine the Median from Ungrouped Data. (Ch. 3)
5. Determine the Median from Grouped Data. (Ch. 3)
6. Compute a Mean from Ungrouped Data. (Ch. 3)
B. Compute a Mean from Grouped Data by the Code Method. (Ch. 3)
7. Determine the Semi-Interquartile Range $(Q)$ from Ungrouped Data. (Ch. 4)
8. Determine the Semi-Interquartile Range $(Q)$ from Grouped Data. (Ch. 4)
9. Compute the Standard Deviation (S.D.) by the Deviation Score Method. (Ch. 4)
10. Compute the Standard Deviation (S.D.) by the Raw Score Method. (Ch. 4)

C. Compute the Standard Deviation (S.D.) by the Code Method. (Ch. 4)

11. Given a Raw Score, Find Its Centile Rank. (Ch. 5)

12. Given a Centile Rank, Find the Corresponding Raw Score. (Ch. 5)

13. Convert Raw Scores to Standard Scores Using a Linear Transformation. (Ch. 5)

14. Convert Standard Scores to Scaled Scores with a Different $M$ and S.D. (Ch. 5)

15. Use the Normal Curve Table to Find Numbers of Cases Expected in Different Parts of the Distribution. (Ch. 5)

D. Convert Raw Scores to Normalized Standard Scores. (Ch. 5)

16. Compute the Correlation Coefficient from Raw Scores Using the Raw Score Formula. (Ch. 6)

17. Develop a Deviation Score Regression Equation. (Ch. 6)

18. Develop a Standard Score Regression Equation and a Raw Score Regression Equation. (Ch. 6)

E. Compute the Correlation Coefficient from a Scatter Diagram Using the Code Method. (Ch. 6)

19. Given a Situation with Two Possible Outcomes, Determine the Probability of a Particular Combination of Outcomes in $n$ Trials. (Ch. 7)

20. Using a Binomial Distribution, Make a Statistical Test of an Hypothesis, Selecting the Appropriate One– or Two–Tailed Test. (Ch. 7)

21. Using the Normal Curve Approximation to the Binomial Distribution, Make a Statistical Test of an Hypothesis. (Ch. 7)

22. Using the Binomial Distribution, Test the Significance of the Difference Between Correlated Frequencies. (Ch. 7)

23. Test the Hypothesis that the Mean of the Population is a Certain Value. (Ch. 8)

24. Establish a Confidence Interval for the True Mean. (Ch. 8)

25. Test the Significance of a Correlation Coefficient. (Ch. 8)

26. Test the Significance of the Difference Between Two Variance Estimates with the $F$ test. (Ch. 8)

27. Test the Significance of the Difference Between Two Independent Random Sample Means Using the $t$ Test. (Ch. 8)

28. Test the Difference Between Two Correlated Means Using Difference Scores. (Ch. 8)

29. Test the Significance of the Difference in Mean Change for Two Independent Random Samples. (Ch. 8)

30. Test the Significance of the Difference in Mean Change for Two Matched Samples. (Ch. 8)

31. Test the Difference Between Means of Several Samples of Equal Size Using One-Way Anova. (Ch. 9)

32.  Test the Difference Between Means of Several Samples of Unequal Size Using One-Way Anova. (Ch. 9)
33.  Test the Effects of Two Factors Simultaneously Using the Two-Way Anova. (Ch. 9)
34.  Test the Difference Between Obtained and Theoretical Frequency Distribution Values. (Ch. 10)
35.  Test the Association Between Two Variables Using Frequencies in a Four-Fold Table. (Ch. 10)
36.  Test the Association Between Two Variables Using Frequencies in Larger Contingency Tables. (Ch. 10)
 F.  Test the Difference Between Obtained and Normal Curve Frequency Distribution Values. (Ch. 10)

Some of the problems are included more for their conceptual importance than for their wide use as applied methods. Problems in Chapters 1 through 6 are concerned with commonly used "descriptive statistics," that is, statistical procedures that describe what a body of data is like. The remaining problems are concerned with "inferential statistics," procedures which permit the investigator to draw certain conclusions based on the data analyzed. There has been an attempt to deemphasize seldom used procedures and obscure theoretical issues to permit greater elaboration with respect to essential concepts and the methods most often used in practice.

It has already been suggested that some or all of the letter series problems, A, B, C, D, E, and F, may be omitted from a given course of instruction in elementary statistics. Some instructors may also wish to leave out the problems in Chapter nine on analysis of variance, Chapter 10 on chi square, or both. The problems selected by a given instructor will depend on the length of the course, the number of credit hours, the pace he wishes to maintain, and his own perception of the relative importance of the methods. Instructors with a heavy experimental orientation will probably wish to include some or all of the problems in Chapter nine on analysis of variance. Instructors interested in sociological research may consider it imperative to include some or all of the problems from Chapter 10 on chi square. A fairly typical one quarter or one semester course might include just the number series problems in chapters one through eight, omitting the letter series problems as well as all of Chapters nine and 10. A two quarter or two semester course would probably include all of the problems.

Exercises are provided at the end of each chapter to give students experience in working examples of each problem. The answers to the odd numbered exercises are provided in Appendix B so the student may check to see that he has worked the problems correctly. The method of working the problem is shown rather fully in many cases. Answers to the even-numbered exercises are not given in case the instructor wishes to assign problems for which the students do not have the answers.

## Supplementary Reading

Students in the behavioral sciences accustomed to reading primarily verbal material tend to expect that they should grasp the meaning of a passage in one reading. If not, they sometimes become dismayed and assume that the material is incomprehensible in general, or at least to them. Most social science students, however, must read mathematical material more than once, often many times, before it really sinks in. The student is urged, therefore, not to be discouraged if the first reading of some passage in this book fails to make everything crystal clear. He should expect, as a matter of course, to read it several times before the pieces all fit neatly together. If it happens the first time, just treat it as an unexpected bonus.

If after several readings the material is still not clear, it is often useful to try a different exposition of the same subject by another author. Where one author's way of explaining a particular point draws a blank, a second author's may hit the mark, particularly when the student already has tried to understand another exposition. Your library will probably have several books on elementary statistics which you can consult. A few titles are listed below. Since there are perhaps 100 published texts on elementary statistics that could be of interest to social science students, the list below obviously will exclude many worthwhile books:

H. L. Alder and E. B. Roessler, *Introduction to Probability and Statistics,* 5th ed. (San Francisco: W. H. Freeman & Co., 1972).

C. I. Chase, *Elementary Statistical Procedures.* (New York: McGraw-Hill Book Co., 1967).

W. J. Dixon and F. J. Massey, *Introduction to Statistical Analysis,* 3d ed. (New York: McGraw-Hill Book Co., 1969).

P. H. DuBois, *An Introduction to Psychological Statistics.* (New York: Harper & Row, Publishers, 1965).

A. L. Edward, *Probability and Statistics.* (New York: Holt, Rinehart, & Winston, Inc., 1971).

A. L. Edwards, *Statistical Methods,* 3d ed. (New York: Holt, Rinehart, & Winston, Inc., 1973).

J. E. Freund, *Modern Elementary Statistics,* 3d ed. (Englewood Cliffs, N.J.: Prentice-Hall, Inc., 1967).

J. P. Guilford and B. Fruchter, *Fundamental Statistics in Psychology and Education,* 5th ed. (New York: McGraw-Hill Book Co., 1973).

P. G. Hoel, *Elementary Statistics.* 3d ed. (New York: John Wiley & Sons, Inc., 1971).

R. H. Kolstoe, *Introduction to Statistics for the Behavioral Sciences,* Rev. ed. (Homewood, Ill.: Dorsey Press, 1973).

W. Mendenhall and M. Ramey, *Statistics for Psychology.* (North Scituate, Mass.: Duxbury Press, 1973).

Q. McNemar, *Psychological Statistics,* 4th ed. (New York: John Wiley & Sons, Inc., 1969).

E. W. Minium, *Statistical Reasoning in Psychology and Education.* (New York: John Wiley & Sons, Inc., 1970).

H. M. Walker and J. Lev, *Elementary Statistical Methods,* 3d ed. (New York: Holt, Rinehart, & Winston, Inc., 1969).

G. H. Weinberg and J. A. Schumaker, *Statistics an Intuitive Approach.* (Belmont, Calif.: Wadsworth Publishing Co., 1962).

The student may also have occasion to consult more advanced books on topics not covered in this or other elementary statistics books. This could occur as a result of reference to some technique in his reading, search for a method of treating available data, or just a desire to increase his knowledge of the subject. A very abbreviated list of references to more advanced topics in statistical analysis is given below. Each of these books in turn will contain many more references. Your librarian should be able to help you locate these or other more advanced books on statistical methods if you should need them. The list of advanced books is as follows:

A. L. Comrey, *A First Course in Factor Analysis.* (New York: Academic Press, Inc., 1973).

A. L. Edwards, *Experimental Design in Psychological Research,* 4th ed. (Holt, Rinehart, & Winston, Inc., 1972).

W. L. Hays, *Statistics for the Social Sciences,* 2d ed. (New York: Holt, Rinehart, & Winston, Inc., 1973).

P. G. Hoel, *Introduction to Mathematical Statistics,* 4th ed. (New York: John Wiley & Sons, Inc., 1971).

R. E. Kirk, *Experimental Design: Procedures for the Behavioral Sciences.* (Belmont, Calif.: Brooks/Cole Publishing Co., 1968).

D. Lewis, *Quantitative Methods in Psychology.* (New York: McGraw-Hill Book Co., 1960).

F. M. Lord and M. R. Novick, *Statistical Theories of Mental Test Scores.* (Reading, Mass.: Addison-Wesley Publishing Co., 1968).

L. A. Marascuilo, *Statistical Methods for Behavioral Science Research.* (New York: McGraw-Hill Book Co., 1971).

J. L. Myers, *Fundamentals of Experimental Design,* 2d ed. (Boston: Allyn and Bacon, Inc., 1972).

C. C. Peters and W. R. Van Voorhis, *Statistical Procedures and Their Mathematical Bases.* (New York: McGraw-Hill Book Co., 1940).

S. Siegel, *Nonparametric Statistics for the Behavioral Sciences.* (New York: McGraw-Hill Book Co., 1956).

M. M. Tatsuoka, *Multivariate Analysis: Techniques for Educational and Psychological Research*. (New York: John Wiley & Sons, Inc., 1971).

B. J. Winer, *Statistical Principles in Experimental Design*, 2d ed. (New York: McGraw-Hill Book Co., 1971).

### Importance of Learning Statistics

Many social science students are enrolled in statistics for one and only one reason—it is a required course! They would prefer to avoid it but must take statistics to get their degree. This is an understandable attitude on the part of students with math shock since the prospect of being exposed to statistics arouses anxiety and most of us try to behave in such a fashion as to reduce the amount of anxiety to which we will be subjected.

This book has been written with sympathy for the plight of these students and with the hope and intention of minimizing the trauma to which they will be subjected while fulfilling their requirements by taking a course in statistics. The student can perhaps help himself to reduce this pain still further, or at least make it more tolerable, by recognizing the value which a knowledge of statistics can have for him. Statistical thinking has become so pervasive in modern civilization that a person cannot really consider himself to be educated unless he comprehends at least the elementary basic statistical ideas and techniques. On almost any job that the student eventually ends up with, circumstances are apt to occur in which he will be able to perform more effectively if he has a knowledge of statistics. It goes without saying, of course, that it is impossible to read the research literature in the social sciences without a knowledge of statistics. It is a truly embarrassing predicament to be a college graduate in a social science field without being able to read even easy articles in that field. The student who does not learn some statistics places himself in that position.

Some students in social science fields try to delude themselves by saying that they are interested in people, not science, and hence do not need to know anything about statistics. To read about what is being done in fields devoted to helping people and especially to participate professionally in those efforts, however, *does* require a knowledge of statistics if the person is to function competently. To try to function in these areas without that knowledge is like trying to practice medicine without having had a course in anatomy. There will always be the need to cover up for ignorance. If the student harbors resentment about having to acquire a knowledge of statistics, therefore, it may help him to be aware that it is an extremely useful skill to have in a wide variety of situations and for many professions, it is an absolute necessity. The author hopes that this text will facilitate appreciably the student's efforts to acquire that skill.

# chapter 2

# Data Representation

The research worker in the social sciences is commonly confronted with a mass of data which he must condense in some fashion if he is to make any sense out of it. A psychologist, for example, may have a dozen or more scores on psychological tests for each of many hundreds of subjects. Poring over page after page of numbers is not a very effective way to obtain a clear picture of what the data are like. One of the most important functions of statistical procedures is to provide effective methods for describing masses of data. In this chapter, three elementary statistical procedures of this kind will be explained and illustrated. The first of these, the frequency distribution, presents data in tabular form. The other two procedures, the frequency polygon and the histogram, present data in pictorial form. The frequency distribution and the frequency polygon are treated in Problems 1 and 2 of the number series of problems. These are the more commonly used procedures. The histogram is less commonly used so it is treated in Problem A of the letter series of problems. Letter series problems may be omitted without disturbing the study of the number series problems.

## Problem 1
### PREPARE A FREQUENCY DISTRIBUTION

This procedure is applied in those situations where there are so many measurements or scores available in the data that some way is needed to summarize the information in a more comprehensible form. A frequency distribution is also prepared in many instances as a preliminary step to the computation of other statistics to be described later, e.g., the mean, median, standard deviation, and the semi-interquartile range.

**Preparing a Frequency Distribution**

To illustrate the preparation of a frequency distribution, consider Data A in Table 2–1. A count shows that there are 50 scores in Data A,

TABLE 2–1
Data A

| 90 | 82 | 76 | 135 | 82 | 68 | 77 | 75 | 124 | 89 |
|----|-----|-----|-----|----|----|-----|-----|-----|----|
| 63 | 38 | 65 | 76 | 98 | 68 | 72 | 50 | 80 | 85 |
| 58 | 111 | 94 | 58 | 63 | 91 | 113 | 12 | 54 | 74 |
| 87 | 70 | 113 | 75 | 46 | 79 | 60 | 117 | 42 | 90 |
| 84 | 84 | 48 | 74 | 96 | 59 | 130 | 33 | 69 | 78 |

10 columns of 5 scores each. Although 50 is not a large number of scores, it is still difficult to assimilate the meaning of Data A in the form shown in Table 2–1. Preparation of a frequency distribution from Data A will be helpful in providing a better impression of what the data are like.

*Finding the Range.* The first step in preparing a frequency distribution from Data A is to locate the highest and lowest scores and compute the range using these two scores. The highest score, or measurement, in Data A is 135 and the lowest score is 12. The range is the difference between these two scores plus 1 point, i.e., $135 - 12 + 1 = 124$.

Why is one point added to the difference between the highest and lowest score to find the range? The reason is that each score covers a full one point range, from one half a point below the score to one half a point above the score. The highest score in Data A, therefore, goes from 134.5 to 135.5 and the lowest score goes from 11.5 to 12.5. The range, then, is really from 11.5 to 135.5, rather than just from 12.0 to 135.0. Adding the extra one point to the difference between the highest and lowest scores takes care of the two half points on either end of the scale needed to reach the exact limits of the highest and lowest scores. Thus, 134.5 and 135.5 are the exact limits of the score 135. A few authors use the convention of treating a score of 135 as going from 135.00 to 135.99. This is not incorrect but the more common procedure, and the one to be followed throughout this book, is to treat the score as extending from one half a point below the score integer (a whole number) to one half a point above the score integer.

*Finding the Class Interval, i.* The next step after computing the range in preparing a frequency distribution is to determine the size of the class interval, $i$. The total range will be broken up into a number of equal steps, called class intervals. Each of these class intervals is of width $i$, where $i$ will usually be a number selected from the following possibilities: 1, 2, 3, 5, 10, or a multiple of 10. Only in case the scores are decimal fractions will class intervals of other sizes be employed. Using one of these preferred class intervals is only a matter of convention or standard practice. Theoretically, there is no reason why the class interval, $i$, could not be 4, 6, 7, 8, 9, 11, or some other number not in the preferred list, but common usage does not include these choices. The choices of 5 and 10 are preferred ones because of their convenience in the base 10 number system used in our culture. The values 1, 2, and 3 are also admitted because $i = 5$ may be too large; $i = 4$ can always be avoided.

Which of the preferred values of $i$ should be selected for use with Data A? This depends upon the range. The commonly accepted rule is to select a value of $i$ that will produce between 10 and 20 class intervals over the entire range of scores. Thus, each potentially acceptable value of $i$ can be divided into the range to see if it gives a result between 10 and 20. If so, it is an acceptable class interval. If the value of $i$ goes into the range more than 20 times or less than 10 times, it is not an acceptable class interval for use with the problem at hand.

With Data A, for example, dividing 124, the range, by the various recommended values of $i$, e.g., 1, 2, 3, 5, 10, and so on, it can be seen that only $i = 10$ is acceptable. With $i = 10$, there are $124/10$ or 12 and a fraction class intervals. The fraction, no matter how small, adds another full interval, so there will be 13 class intervals for Data A with $i = 10$. The next smaller class interval would give $124/5$ or 24 and a fraction, i.e., 25, class intervals. This is more than 20, so $i = 5$ is not acceptable. The next larger possible class interval would be $i = 20$ which would give $124/20$, or 6 and a fraction, i.e., 7 class intervals. This is less than 10 class intervals, hence $i = 20$ is not acceptable in this example. Often this process will yield two values of $i$ that are acceptable. In the event that there are two values of $i$ that are acceptable, either could be used. Typically, the smaller one will be chosen if there are many scores, e.g., hundreds, and the larger value of $i$ will be chosen if there are few scores, e.g., dozens.

Why should the number of class intervals be between 10 and 20? This rule is somewhat arbitrary to be sure but it is based on sound reasoning. When a score of 17 is tallied in the class interval of 10–19, for example, some information is lost because the score is then treated as though it were at the midpoint of the interval, i.e., at 14.5, not 17.0. This represents an error of 2.5 points. Other scores tallied in this interval also have their associated errors resulting from being treated as though they fell at the midpoint of the interval. These errors average out to some extent since some of the errors are positive and some are negative, but they usually do not average exactly to zero. The larger the size of the class interval, the greater this residual error is apt to be. When there are fewer than 10 class intervals, this residual error may be too large to be tolerated since with only a few class intervals $i$ must be large if the entire range is to be covered. On the other hand, if more than

20 class intervals are used, this residual grouping error is reduced to negligible proportions because the intervals are small. The economy and convenience derived from the grouping process, however, is diminished when a large number of class intervals is used. Keeping the number of class intervals between 10 and 20 gives an important dividend in economy without giving up too much in the way of accuracy.

*Locating the Bottom Class Interval.* After determining the size of the class interval, $i$, the next step is to decide where to start the bottom class interval. The rule that will be followed in this book is to start the bottom class interval on the even multiple of $i$ next below the lowest score in the data. In Data A, $i = 10$ and the lowest score is 12. The bottom class interval will not begin on 12, the lowest score, but will begin on 10 which is the first even multiple of 10 below 12. If the lowest score in the data had been 8, with $i = 10$ the lowest class interval would start at 0. If the lowest score had been 27, the lowest class interval would start at 20. When the value of $i$ is 3, the lowest class interval must start on a value divisible by 3. If $i = 5$, the lowest class interval must start on a value divisible by 5, and so on. The reason for this rule is that frequency distributions look better in the decimal system when class intervals begin with multiples of 10. There is less reason for the rule with values of $i = 2$ or $i = 3$, but to maintain consistency, the same rule is followed in these cases, too.

It should be noted that starting the lowest class interval on a multiple of $i$ effectively increases the range to be covered if the lowest score is not divisible exactly by $i$. With Data A, the range to be covered was increased from 124 to 126 because the lowest class interval is started at 10 instead of at 12, the lowest score. The range of 126 divided by 10, the value of $i$ previously selected, still gives the same number of class intervals as 124 divided by 10. The number of class intervals needed was not increased in this instance, therefore, by starting the lowest class interval at a point below the lowest score. In other cases, however, this increment to the effective range may increase the number of class intervals needed by one, thereby possibly affecting the decision about which value of $i$ to choose for the frequency distribution. It might permit the use of a value of $i$ that would only give nine class intervals if the lowest class interval were started at the lowest score. Starting the lowest class interval on a lower value that is a multiple of $i$ could increase the number of class intervals to 10, making the value of $i$ acceptable. By the same token, increasing the number of class intervals might push the number from 20 to 21 which would render a previously acceptable class interval unacceptable when the starting point is lowered. In determining the number of class intervals and the value of $i$ to be selected, it is always necessary to take into account any increment to the range produced by lowering the starting point for the lowest class interval to a point below the lowest score.

*Tallying the Scores.* Once the class interval size, $i$, and the starting point for the bottom class interval have been determined, all the class intervals are listed as in column $X$, Table 2–2. In this example, the lowest class interval runs from 10 to 19, the next one runs from 20 to 29, and so on, up to 130–139 for the highest class interval. After listing the class intervals in a column headed by $X$, the scores are tallied in the Tally column as shown in Table 2–2. Each score in Data A from Table 2–1 results in a tally mark in the Tally column of Table 2–2.

TABLE 2–2
Frequency Distribution

| $X$ | Tally | $f$ |
|---|---|---|
| 130–139 | // | 2 |
| 120–129 | / | 1 |
| 110–119 | //// | 4 |
| 100–109 |  | 0 |
| 90– 99 | ////// | 6 |
| 80– 89 | //////// | 8 |
| 70– 79 | /////////// | 11 |
| 60– 69 | /////// | 7 |
| 50– 59 | ///// | 5 |
| 40– 49 | /// | 3 |
| 30– 39 | // | 2 |
| 20– 29 |  | 0 |
| 10– 19 | / | 1 |
| $N$ |  | 50 |

When the tallying has been completed, the number of tallies in each row is entered in the column headed by $f$, which stands for score frequency. If there are no scores in a class interval, and hence no tallies for that row, a frequency of zero is entered on that row in the frequency column, e.g., for class intervals 20–29 and 100–109.

The sum of the frequencies is entered at the bottom of column $f$ in Table 2–2. This is the total number of scores, $N$, in the distribution. The letter $N$ is shown on the same row as the numerical total of the frequencies. In Table 2–2, the letter $N$ appears at the bottom of the $X$ column, opposite 50, the number of scores in the distribution. The sum of the frequencies, $N$, should be checked by counting the number of scores in the original data. To be sure there is no error, however, it would be necessary to do the entire frequency distribution over again, checking it line by line with the original.

Note that the interval widths appear to be only 9 points, i.e., 10–19, 130–139, and so on. The exact limits of these intervals, however, are not as shown in Table 2–2. The bottom class interval has exact limits that run from 9.5 to 19.5, an interval of 10 points as called for by the choice of $i = 10$ for the class interval size. The extra point comes from the fact that the score 10 goes from 9.5 to 10.5 and the score 19 goes from 18.5 to 19.5, giving a half point below 10.0 and a half point above 19.0. The written class interval limits in the frequency distribution are whole numbers rather than the exact interval limits. This not only gives a neater frequency distribution as far as appearances are concerned but also helps to avoid errors in tallying.

If the bottom two intervals were written as 9.5–19.5 and 19.5–29.5, respectively, in tallying a score of 19 it would be easy to mistakenly tally a score of 19 in the interval 19.5–29.5 instead of in the interval 9.5–19.5 where it belongs. The appearance of 19 as part of the exact lower limit of the interval 19.5–29.5 could trigger this type of error. The only problem with using integers or whole numbers as the written class interval limits in the frequency distribution is that the reader must remember to add one point to the difference between the upper and lower class interval written limits to obtain the proper class interval size, $i$.

### Summary of Steps for Preparing a Frequency Distribution

The procedures described above for preparing a frequency distribution from a set of raw scores or measurements may be summarized briefly as follows:

1. Find the range of scores by adding one point to the difference between the highest and the lowest score in the body of data.
2. Select a class interval, $i$, from the recommended values of $i$ (e.g., 1, 2, 3, 5, 10, or a multiple of 10) such that the range divided by $i$ will yield at least 10 class intervals but not more than 20 class intervals.
3. Start the lowest class interval on the first multiple of $i$ immediately below the lowest score in the data set. Check to see if this increase in the effective range will affect the value of $i$ selected and readjust if necessary.
4. List the class intervals in a column, the lowest at the bottom and the highest at the top, and tally the scores in the Tally column.
5. Enter the sum of the row tallies in the frequency column, $f$, and sum up the frequencies to find $N$.
6. Check to see that $N$ agrees with the number of scores in the data.

### Problem 2
### PREPARE A FREQUENCY POLYGON

It is often convenient to display the information from a frequency distribution in pictorial form rather than in tabluar form, as shown in the frequency distribution in Table 2–2. Two common methods for presenting results pictorially are the frequency polygon and the histogram which will be treated in Problem A. The frequency polygon contains no information beyond that found in the frequency distribution. The information merely is presented in a different form. Thus, given a frequency distribution for a body of data, the frequency polygon can be constructed without additional information.

### Preparing a Frequency Polygon

The frequency polygon is a graph with the score dimension laid out along the horizontal axis (the abcissa) and the frequency dimension laid out along the vertical axis (the ordinate). One point is located on the graph for each class interval in the frequency distribution. The point for a given interval is located above the midpoint of the interval at a height corresponding to the frequency for that interval.

The frequency polygon for Data A which can be prepared from the information contained in the frequency distribution for Data A shown in Table 2–2 is shown in Figure 2–1. Note that the point for the class

FIGURE 2–1
Frequency Polygon

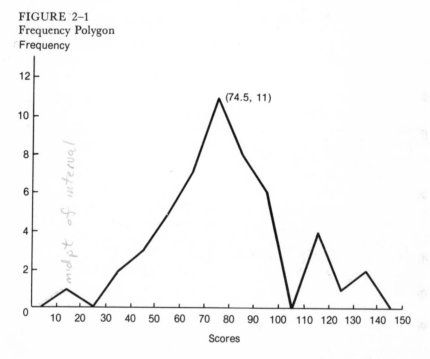

interval 10–19 in Table 2–2 is plotted above the point 14.5 on the score axis (abcissa) and opposite one on the frequency axis (ordinate). This is because 14.5 is the midpoint for this class interval, i.e., halfway between

9.5 and 19.5, and one is the frequency for this class interval in the frequency distribution for Data A (Table 2–2). Each of the other class intervals is represented by a point determined by the class interval midpoint, which fixes the horizontal axis coordinate, and the class interval frequency, which fixes the vertical axis coordinate.

It should be noted that the points for the intervals 20–29 and 100–110 fall on the horizontal score axis, i.e., at 0 on the vertical axis. This is due to the fact that there are no scores in these intervals.

In preparing a frequency polygon, it is customary to add an interval below the bottom interval containing scores and give it a zero frequency. Another interval with zero frequency is added above the top interval containing scores. Since these added intervals have zero frequencies, their points in the frequency polygon are plotted on the base line, the horizontal axis, at the midpoints of these intervals. This has the effect of bringing the graph back down to the base line on either side of the figure to provide a more pleasing effect. Without these two end points, the figure would be left hanging in the air. In Figure 2–1, these two end points are at 4.5 and 144.5, respectively.

After all the points have been plotted on the graph, they are connected with straight lines as shown in Figure 2–1.

*Proportions of the Figure.* The horizontal and vertical axes in Figure 2–1 are on different scales so that the height of the frequency polygon bears a reasonable proportional relationship to the width. If the same scale were used for both the score and the frequency axes, the figure in this instance would be extremely flat. In other instances, it might be extremely tall and narrow. By choosing the right scales for the two axes, a figure may be obtained that is pleasing to the eye and which effectively displays important features of the data for visual inspection. This is important because the main reason for using a frequency polygon is to make the information easier for the reader to examine and assimilate than it is in the frequency distribution itself. For individuals accustomed to reading frequency distributions, there is little point in preparing a frequency polygon. Many people in lay audiences, however, are repelled by tabular presentations while showing more tolerance for the same information presented in pictorial form.

*Labeling the Numerical Axes.* Even if the class interval, $i$, is a number like 2 or 3, which produces class intervals beginning with numbers other than multiples of 10, the numbers listed below the horizontal axis and beside the vertical axis to mark off the scales will often be multiples of 10, depending on the range to be covered. If the horizontal score axis is to cover 100 points, for example, it is likely that only the numbers at multiples of 10 would actually be written below the axis. The points in between might be marked by dots or small vertical lines touching the horizontal axis but the actual numbers represented by those dots would be omitted to give a more pleasing figure. Only in the case of a very narrow range of scores would all the integers in the entire range be written below the score axis. The same rules are applied to the vertical frequency axis. In Figure 2–1, for example, since the maximum frequency is small, every other integer has been listed. If the maximum frequency had been 45, probably only the numbers 5, 10, 15, and so on, up to 45 would have been used to label the frequency axis. With a maximum frequency of 100, probably only the values 10, 20, 30, and so on, up to 100 would have been listed with small marks at the points half way between these values. In some cases, additional dots at the single integer divisions might be included if a high degree of precision is needed.

## Summary of Steps for Preparing a Frequency Polygon

1. Lay out a chart or graph with the score axis along the horizontal dimension and the frequency axis along the vertical dimension. Choose scales that will accommodate the full range of scores plus at least one extra class interval and the full range of frequencies.
2. Label the axes numerically, using numbers that are multiples of 10 if the range of values is large enough, numbers that are multiples of 5 as the next best choice, and finally label with individual integers if necessary because the range of values is too small for anything else.
3. Plot one point on the graph for each class interval, placing the point above the exact midpoint of the interval and at a height equal to the frequency for that interval.
4. Place a point on the score axis, i.e., with zero frequency, at the midpoint of the interval next below the lowest class interval and at the midpoint of the interval next above the highest class interval.
5. Connect all these points by drawing straight lines between the points for adjacent class intervals.

*Smoothing the Frequency Polygon.* A procedure sometimes followed in preparing frequency polygons is to "smooth" the curve, thereby eliminating excessive irregularity in the figure. The frequency for each class interval is averaged with the frequencies of the intervals above and below it. In the example given in Table 2–2, averaging the frequencies for the lowest class interval, i.e., 10–19, would give one third of the sum of the frequencies for the interval 10–19 plus the frequencies for the two intervals on either side, e.g., $\frac{1}{3}(0 + 1 + 0)$ which equals $\frac{1}{3}$. For the interval 20–29, it would be $\frac{1}{3}(1 + 0 + 2)$ which equals 1. For the interval 100–109, the smoothed frequency would be $\frac{1}{3}(6 + 0 + 4)$ or $3\frac{1}{3}$. The smoothing process here would reduce the big dip in the frequency polygon at 104.5 in Figure 2–1.

The smoothing process may also be carried out giving double weight to the actual frequency. Thus, for the interval 10–19 in this problem, the smoothed frequency would be $\frac{1}{4}(0+1+1+0) = \frac{1}{2}$; for the interval 20–29, it would be $\frac{1}{4}(1+0+0+2) = \frac{3}{4}$. Although smoothing the curve does yield a more pleasing figure in many instances, the process is not recommended by this author because it distorts reality. The original frequency distribution can no longer be recovered from the frequency polygon after smoothing. Information has been lost in the smoothing process and the smoothed curve may or may not be a more accurate representation of the real situation. The benefit derived from this loss of information in the smoothing process is minimal. The reader may mentally supply a smoothed curve given the unsmoothed curve. He cannot, however, mentally supply the unsmoothed curve given only the smoothed curve.

## Problem A
## PREPARE A HISTOGRAM

Problem A is the first of the supplementary letter series problems. These problems may be omitted by the reader without interfering with his understanding of the numbered problems. Problems in the number series are concerned with more widely used statistical procedures than those in the letter series.

The histogram is merely an alternate type of figure to the frequency polygon for presenting pictorially the information contained in the frequency distribution. The histogram can be prepared from the frequency polygon or from the frequency distribution just as the frequency distribution and the frequency polygon can be reproduced from the histogram.

### Preparing the Histogram

The histogram for Data A from Table 2–1 is shown in Figure 2–2. The histogram basically consists of a series of bars touching each other (usually with the lines where they touch being removed) to give a "New York skyline" effect. Each class interval has a bar above it that has a width equal to $i$, the class interval size. The base of the bar runs from the exact lower limit of the interval to the exact upper limit of the interval. The height of the bar corresponds to the frequency for that class interval. Since the bars extend to the exact interval limits, it will be noted in Figure 2–2 that the bars begin at 9.5, 19.5, 29.5, and so on, not at 10, 20, 30, the written limits. The bars end at 19.5, 29.5, 39.5, and so on, not at 19, 29, and 39, the written limits.

The same rules for numerical labelling of the coordinate axes used with the frequency polygon are also appropriate with the histogram.

FIGURE 2-2
Histogram

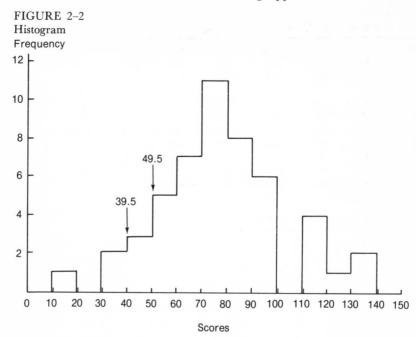

That is, try to write values on the score and frequency axes that are multiples of 5 or preferably 10. With a small range of possible values on the axis, however, it will be necessary to label with numbers other than multiples of 5 and 10. This is the case with the frequency axis in both Figures 2–1 and 2–2.

### Summary of Steps for Preparing a Histogram

1. Lay out a chart or graph with the score axis along the horizontal dimension and the frequency axis along the vertical dimension.
2. Label the axes appropriately to cover the range of possible scores and frequencies, using multiples of 5 or 10 to mark the division points on the scale wherever possible.
3. Draw a bar the full width of the interval, from exact lower limit to exact upper limit, for each interval in the frequency distribution. The points for the frequency polygon would fall at the centers of these bars if they were entered on the same graph.
4. Drop perpendicular lines from the outside ends of the bars for the two extreme class intervals until the lines touch the horizontal axis.
5. Using vertical lines perpendicular to the horizontal axis, connect the ends of the bars for adjacent class intervals.

TABLE 2–3
Data B (a sample of I.Q. scores)

| | | | | | | | |
|---|---|---|---|---|---|---|---|
| 100 | 116 | 84 | 95 | 102 | 110 | 96 | 90 |
| 117 | 135 | 97 | 126 | 75 | 133 | 113 | 72 |
| 92 | 91 | 104 | 109 | 106 | 126 | 82 | 145 |
| 88 | 103 | 124 | 105 | 81 | 85 | 76 | 132 |
| 128 | 58 | 94 | 64 | 77 | 87 | 114 | 101 |
| 101 | 92 | 107 | 119 | 82 | 109 | 111 | 98 |
| 94 | 105 | 112 | 99 | 110 | 93 | 91 | 102 |
| 107 | 113 | 90 | 115 | 90 | 100 | 97 | 101 |
| 107 | 89 | 97 | 95 | 117 | 86 | 107 | 85 |
| 68 | 94 | 106 | 112 | 109 | 89 | 100 | 114 |

TABLE 2–4
Data C (class examination scores)

| | | | |
|---|---|---|---|
| 75 | 84 | 69 | 77 |
| 85 | 69 | 80 | 71 |
| 79 | 95 | 50 | 49 |
| 76 | 61 | 64 | 90 |
| 83 | 58 | 70 | 83 |
| 91 | 72 | 41 | 74 |
| 82 | 63 | 37 | 66 |
| 76 | 79 | 76 | 88 |
| 87 | 58 | 70 | 83 |
| 70 | 58 | 66 | 75 |

TABLE 2–5
Data D (number of bar presses in a learning experiment)

| | | | | | |
|---|---|---|---|---|---|
| 4 | 8 | 3 | 18 | 33 | 11 |
| 20 | 7 | 5 | 11 | 10 | 17 |
| 2 | 10 | 9 | 8 | 6 | 12 |
| 9 | 30 | 7 | 6 | 14 | 9 |
| 14 | 28 | 14 | 9 | 16 | 8 |

TABLE 2–6
Data E (reported hours of sleep for previous night)

| | | | | |
|---|---|---|---|---|
| 8 | 7 | 8 | 6 | 8 |
| 5 | 7 | 7 | 8 | 10 |
| 9 | 6 | 9 | 9 | 7 |
| 7 | 2 | 4 | 5 | 9 |
| 8 | 8 | 7 | 6 | 8 |
| 8 | 9 | 6 | 7 | 8 |
| 9 | 12 | 8 | 8 | 7 |
| 7 | 10 | 8 | 9 | 6 |

TABLE 2–7
Data F (typing errors per subject in a speed trial)

| | | | | |
|---|---|---|---|---|
| 4 | 8 | 7 | 15 | 10 |
| 7 | 10 | 13 | 5 | 6 |
| 20 | 16 | 3 | 19 | 17 |
| 11 | 5 | 12 | 6 | 8 |
| 6 | 9 | 23 | 10 | 7 |
| 10 | 8 | 10 | 15 | 9 |
| 9 | 10 | 7 | 25 | 5 |
| 7 | 22 | 13 | 4 | 11 |
| 6 | 8 | 7 | 12 | 8 |
| 4 | 6 | 9 | 10 | 3 |

TABLE 2–8
Data G (100 yard dash times of
college sprinters)

TABLE 2–9
(a sample of grade point averages)

| 9.8 | 9.8 | 9.9 | 9.4 | 2.0 | 2.5 | 2.4 | 2.9 |
| 9.9 | 9.5 | 9.7 | 9.6 | 2.2 | 2.8 | 2.7 | 3.1 |
| 9.7 | 9.6 | 9.5 | 9.8 | 2.8 | 2.7 | 2.5 | 2.1 |
| 9.8 | 9.3 | 9.9 | 9.7 | 3.1 | 3.5 | 3.0 | 2.4 |
| 9.6 | 10.1 | 9.4 | 9.8 | 3.2 | 1.9 | 2.9 | 2.8 |
| 10.0 | 9.7 | 9.9 | 9.7 | 2.5 | 2.4 | 3.7 | |
| 9.7 | 9.6 | 9.7 | 9.7 | 2.4 | 2.3 | 2.2 | |
| 9.5 | 9.5 | 10.0 | 9.6 | 2.2 | 2.8 | 2.1 | |
| 9.6 | 10.2 | 9.2 | 9.8 | 2.7 | 2.7 | 2.5 | |
| 9.4 | 9.6 | 9.9 | 9.7 | 3.0 | 3.3 | 2.5 | |

TABLE 2–10
Data I (a sample of reaction times in milliseconds)

| 300 | 420 | 260 | 420 | 425 |
| 320 | 480 | 250 | 465 | 275 |
| 305 | 490 | 285 | 430 | 335 |
| 225 | 325 | 325 | 340 | 350 |
| 285 | 250 | 300 | 360 | 360 |
| 355 | 265 | 330 | 290 | 375 |
| 455 | 230 | 345 | 285 | 330 |
| 500 | 240 | 360 | 290 | 315 |
| 520 | 310 | 375 | 300 | 290 |
| 475 | 305 | 380 | 305 | 285 |

## Exercises

2–1. Prepare a frequency distribution, a frequency polygon, and a histogram (if Problem A is assigned) for Data B, Table 2–3.

2–2. Prepare a frequency distribution, a frequency polygon, and a histogram (if Problem A is assigned) for Data C, Table 2–4.

2–3. Prepare a frequency distribution, a frequency polygon, and a histogram (if Problem A is assigned) for Data D, Table 2–5.

2–4. Prepare a frequency distribution, a frequency polygon, and a histogram (if Problem A is assigned) for Data E, Table 2–6.

2–5. Prepare a frequency distribution, a frequency polygon, and a histogram (if Problem A is assigned) for Data F, Table 2–7.

2–6. Prepare a frequency distribution, a frequency polygon, and a histogram (if Problem A is assigned) for Data G, Table 2–8.

2–7. Prepare a frequency distribution, a frequency polygon, and a histogram (if Problem A is assigned) for Data H, Table 2–9.

2–8. Prepare a frequency distribution, a frequency polygon, and a histogram (if Problem A is assigned) for Data I, Table 2–10.

chapter **3**

# Measures of Central Tendency

The frequency distribution and its graphical representations, the frequency polygon and histogram, provide the research worker with a much better impression of his data than he can get from inspection of mere columns of randomly arranged numbers. These methods of data representation do not provide, however, a single precise numerical figure to represent the entire distribution of scores as far as overall size is concerned. Measures of central tendency are designed to provide a single numerical index that will tell in general how high or how low the scores are on the average for the entire distribution.

The measure of central tendency packs a great deal of information about a distribution of scores into a single number. To be sure, it omits a great deal of information, too, but it does convey a lot very quickly. One of the things not described by a measure of central tendency, however, is how spread out the scores are in the distribution. Measures of dispersion, to be described in the next chapter, are used to compare distributions on the relative extent to which the scores are very close together, widely spread apart, or somewhere in between. Two single numbers, a measure of central tendency, and a measure of dispersion, will go a long way toward describing what really needs to be known about an entire distribution involving perhaps hundreds of individual scores.

The three measures of central tendency to be described in this chapter are the mode, the median, and the mean. Methods for computing these statistics will be described and illustrated.

## Problem 3
## DETERMINE THE MODE OF A DISTRIBUTION

The mode is the easiest of all measures of central tendency to obtain. It is defined as the most frequent score. If data are not grouped in class intervals, merely find the score that appears most often. Or, if the data are grouped in class intervals of size 1, the mode is the score with the greatest frequency. In case the data are grouped in class intervals and the class interval, $i$, is greater than 1, the mode is the midpoint of the interval with the greatest frequency.

Table 3–1 represents a frequency distribution where the data have been grouped into class intervals with a class interval width, $i$, of 1 point.

TABLE 3–1
Frequency Distribution

| X | f | cf |
|---|---|-----|
| 95 | 3 | 209 |
| 94 | 5 | 206 |
| 93 | 7 | 201 |
| 92 | 10 | 194 |
| 91 | 13 | 184 |
| 90 | 15 | 171 |
| 89 | 16 | 156 |
| 88 | 20 | 140 |
| 87 | 25 | 120 |
| 86 | 22 | 95 |
| 85 | 19 | 73 |
| 84 | 16 | 54 |
| 83 | 14 | 38 |
| 82 | 11 | 24 |
| 81 | 8 | 13 |
| 80 | 3 | 5 |
| 79 | 2 | 2 |
| N | 209 | |

Ignore the "*cf*" column in Table 3–1 as far as Problem 3 is concerned; it will be used with a later problem. The mode of the distribution in Table 3–1 is 87, since the frequency for the score of 87 is 25 and 25 is higher than any other frequency in the distribution. Table 3–2 represents a frequency distribution where the data have been grouped into class intervals with $i = 5$. The mode of the distribution in Table 3–2 is 47.0, the midpoint of the interval with the largest frequency. This is the interval 45–49 with exact interval limits of 44.5 to 49.5 and a frequency of 25. The mode, 47.0, lies halfway between the exact limits, i.e., $44.5 + \frac{1}{2}(5)$, or halfway between 44.5 and 49.5. It is also halfway between the class interval written limits, 45 and 49.

TABLE 3–2
Frequency Distribution

| X | f | cf | x' | fx' |
|---|---|---|---|---|
| 60–64 | 3 | 160 | 5 | 15 |
| 55–59 | 8 | 157 | 4 | 32 |
| 50–54 | 15 | 149 | 3 | 45 |
| 45–49 | 25 | 134 | 2 | 50 |
| 40–44 | 12 | 109 | 1 | 12 |
| 35–39 | 10 | 97 | 0 | 0 |
| 30–34 | 13 | 87 | -1 | -13 |
| 25–29 | 18 | 74 | -2 | -36 |
| 20–24 | 23 | 56 | -3 | -69 |
| 15–19 | 16 | 33 | -4 | -64 |
| 10–14 | 10 | 17 | -5 | -50 |
| 5– 9 | 5 | 7 | -6 | -30 |
| 0– 4 | 2 | 2 | -7 | -14 |
| N | 160 | | | |

Table 3–2 is an unusual frequency distribution in that there are two distinct modes rather than just one. According to the definition of the mode as the midpoint of the interval with the largest number of cases, the mode is 47.0 and technically there is only one mode. On the other hand, there is another large frequency, for the interval 20–24, which is only two scores less than the frequency for the interval containing the mode. Furthermore, the frequencies fall off in magnitude on either side of both these peak frequency intervals. Such a distribution is referred to as a "bimodal" distribution, i.e., with two modes, even though the two peak frequencies are not identical. The frequency polygon for the frequency distribution in Table 3–2 is shown in Figure 3–1. The two separate peaks in the curve clearly reveal the bimodal character of the distribution. In case there are two definite and distinct modes, both should be reported.

In searching for the mode of a given distribution, it may occur that two frequencies for adjacent intervals are equal and larger than any of the other frequencies. In this case, the mode is taken to be at the point halfway between the midpoints of these adjacent intervals, i.e., at the exact interval limit separating the two intervals.

## When to Use the Mode

The research worker can, at his discretion, choose any of the measures of central tendency to describe his particular data. This choice, however, is ordinarily dictated by the nature of the situation that he is dealing with. The mode, for example, is chosen in those cases where the most common score is desired. A manufacturer who is deciding how many pairs of shoes to produce of a certain size is interested in what sizes

FIGURE 3–1
A Bimodal Distribution Frequency Polygon

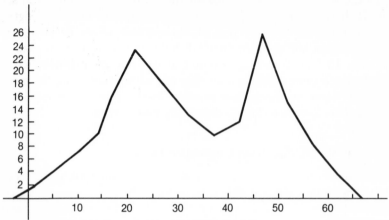

occur most frequently. If there are more size nines than any other size, he wants to make more size nine shoes than any other size. Any measure of central tendency that failed to yield a value closer to size nine than to any other shoe size would be giving him misleading information.

Another type of situation where the mode might be chosen as the measure of central tendency would be in finding the class size a student could expect to encounter at the state university. In this case, many classes are small but there are some very large classes. The most likely class to be encountered, if picking a class at random, would be a small class. Using the modal class size would be a good measure of central tendency to provide an indication of what the student is most likely to find. Choosing a measure of central tendency such as the mean could result in a class size figure that would never be encountered in practice.

The mode is also used to obtain a very quick estimate of the central tendency of the distribution because it can be determined without computation merely by inspection. It is not a very stable measure, however, since it can change considerably with the addition or deletion of a few scores.

## Problem 4
### DETERMINE THE MEDIAN FROM UNGROUPED DATA

The median of a distribution of scores is defined as the point on the score scale below which 50 percent of the scores fall. It follows logically that it is also the point on the score scale above which 50 percent of the scores fall. Roughly speaking, it is the middle score in a collection of scores. This definition has to be interpreted carefully, however, because the middle score is not necessarily in the center of the numerical score range. If there are some extreme scores, either high or low, this lengthens the score range in the direction of the extreme scores in such a way that the median will probably not fall halfway between the bottom and top scores. In fact, the extreme scores at one end of a distribution can be made much more extreme than they are without affecting the median at all. The score in the middle remains in the same place even if the extreme scores are altered to be drastically more extreme.

## Computing the Median from Ungrouped Data

When scores have not been grouped into class intervals and a median is to be computed, the first step is to arrange them in order of size from lowest to highest. Thus, the scores 10, 9, 7, 7, 11, 14, 14, 12, and 6 would be rearranged as follows: 6, 7, 7, 9, 10, 11, 12, 14, and 14. Counting the scores shows that $N = 9$. This is an odd number of scores so it is possible to have a middle score and indeed 10 is the middle score, hence the median here is 10.0. This is true in this distribution of scores because there are no other scores of size 10 except the one score. If there had been another score of 10, e.g., no 9 but two 10's, then 10.0 would not be the median. An example of this more complicated situation will be given later.

How does a median of 10.0 for these nine scores fit the original definition of the median given at the start of Problem 4? Half the scores, i.e., $\frac{9}{2}$, or $4\frac{1}{2}$ scores, should be below 10.0 and $4\frac{1}{2}$ should be above 10.0. Below the score of 10, there are the scores 6, 7, 7, and 9, or four scores instead of $4\frac{1}{2}$ scores. Also, above the score of 10, there are the scores 11, 12, 14, and 14, again four scores instead of $4\frac{1}{2}$ scores.

It must be noted that the score of 10 falls in an interval that runs from 9.5 to 10.5. It is not considered to fall precisely at 10.0. This single score in the interval 9.5 to 10.5 is treated as though it were evenly spread throughout the entire interval from 9.5 to 10.5. This means that half of the score lies between 9.5 and 10.0 and half lies between 10.0 and 10.5. If we take the median to be 10.0, then not only do the four scores of 6, 7, 7, and 9 fall below the median but also the half score that is the portion of the score 10 that falls between 9.5 and 10.0. This gives the total of $4\frac{1}{2}$ scores that should fall below the median as demanded by the definition. At the same time, above 10.0 will be the scores of 11, 12, 14, and 14 plus the half score which is the part of the score 10 falling between 10.0 and 10.5. This gives the total of $4\frac{1}{2}$ scores that should lie above the median in this example.

*An Example with an Even Number of Scores.* Consider the scores 3, 8, 10, 10, 12, 13, 17, 25, 26, and 60, already arranged in order from lowest to highest. Since there is an even number of scores, $N = 10$, there can be no middle score. The two single scores 12 and 13 fall in the middle. In this case, the median falls at 12.5, the exact interval limit between the intervals for the scores of 12 and 13. Below 12.5, there are $1\%$ or 5 scores, 3, 8, 10, 10, and 12, since all of the score 12 falls below the upper limit of the interval for 12, namely, 12.5. Above 12.5, the 5 scores 13, 17, 25, 26, and 50 fall. Thus 12.5 fits the definition of the median that 50 percent of the scores will be above it and 50 percent will be below it.

*When There Are Multiple Scores in the Region of the Median.*    In the two previous examples, the median was either equal to a single middle score with an odd number of scores, or between two single middle scores with an even number of scores. This is unusual. More common is to find that there is no single middle score with an odd number of scores or two single middle scores with an even number of scores. Consider the following scores already arranged in order of size from lowest to highest: 3, 7, 12, 15, 15, 25, 50, 100. In this case, $N = 8$, so four scores must fall below the median and four scores must fall above the median. The value 15.0 is the only point in the distribution for which this is true, hence 15.0 is the median of these scores. Note that there are two scores in the interval 14.5 to 15.5 which contains the median. Half of these two cases, i.e., one score, falls between 14.5 and 15.0 and the other falls between 15.0 and 15.5. Hence, below 15.0 is the one case, between 14.5 and 15.0, plus the three scores 3, 7, and 12.

The determination of the median becomes still more complicated if there are several identical scores for the one-point interval in which the median must fall. Consider the following example of scores already arranged in order of increasing size: 3, 7, 12, 15, 15, 15, 15, 25, 50, 75, 90, and 100. In this example, $N = 12$, hence six scores fall below the median and six scores fall above the median. Counting up from the bottom shows that at 14.5 three scores have been counted or passed on the way up and at 15.5 a total of seven scores have been covered, the three below the score interval 14.5 to 15.5 plus the four scores of 15 contained in the score interval 14.5 to 15.5. Since three scores is less than the needed six, the median must be above 14.5. Also, since seven is more than the needed six, the median must be below 15.5. This places the median somewhere between 14.5 and 15.5.

There are three scores below 14.5 and six are needed before coming up to the median point since the median has six cases below it in this example. Since there are four cases in the interval 14.5 to 15.5, this means that three quarters of the cases in that interval must fall below the median and only one quarter above it. Since the cases are spread evenly throughout the interval, it is necessary to rise above 14.5 a distance equal to three quarters of the interval before coming to the point below which three of the cases in the interval will fall, plus the three below this interval, making a total of six. The interval from 14.5 to 15.5 is one score point, hence the median is three quarters of one point, or 0.75, above 14.5 or the median equals $14.5 + 0.75 = 15.25$. Thus, the median is at 15.25 because exactly six cases fall below 15.25 and six cases fall above 15.25, as required by the definition of the median.

*Rounding Off the Median to One Decimal Place.* The median of 15.25 would be rounded off to one decimal place, i.e., to 15.2. There is usually little reason to report measures of central tendency to a greater level of precision than one decimal place where the scores are integers. In rounding, the rule is to round up if the second and third digits to the right of the decimal place give a number larger than 50; round down if they give a number less than 50. Thus, 6.651 becomes 6.7 because the 51 is greater than 50; 6.649 becomes 6.6 because 49 is less than 50; 100.104 becomes 100.1; 20.380 becomes 20.4; 8.92 becomes 8.9; 4.29 becomes 4.3; 9.97 becomes 10.0; 0.02 becomes 0.0, and so on. In the case that the second and third digits give a number exactly equal to 50, and any additional digits to the right are also zero, the rule given above does not apply. To determine the direction of rounding in this case, round always to the even number rather than to an odd number. Thus, 13.450 becomes 13.4 because the second digit to the right is 5 and all digits to the right of that are zero plus the fact that 4 is an even number; 107.75 becomes 107.8 because the second digit to the right is 5 and 7 is an odd number so the rounding is up to the even digit, 8; 9.95 becomes 10.0; 6.55 becomes 6.6; 8.850006 still rounds up to 8.9, however, since 50006 is greater than 50000. This last case does not fit the requirements for the immediately preceding examples because all the digits to the right of the 5 are not zeroes. If the number had been 8.850000, the rounded number would have been 8.8.

*When Several Medians Exist.* When the median falls at the exact limit between two intervals, such that one or more score intervals on one side or the other are empty, many possible score points could satisfy the requirements for the median. Consider the following scores, already arranged in order of increasing size: 3, 4, 8, 9, 9, 11, 15, 25, 50, 100. With $N = 10$, there must be five scores below the median. Counting up from the bottom shows that exactly five scores have been covered by the time the point 9.5 has been reached. It might be concluded, therefore, that 9.5 is the median, especially since there are also five scores above 9.5. It may be noted, however, that any point between 9.5 and 10.5 will also have five scores above it and five scores below it. This is because there are no scores of 10 in this distribution and hence no additional scores are added to the five already passed in moving upward from 9.5 up to 10.5. Where a range of such values exists, all of which will satisfy the definition of the median, choose the median to be the midpoint of this range. In this example, therefore, the median would be placed at 10.0, halfway between 9.5 and 10.5, the range of values all of which meet the requirements for the median. Sometimes this range of points all of which satisfy the median will extend over an interval of several score points, or even over more than one interval.

## Summary of Steps for Computing the Median from Ungrouped Data

1. Arrange the scores in order of increasing size from lowest to highest, repeating equal scores.
2. Count the number of scores in the distribution, $N$, and find $N/2$.
3. Count up scores from the bottom until a score interval is reached that has $N/2$ scores or less below its exact lower limit and $N/2$ scores or more below its exact upper limit.
4. Take the difference between $N/2$ and the number of scores below the exact lower limit of the score interval containing the median, i.e., the interval located in Step 3. Divide this difference by the number of scores in this score interval. This is the proportion of cases from this interval that will be below the median or the number of scores that must be passed in this interval before reaching the median moving up from the bottom of the distribution.
5. Multiply the proportion of scores in the interval below the median by the value of $i$, the class interval size; this is always 1 for ungrouped data. Add this product to the exact lower limit of the interval containing the median to obtain the median itself.
6. If there is a range of points that satisfy the definition of the median, place the median at the midpoint of this range.

*When to Use the Median.* Since the median is the point in the score distribution below which (and above which) 50 percent of the cases fall, it is determined by the location of the scores in the middle of the distribution, not by the scores at the extremes of the distribution. It follows from this fact that the extreme scores in a distribution could be altered drastically without affecting the median at all. The highest score in a distribution could be increased by 1000 points, or a million, without changing the median of the distribution at all.

The median is a good measure to use where there are some very extreme scores at one end of the distribution or the other which should not be allowed to influence the central tendency value obtained. One example of this would be in determining a measure of central tendency for reaction times. Some of these tend to be abnormally long because the subject was not paying attention or did not hear or see the signal. These extreme measures are not really valid indications of his reaction time but rather are due to extraneous factors. By selecting the median as the measure of central tendency, these extreme scores are not allowed to raise spuriously the summary reaction time figure obtained.

Another situation in which the median is the preferred measure of central tendency is that where the distribution is truncated, i.e., when the scores at one end of the distribution are not actually precisely known. For example, if subjects are given a task of doing arithmetic problems under speeded conditions and the time alloted is too generous, some subjects may come to the end of the problems before the time runs out. These subjects could have earned higher scores if more problems had been included. As it is, the only information available is that their scores should have been higher than they actually are. Since these individuals will almost always have scores well above the median, it is possible to calculate the median even though some of the highest scores in the distribution have not been determined exactly.

## Problem 5
## DETERMINE THE MEDIAN FROM GROUPED DATA

In most real-life data collection situations, the number of scores available will be too large to permit the calculation of the median by the procedures described in Problem 4 above. In such cases, scores are first arranged in a frequency distribution using the methods described in Problem 1. Whether the median is to be calculated from ungrouped data, as in Problem 4, or from grouped data, as in this·problem, however, the basic principles used are the same. In both cases the median is defined as the point below which 50 percent of the cases fall and in both cases it is assumed that scores in a given interval are spread evenly throughout that interval for purposes of determining a precise location for the median. Perhaps the major difference between the procedures in Problem 4 and those in Problem 5 is that in Problem 4 scores are spread over intervals of only one point whereas in Problem 5, scores are spread evenly over class intervals with widths equal to $i$ points, where $i$ is the class interval size.

Some accuracy is lost in computing the median from grouped data because grouping the data into intervals loses some information. Scores in the interval 5 to 9, for instance, can be fives, sixes, sevens, eights, or nines, or any combination of these scores. Since the particular distribution of scores in this interval is not known in grouped data, the scores are treated as though they are spread evenly over the entire interval, from 4.5 to 9.5, the exact limits of the interval.

### Computing the Median from Grouped Data

To illustrate the procedures for calculating a median from grouped data, some examples will be worked out in detail. Consider the frequency distribution in Table 3–1. This situation is very similar to that for calculating the median from ungrouped data, Problem 4, since the class interval size is $i = 1$. The only difference is that instead of stringing the scores out horizontally from smallest to largest, writing the identical scores as many times as they are repeated, in Table 3–1 the scores are listed vertically and only once, with a frequency to indicate how many times that particular score is repeated. With so many scores, it is much easier to compute the median from a frequency distribution like that in Table 3–1 than to compute it by writing out the scores in order of size, as in Problem 4.

The median is the point below which 50 percent of the scores fall so 50 percent of $N = 209$ is determined first to be $.50 \times 209 = {}^{209}\!/_2 = 104.5$. So, the median will be the point in the distribution below which there are exactly 104.5 cases. This emphasizes the point that scores are divided up into fractions and spread evenly over intervals in calculating the median. Scores are not treated as individual indivisible units.

The last column in Table 3–1, headed by $cf$, gives the cumulative frequencies for the frequency distribution. These are obtained by starting with the lowest interval and placing its frequency, shown in column $f$, as the lowest cumulative frequency in column $cf$. This value represents the number of scores in the distribution below the exact *upper* limit of that interval. In this case, in Table 3–1, the cumulative frequency, $cf$, for the bottom interval is 2, indicating that there are 2 cases below 79.5, the exact upper limit of the interval 78.5 to 79.5. Since there are no intervals below this one containing nonzero frequencies, the frequency from column $f$ and the cumulative frequency from column $cf$ are identical.

To obtain the cumulative frequency for the next higher interval in Table 3–1, i.e., for the interval corresponding to the score 80, add the frequency for this interval, 3, from column $f$, to the cumulative frequency, 2, for the next lower interval, from column $cf$, and place the result, $3 + 2 = 5$, as the cumulative frequency in column $cf$ for the interval corresponding to the score 80. For the next higher cumulative frequency, add the frequency for the next higher interval, 8, to 5, the cumulative frequency for 80, to get the cumulative frequency, 13, for the score 81. The cumulative frequency for score 82 is $11 + 13 = 24$; for 83, it is $14 + 24 = 38$, and so on, up to the top interval where the cumulative frequency for the score 95 is $3 + 206 = 209$. Note that the top cumulative frequency equals $N$, the number of cases. This is to be expected since all the scores in the distribution must fall below the upper limit of the highest interval. To avoid errors in calculating the median, it is essential to note that the cumulative frequency represents the number of cases below the exact *upper* limit of the interval, not the midpoint or the lower limit.

Having determined that there are $N/2 = 104.5$ scores below the median for the frequency distribution in Table 3–1, the cumulative frequencies in the *cf* column are inspected to locate the interval that must contain the median. This must be the interval 86.5 to 87.5, corresponding to the score 87, since below 86.5 there are only 95 cases, which is less than 104.5, and below 87.5 there are 120 cases, which is more than 104.5. It follows that the point below which there are 104.5 cases must lie somewhere between 86.5 and 87.5.

There are 25 scores in the interval 86.5 to 87.5, the frequency from column *f* for the score 87. These 25 scores are considered to be spread evenly throughout the interval. A portion of these 25 scores must fall below the median since there are 104.5 scores altogether below the median and there are only 95 scores below 86.5. The difference, $104.5 - 95.0 = 9.5$ scores, or 9.5 scores of the 25 scores in the interval 86.5 to 87.5 must fall below the median. In moving up from 86.5 toward the median, therefore, a total of 9.5 of the 25 scores in the interval for the score 87 must be passed before the median is reached.

Since the 25 scores in the interval 86.5 to 87.5 are considered to be spread evenly over the interval, and 9.5 scores must be passed before the median is reached, the distance between 86.5, the lower limit of the interval, and the median must be 9.5/25 of the total interval width, which is 1 point in this case. So, the distance between 86.5 and the median is equal to $(9.5/25)(1) = 9.5/25$. The calculations to obtain the median for the frequency distribution in Table 3–1 may be summarized as follows:

$$\text{Mdn} = 86.5 + \frac{104.5 - 95}{25} \times (1)$$

This interpolative process may be expressed in words as follows:

$$\text{Median} = \left[ \begin{array}{c} \text{Exact lower limit of interval} \\ \text{containing the median} \end{array} \right] + \frac{\dfrac{N}{2} - \left[ \begin{array}{c} \text{Number of scores below lower limit} \\ \text{of interval holding Mdn} \end{array} \right]}{\left[ \begin{array}{c} \text{Number of cases in the interval that} \\ \text{contains the median} \end{array} \right]} \times (i)$$

Where

$N$ = number of scores in the distribution

$i$ = class interval size

Completing the computations for the median of the frequency distribution in Table 3–1 gives:

$$\text{Mdn} = 86.5 + \frac{9.5}{25} \times (1) = 86.5 + 0.38 = 86.88 = 86.9$$

Note that the median is rounded off to one decimal place.

*A Second Example.* To illustrate the computation of a median with grouped data where $i = 5$, the median will be calculated for the frequency distribution in Table 3–2. There are 160 cases in the frequency distribution in Table 3–2, so $N/2 = 160/2 = 80$ scores must fall below the median. The point on the score scale is sought such that there will be 80 scores below it and 80 scores above it. Examining the cumulative frequencies in column *cf* of Table 3–2, it will be noted that there are 74 scores below 29.5 and 87 scores below 34.5, so the median must lie in this interval. Using the same principles that were applied in the previous example for Table 3–1, the median for the data in Table 3–2 may be calculated as follows:

$$\text{Mdn} = 29.5 + \frac{80 - 74}{13} \times (5)$$

$$= 29.5 + \frac{6}{13}(5) = 29.5 + \frac{30}{13}$$

$$= 29.5 + 3.08 = 32.58 = 32.6$$

In the example above, 29.5 is the exact lower limit of the interval found to contain the median; 80 is the number of cases below the median; 74 is the number of cases found below the interval containing the median; 13 is the number of cases in the interval containing the median; and 5 is $i$, the class interval size. The numerator of the fraction, $80 - 74 = 6$, represents the number of cases needed from this interval to reach the median. Dividing this figure by the frequency for the interval gives the proportion of the class interval width that must be added to the lower limit of the interval containing the median to reach the median itself.

## Summary of Steps for Calculating the Median from Grouped Data

1. Find $N/2$, or 50 percent of $N$, the number of scores or cases falling below the median.
2. Write down the cumulative frequencies for the frequency distribution and using these locate the interval that contains the median, the point with $N/2$ scores below it.
3. Determine the fraction of the class interval width that must be covered to reach the median by taking the ratio of the number of cases needed from the interval to the number of cases in the interval containing the median.
4. Multiply this fraction by the class interval width, $i$, and add the result to the exact lower limit of the interval containing the median to obtain the value of the median itself.
5. Round off to one decimal place.

## Problem 6
### COMPUTE A MEAN FROM UNGROUPED DATA

The mean, $M$, for a set of scores is merely the average of those scores. That is, the scores are added up and the total is divided by the number of scores added. For the scores 3, 8, 8, 10, 12, 5, and 1, for example, the sum is 47 and dividing 47 by 7, the number of scores, $N$, gives $\frac{47}{7} = 6.71 = 6.7$. The mean is rounded off to one decimal point in most situations, as with the other measures of central tendency already discussed.

It will be noted here that there is no need to arrange the ungrouped raw scores in order of increasing size since the scores are merely summed as is. The order of addition of a set of scores does not affect the total. With ungrouped data, this gives the mean an advantage over the median as far as ease of calculation is concerned. This advantage will not be very important, however, unless there are many scores in the frequency distribution. This particular method of calculating the mean is commonly used when $N$ is small. It is also used even where $N$ is large when electronic or mechanical calculating equipment is available. With $N$ large, it is often preferable to break the scores up into groups of smaller size so the sums of scores can be obtained and checked within each group separately. Scores are broken into groups this way for adding because it is difficult to add a very long column of numbers without an error, even using a machine. After the sums of the separate groups have been obtained, they may be added to obtain an overall sum. This overall sum is divided by the total number of scores, $N$, to obtain the final mean value.

### Raw Score Formula for the Mean

The process used above in calculating the mean from ungrouped data may be represented symbolically by the following raw score formula for the mean:

$$M = \frac{\Sigma X}{N} \qquad (3\text{--}1)$$

Where

$$M = \text{mean}$$

$$\Sigma X = \text{sum of the raw scores}$$

$$N = \text{number of scores in the distribution}$$

The capital letter $X$ has already been used in earlier tables to represent raw scores and $N$ has been used to represent the number of cases. The new symbol on the right of Formula 3–1 is the capital Greek letter sigma, $\Sigma$, which is used as a summation sign. This symbol means to add up the elements that appear to the right of the sign. In this formula, only $X$ appears to the right and $X$ stands for a raw score from the distribution, so $\Sigma X$ means to add up all the $X$ scores in the distribution.

*The Summation Sign*, $\Sigma$.  Given that the ungrouped raw scores in a distribution are 4, 7, 10, 20, 16, 7, 5, and 2, what does $\Sigma X$ for this distribution stand for? The result of this operation may be shown as follows:

$$\Sigma X = 4 + 7 + 10 + 20 + 16 + 7 + 5 + 2 = 71$$

The symbols $\Sigma X$ are actually shorthand notation for a more complete symbolic representation shown by the following:

$$\Sigma X = \sum_{j=1}^{N} X_j = X_1 + X_2 + X_3 + \cdots + X_N$$

In this more complete notation, $X_j$ represents an indefinite raw score, i.e., score $j$ where score $j$ is any one of the scores $X_1, X_2, X_3, \ldots, X_N$. In the example given in this section, above, $X_1 = 4$, $X_2 = 7$, $X_3 = 10$, $X_4 = 20$, $X_5 = 16$, $X_6 = 7$, $X_7 = 5$, and $X_8 = 2$ with $N = 8$. $X_j$ can stand for any of these eight scores. The symbols $\sum_{j=1}^{N} X_j$ are interpreted verbally as follows: sum the scores for which the general term is $X_j$, starting with $j = 1$ and going up to $j = N$. In this particular example, it would be $\sum_{j=1}^{8} X_j$, with the last term in the summation being the eighth term because $N = 8$. Shorthand notation for formulas involving summation signs, i.e., without putting the limits on the summation signs and subscripts on the variable terms, will be used in this book wherever it is possible to do so without ambiguity in order to make the formulas less forbidding in appearance. It is to be understood that the subscripts and summation limits are implied even if they have not been actually included in the formulas. If the subscript were not implied, $X$ would be merely one score, albeit an unknown one, and the symbol $\Sigma X$ would mean:

$$X + X + X + \cdots + X = NX$$

where $N$ values, all equal to $X$, are added up. The implied subscript on the $X$ makes it $X_j$ which represents a *variable*, not a constant. The sum of a variable term gives:

$$\sum_{j=1}^{N} X_j = \Sigma X = X_1 + X_2 + X_3 + \cdots + X_N$$

where at least some of the values of $X_j$ are not equal to each other. If they were all equal to each other and to the number symbolized by $X$ (without a subscript), then the result of the summation would be merely to add up the same number, a constant, $N$ times to get $N \times X$, or $NX$. The $\Sigma X$, therefore, means a sum of different terms only where $X$ is a variable, not a constant. This implies that $X$ really stands for $X_j$ but the subscript $j$ has been left off for convenience. In this book, $X$ will always be used to represent a variable term rather than a constant term.

The summation of a constant term could be indicated by the symbols $\Sigma a$, where $a$ is merely a number, e.g., 5. If the summation is over $N$ terms, then $\Sigma a$ really stands for $\sum\limits_{j=1}^{N} a$. There is no subscript on the constant term, $a$, because it is just a number, i.e., 5 in this case. The result then, is:

$$\sum a = \sum_{j=1}^{N} a = a + a + a + a + \cdots + a = N \times 5 = 5N$$

If $N$ happens to be 10, then the final result of the summation would be 50. The difference between summations of constant terms, variable terms, and mixtures of the two will be important in later problems so it is essential to understand the meaning of the summation sign and its use before proceeding further.

## Computing the Mean from Grouped Data

If the data have already been grouped into class intervals, Formula 3–1 can still be used to calculate the mean. The midpoint of each interval is multiplied by the frequency for that interval. These products are summed to obtain $\Sigma X$. This sum is divided by $N$, the sum of the frequencies to obtain the mean. A method for obtaining the mean of grouped data with less labor where no calculating equipment is available is described in Problem $B$, the last problem in this chapter. The mean computed from grouped data will usually be slightly different from the mean computed with ungrouped data. With grouped data, all scores are treated as though they were equal to the midpoint of the interval. This usually introduces some error into the final result. In most cases, however, the magnitude of this error is too small to be of any practical consequence.

## When to Use the Mean

The mean is the most stable of the measures of central tendency because it is influenced by every score in the distribution. It is a good measure of central tendency to use if accuracy and stability are important. If the distribution has many extreme scores at one end, however, the mean is pulled in the direction of these scores, giving a misleading indication of the typical score. In computing a mean for reaction time scores, for example, very long reaction times due to inattention would add substantially to the central tendency value obtained when in fact they should not be allowed to do so because they are not representative reaction times.

### Comparison of the Mean, Median, and Mode

The mean is by far the most accurate, stable, and the most commonly used measure of central tendency. It is ordinarily the one to be chosen, especially if certain additional statistical computations are to follow, e.g., standard deviation, correlation coefficient, or tests of significance. If the mean is not chosen, it is likely to be for one of the following reasons:

1. Only a very crude and quickly determined measure of central tendency is desired in which case the mode, or possibly the median, will be preferred.
2. Because some scores are not exactly known, but only identified as either very large or very small, the mean cannot be computed. In this case, the median will usually be chosen, but occasionally the mode will be preferred.
3. If the distribution of scores is badly *skewed*, that is, with a frequency polygon showing a long tail either to the left or to the right, the median may be preferred to give a more realistic figure for the central tendency of the distribution of scores. This would be particularly true if the accuracy of the extreme scores in the tail of the distribution is suspect.
4. If only the most common score is desired, the mode will be chosen as the preferred measure of central tendency.

### Problem B
### COMPUTE A MEAN FROM GROUPED DATA BY THE CODE METHOD

The main reason for computing a mean from grouped data by the code method is to save labor. Since small electronic calculators are becoming so generally available, code methods for computing means and other statistics are not as widely used as they once were. For this reason, computation of the mean by the code method has been placed in the letter series of supplementary problems. If this problem is omitted, however, it will also be necessary to omit Problems C and E where the code method is used to compute the standard deviation and the correlation coefficient.

Once a frequency distribution has been prepared for a body of data, computing the mean by the code method is relatively painless, even without computing equipment. If a desk calculator, or even an adding machine, is available and no frequency distribution is needed for other purposes, it may be less time consuming to compute the mean by the method of Problem 6 rather than by the method to be described here. In the absence of computing equipment, however, the method to be described here involves less work for larger samples than computing the mean by the raw score formula used in Problem 6.

**Formula for Computing the Mean by the Code Method**

$$M = M_G + \frac{\Sigma fx'}{N} \cdot i \tag{3-2}$$

Where

$M$ = mean

$M_G$ = guessed mean

$f$ = class interval frequency

$x'$ = deviation of the midpoint of a class interval from the guessed mean divided by the class interval size

$i$ = class interval size

The mathematical derivation of Formula 3–2 is given in the Mathematical Notes section below for those students who wish to concern themselves with such matters. Students interested only in applying the methods without reference to the underlying logic may skip the mathematical notes sections here and throughout the book.

**Computing the Mean by the Code Method Formula**

Application of Formula 3–2 to compute the mean of grouped data by the code method is best understood with reference to an actual example. A frequency distribution is shown in Table 3–3 with the additional columns that are needed in calculating the mean by the code method. The actual computations are also shown in Table 3–3.

The guessed mean, $M_G$, is chosen to be the midpoint of one of the intervals in the distribution. Any interval could be chosen without affecting the obtained mean value but $M_G$ is usually placed at the midpoint of an interval near the center of the distribution. If it were placed at the

TABLE 3–3
Computation of the Mean by the Code Method

| X | f | x' | fx' |
|---|---|---|---|
| 110–119 | 3 | 4 | 12 |
| 100–109 | 8 | 3 | 24 |
| 90– 99 | 12 | 2 | 24 |
| 80– 89 | 20 | 1 | 20 |
| 70– 79 | 36 | 0 | 0 |
| 60– 69 | 33 | −1 | −33 |
| 50– 59 | 24 | −2 | −48 |
| 40– 49 | 16 | −3 | −48 |
| 30– 39 | 7 | −4 | −28 |
| 20– 29 | 2 | −5 | −10 |
| Σ | 161 | | −87 |

$$M = M_G + \frac{\Sigma fx'}{N} \cdot i \qquad\qquad (3\text{–}2)$$

$$M = 74.5 + \frac{(-87)}{161} \times (10) = 74.5 - \frac{870}{161}$$

$$M = 74.5 - 5.404 = 69.096$$

$$M = 69.1$$

midpoint of the bottom interval, the same mean would be obtained but at a cost of more computational work. In Table 3–3, the guessed mean, $M_G$, was taken to be the midpoint of the interval 70–79, i.e., 74.5. The guessed mean is never placed at any point other than the midpoint of an interval.

The values in the column of Table 3–3 headed by x' represent deviations of the midpoints of the class intervals from the guessed mean in terms of class interval units. The value of x' for the interval 70–79 is zero because the midpoint of this interval does not deviate from $M_G$ at all. It *is* $M_G$. The midpoint of the interval 80–89 is 84.5, which is 10 points higher than 74.5, the value of $M_G$. Dividing this value of 10 by $i$, the class interval size, gives $\frac{10}{10}$ or 1 for the value of x' corresponding to the interval 80–89. In other words, the midpoint of interval 80–89 is one class interval above $M_G$. The value of x' for the interval 90–99 is 2, because 94.5 is 20 points or two class intervals above 74.5, the value of $M_G$. The next two class intervals above have 3 and 4, respectively, for their x' values because their midpoints are three and four class intervals above $M_G$.

The values of x' for class intervals below $M_G$, however, will be negative, since subtracting the value of $M_G$ from their midpoints gives negative numbers. For the interval 60–69, x' is given by $\frac{64.5 - 74.5}{10} = \frac{-10}{10}$

$= -1$. For the interval 50–59, $x' = \dfrac{54.5 - 74.5}{10} = \dfrac{-20}{10} = -2$. For the bottom interval, $x' = \dfrac{24.5 - 74.5}{10} = \dfrac{-50}{10} = -5$.

In practice, there is no necessity to compute these values of $x'$. A zero is placed in column $x'$ opposite some selected interval, usually near the middle of the distribution. Then, the consecutive intervals above this one are given values of $x'$ equal to 1, 2, 3, and so on, until the top interval is reached. The intervals below the guessed mean interval receive consecutive values of $x'$ equal to $-1$, $-2$, $-3$, and so on, down to the bottom interval.

The column in Table 3–3 headed by $fx'$ is obtained by multiplying the value of $f$ (frequency) in a given row times the value of $x'$ on the same row. Thus, for the top interval, $fx' = 3 \times 4 = 12$. For the next interval down, $fx' = 8 \times 3 = 24$. The value of $fx'$ for the interval containing the guessed mean is always zero. The values of $fx'$ will always be negative for any intervals below the interval containing the guessed mean since the frequencies are being multiplied by negative numbers. For example, $fx'$ for the interval 40–49 is $16 \times (-3) = -48$. Some statisticians will actually place the guessed mean in the bottom interval to avoid any negative $fx'$ values. They would rather multiply larger numbers than deal with negative $fx'$ values. The practice of putting $M_G$ in the bottom interval is most likely to be helpful if a desk calculator is being used to carry out the multiplications. In this case, multiplication of larger numbers is no burden.

The bottom row in Table 3–3 shows the sums ($\Sigma$) for the frequency ($f$) column and for the $fx'$ column. The sum of the frequency column, symbolized by $\Sigma f$, equals 161, the value of $N$. The sum of the $fx'$ column is symbolized by $\Sigma fx'$. The numerical value of this sum for the example in Table 3–3 is $-87$, the algebraic sum of the numbers in the $fx'$ column. The positive values above the guessed mean interval are added to obtain a total of $+80$. The sum of the negative $fx'$ values is $-167$. The positive sum plus the negative sum gives $-87$, the algebraic sum of the entries in the $fx'$ column.

All the figures have now been obtained that are needed for substitution in Formula 3–2, the code method formula for the mean. These substitutions and subsequent calculations are shown at the bottom of Table 3–3. The value 74.5, the midpoint of the interval with $x' = 0$, is substituted for $M_G$; the sum of the $fx'$ column is substituted for $\Sigma fx'$; the sum of the frequency column is substituted for $N$; finally, the class interval width, $i = 10$, is substituted for $i$. The numerical calculations are carried out, remembering to carry the negative sign through for $\Sigma fx'$. The final value of the mean is rounded off to one decimal place.

*Another Example of the Mean by the Code Method.* A second example of computing the mean of grouped data by the code method is shown in Table 3–4. In this example, the guessed mean, $M_G$, is at 10.5,

TABLE 3–4
Computation of the Mean by the Code Method

| $X$ | $f$ | $x'$ | $fx'$ |
|-----|-----|------|-------|
| 18–19 | 1 | $4^9$ | $4^9$ |
| 16–17 | 6 | $3^8$ | $18^{48}$ |
| 14–15 | 10 | $2^7$ | $20^{70}$ |
| 12–13 | 15 | $1^6$ | $15^{90}$ |
| 10–11 | 18 | $0^5$ | $0^{90}$ |
| 8– 9 | 12 | $-1^4$ | $-12^{48}$ |
| 6– 7 | 9 | $-2^3$ | $-18^{27}$ |
| 4– 5 | 6 | $-3^2$ | $-18^{12}$ |
| 2– 3 | 3 | $-4^1$ | $-12^3$ |
| 0– 1 | 2 | $-5^0$ | $-10^0$ |
| $\Sigma$ | 82 | | $-13$ |

$$M = M_G + \frac{\Sigma fx'}{N} \cdot i \qquad\qquad (3\text{-}2)$$

$$M = 10.5 + \frac{(-13)}{82} \times (2) = 10.5 - \frac{26}{82}$$

$$M = 10.5 - 0.32 = 10.18$$

$$M = 10.2$$

the midpoint of the interval where the zero value of $x'$ was placed. The zero here was placed opposite the largest frequency. This choice for $M_G$ results in a negative total for the sum of the $fx'$ column, as in the example in Table 3–3. Had $M_G$ been placed at 8.5, one class interval lower, the sum of the positive $fx'$ values would have been larger than the sum of the negative $fx'$ values and the algebraic sum of the two would have been positive. It is desirable to have a positive sum of the $fx'$ column to reduce the probability of error due to a failure to carry the negative sign through in the formula calculations when $\Sigma fx'$ is negative.

## Summary of Steps for Calculating the Mean by the Code Method

1. Prepare the frequency distribution and add column headings $x'$ and $fx'$ to the right of the $X$ and $f$ headings.
2. Place a zero in the $x'$ column opposite some interval with a large frequency, near the center of the distribution.
3. Place the consecutive numbers 1, 2, 3, and so on, in the $x'$ column above the zero until the top class interval is reached. Place the num-

bers −1, −2, −3, and so on, in the $x'$ column below the zero until the bottom class interval is reached.

4. For each row, multiply the $f$ value by the $x'$ value and place the product in the $fx'$ column.

5. Sum the frequency column to get $N$ and take the algebraic sum of the $fx'$ column, i.e., taking into account the signs on the $fx'$ values.

6. Multiply the sum of the $fx'$ column by $i$, the class interval size, and divide this product by $N$.

7. Add the result obtained in step 6 algebraically to $M_G$, the midpoint of the interval with the zero in the $x'$ column.

## Mathematical Notes

This section may be omitted by the student who is willing to take Formula 3–2 on faith. The problems can be worked without reading this material which is devoted to a mathematical derivation for Formula 3–2.

When scores are placed in a frequency distribution, they lose their original identity, so to speak, and become lumped as though they were equal to each other score in their own class interval. The calculations for the mean in this case produce a value of the mean which is consistent with the assumption that the mean of all the scores in a class interval is the midpoint of the interval. If every score can be treated as though it were at the midpoint of its interval, then any score can be expressed as follows:

$$X = M_G + ix' \tag{3–3}$$

In Table 3–4, for example, the scores in the first interval are at 0.5, so, using Formula 3–3, $0.5 = 10.5 + (2)(−5) = 10.5 − 10 = 0.5$, as expected. Scores for the next interval are at 2.5, so, using Formula 3–3 again, $2.5 = 10.5 + (2)(−4) = 10.5 − 8 = 2.5$. Each other interval can be treated in the same way up to and including the top interval which gives $18.5 = 10.5 + (2)(4) = 10.5 + 8 = 18.5$. Since every score in grouped data may be treated as being at the midpoint of the interval within which it is contained, Formula 3–3 above can be substituted into Formula 3–1 as follows:

$$M = \sum \frac{X}{N} \tag{3–1}$$

Substituting Formula 3–3 in 3–1 gives:

$$M = \sum \frac{M_G + ix'}{N} \tag{3–4}$$

Since the sum of a sum of two terms equals the sum of the separate terms, Formula 3–4 becomes:

$$M = \frac{\Sigma M_G + \Sigma ix'}{N}$$

(3–5)

It has already been pointed out that the sum of a constant is just the sum of that constant $N$ times if the summation is from 1 to $N$. This means that $\Sigma M_G$ is equal to $M_G$ times $N$, since $M_G$ is just a constant, i.e., the guessed mean. Eq. 3–5 can be rewritten, therefore, as:

$$M = \frac{N M_G}{N} + \frac{\Sigma ix'}{N}$$

(3–6)

The first term on the right of Eq. 3–6 is just $M_G$, since the $N$ in the numerator cancels the $N$ in the denominator. The term on the right has the following equivalent forms:

$$\frac{\Sigma ix'}{N} = i\left(\frac{\Sigma x'}{N}\right) = i\left(\frac{\sum\limits_{j=1}^{N} x'_j}{N}\right)$$

$$= i\left(\frac{\sum\limits_{k=1}^{n} f_k x'_k}{N}\right) = i\left(\frac{\Sigma fx'}{N}\right) = \left(\frac{\Sigma fx'}{N}\right) \cdot i$$

(3–7)

Substituting the last term on the right of Eq. 3–7 into Eq. 3–6 and cancelling the $N$'s gives:

$$M = M_G + \frac{\Sigma fx'}{N} \cdot i$$

(3–2)

This is the formula for the mean by the code method which was to be derived in this section. An important part of the derivation is the proof that the various terms in Formula 3–7 are equivalent mathematically. In fact, the different terms in Formula 3–7 are merely different ways of expressing the same result. The term on the left in Formula 3–7, for example, indicates that a value of $x'$ for each person is to be multiplied by $i$; these values are to be summed and the total divided by $N$. No subscripts are shown but the summation from 1 to $N$ is implied. The first term on the right of Formula 3–7 is different from the one on the left only in that the constant, $i$, is moved outside the summation sign. If $i$ were 5, and the only values of $x'$ were: $-1$, 0, 0, 1, 1, 1, and 2, then the expression on the left would be:

$$(1/7) \cdot [5(-1) + 5(0) + 5(0) + 5(1) + 5(1) + 5(1) + 5(2)] = 20/7$$

The first expression on the right of Formula 3.7 for the same scores would be:

$$5 \cdot [(1/7)(-1 + 0 + 0 + 1 + 1 + 1 + 2)] = 5 \times (4/7) = 20/7$$

The second expression on the right of Formula 3–7 differs from the one immediately before it only in that the limits are placed on the summation sign and the subscript $j$ is placed on the $x'$ to indicate that it is a variable term taking on successive values for the various terms 1 through $N$.

The third term on the right of Formula 3–7 sums the terms in a different way from the preceding expressions. All the values of $x'$ that are equal, i.e., all those that are from the same class interval, are collected together and summed by multiplying their common numerical value by the frequency, $f_k$, the number of cases with the same value of $x'$, instead of summing all the $x'$ values separately regardless of their size.

For the few scores shown above, this new way of summing terms would give:

$$5 \cdot ((\tfrac{1}{7}) \cdot [1(-1) + 2(0) + 3(1) + 1(2)]) = {}^{20}\!\!/_{7}$$

There is only one value of $x' = -1$, so it is multiplied by a frequency of 1; there are two zeroes, so $x' = 0$ is multiplied by 2; there are three values of 1, so $x' = 1$ is multiplied by 3; finally, there is only one $x' = 2$, so 2 is multiplied by 1. The result is still the same, ${}^{20}\!\!/_{7}$.

Note that the limits on the summation for the third expression on the right of Formula 3–7 are from $k = 1$ to $k = n$. This is a small $n$, not a large $N$. The small $n$ here represents the number of class intervals. Thus, the number of terms in this summation equals the number of class intervals, i.e., the number of *different* values of $x'$. All the values of $x'$ for everyone in the same class interval are equal and the sum of $x'$ for cases in that interval is obtained merely by multiplying $x'$ for that interval by the interval frequency. The next to the last expression on the right of Formula 3–7 is identical to the previous one except that it merely drops the subscript on the $x'$ variable and the limits on the summation sign, leaving them implied rather than explicitly stated. The presence of the $f$ in the formula before the $x'$ indicates that the summation will be over $n$, the number of class intervals, not over $N$, the number of cases. Finally, the last expression on the right of Formula 3–7 differs from the immediately preceding one only in that $i$ has been moved from in front of the expression to the right-hand side, to coincide with the way it appears in the main formula, 3–2, for calculating the mean by the code method. The placement of the constant in a product of terms is arbitrary since multiplication of numbers is commutative, i.e., $a \times b = b \times a$.

## Exercises

3–1.  What is the mode of the frequency distribution in Table 3–3?

3–2.  What is the mode of the frequency distribution in Table 3–4?

3–3.  Find the median of the frequency distribution in Table 3–3.

3–4.  Find the median of the frequency distribution in Table 3–4.

3–5.   Find the median of the following scores: 3, 15, 15, 25, 25, 25, 60, 75, 100, and 200.

3–6.   Find the median for the following scores: 20, 21, 23, 23, 23, 26, 26, 29, 30, 31.

3–7.   Find the median for the scores: 400, 800, 35, 30, 35, 900, 200, 20, 20, 35, 14, and 35.

3–8.   Find the median of the scores: −11, −10, −8, −4, −4, 0, 1, 2, 2, 10, and 30.

3–9.   Compute the median for the frequency distribution given as the answer to Exercise 2–1 in the back of the book. Compute the median working up from the bottom and also working down from the top of the distribution.

3–10.   Calculate the median both from the bottom up and from the top down for the frequency distribution given as the answer to Exercise 2–3 in the back of the book.

3–11.   Find the mean of the scores: 4, 8, 10, 8, 7, 6, 12, 2, 9, 7, 8, and 8.

3–12.   Find the mean of the scores: 90, 110, 95, 98, 102, 108, 125, 68, 89, 112, 100, 92, and 94.

3–13.   Find the mean of the frequency distribution in Table 3–2 using the code method and setting the guessed mean in the interval 35–39. (Problem B)

3–14.   Find the mean of the frequency distribution in Table 3–1 using the code method and setting the guessed mean in the interval 87. (Problem B)

3–15.   Calculate the mean of the frequency distribution in Table 3–3 using the code method and setting the guessed mean in the interval 20–29. (Problem B)

3–16.   Calculate the mean of the frequency distribution in Table 3–4 using the code method and setting the guessed mean in the interval 0–1. (Problem B)

3–17.   Compute the mean by the code method for the distribution given as the answer to Exercise 2–1 in Appendix B; place the guessed mean in the interval 100–109. (Problem B)

3–18.   Compute the mean for the distribution given as the answer to Exercise 2–3 in Appendix B, placing the guessed mean in the interval 12–14. (Problem B)

# chapter 4

# Measures of Variability

The measures of central tendency described in the previous chapter will provide an indication of the general level of scores in a body of data but they will not give any indication of how spread out the scores are. Two distributions could both have a mean of 100, for example, but one could have a range of scores from 90 to 110 and the other could have a range of scores from 10 to 1000. It is just as important to have some method for determining the amount of spread in a group of scores as it is to have a method for determining the central tendency. The two most commonly used methods for determining the amount of spread in a body of scores, called "measures of variability," are the semi-interquartile range $(Q)$ and the standard deviation $(S.D.)$. This chapter will be concerned with the calculation of these statistics for both ungrouped and grouped data.

**Problem 7**
**DETERMINE THE SEMI-INTERQUARTILE RANGE $(Q)$**
**FROM UNGROUPED DATA**

Before presenting a formula for the semi-interquartile range $(Q)$ it will be necessary to define the three quartile points, $Q_1$, $Q_2$, and $Q_3$. $Q_2$ is easiest to understand at this point because it is identical to the median, i.e., it is the point below which 50 percent of the cases fall. $Q_1$ is the point below which 25 percent of the cases fall. $Q_1$ and $Q_3$ are needed in calculating $Q$, the semi-interquartile range. Note that $Q$, the semi-interquartile range, has no subscripts. It is a width, or distance, measure. $Q_1$, $Q_2$, and $Q_3$ have subscripts and represent specific points on the score scale, not widths or distances. $Q_1$ and $Q_3$ are just like $Q_2$

in that they are defined as points on the score scale below which a specified percentage of the cases fall. This suggests that $Q_1$ and $Q_3$ should be calculated in a manner very similar to that used for calculating $Q_2$ or the median (Mdn). This is in fact the case. If the methods for calculating the median, as described in Problems 4 and 5, are well understood, calculating $Q$ will be easy; if not, these methods should be reviewed before proceeding.

### Formula for the Semi-Interquartile Range (Q)

$$Q = \frac{Q_3 - Q_1}{2} \qquad (4\text{--}1)$$

Where

$Q$ = the semi-interquartile range

$Q_3$ = the third quartile point, the point below which 75 percent of the scores fall

$Q_1$ = the first quartile point, the point below which 25 percent of the scores fall

Inspection of Formula 4–1 reveals that $Q$ is just one half of the distance between $Q_3$ and $Q_1$, the third and first quartile points, respectively. It is for this reason that it is called the *semi-interquartile* range. $Q_3 - Q_1$ is the interquartile range but this distance is very rarely ever used as a measure of variability.

### Computation of Q for Ungrouped Data

To find $Q$, $Q_3$ and $Q_1$ must be found first and substituted into Formula 4–1. Since $Q_3$ and $Q_1$ are like $Q_2$, the median, in their definition, the same general method used for calculating the median of ungrouped data (Problem 4) will be used for finding $Q_3$ and $Q_1$ except that 75 percent and 25 percent, respectively, will play the same role in calculating $Q_3$ and $Q_1$ that 50 percent played in calculating the median, $Q_2$. It will be recalled that in calculating the Mdn, the first step was to find $N/2$, or 50 percent of the total number of cases, $N$. The remaining steps were devoted to finding the point with this many cases below it. In calculating $Q_3$ and $Q_1$, the first step will be to find 75 percent and 25 percent of the total number of cases, respectively. The remaining steps will be the same as in calculating the median, i.e. finding the point with that many cases below it in the distribution of scores.

*An Example.* Consider the following ungrouped scores, already arranged in order of increasing size:

1, 10, 25, 25, 30, 35, 35, 60, 60, 60, 75, 100.

The total number of cases is $N = 12$. If the value of the median, which is equal to $Q_2$, were being sought, it would be obtained as follows:

$$12 \times 0.50 = 6 \text{ (50 percent of } N)$$

$$Q_2 = \text{Mdn} = 34.5 + \frac{6-5}{2} \cdot (1)$$

$$= 34.5 + 0.50 = 35.00$$

The lower limit of the interval containing $Q_2$, the point below which 6 cases fall, was 34.5. Below 34.5 there were 5 cases, so $6 - 5 = 1$ represents the number of cases still needed from the interval. There are two cases in the interval so this value of 1 is divided by 2 to get the proportion of the score interval that must be counted to get up to $Q_2$. Since the score interval is only one point, this fraction of $\frac{1}{2}$ is multiplied by 1 and added to 34.5, the lower limit of the interval to get the value of $Q_2$. The only difference between calculating $Q_2$ and the median is that two decimal places are retained for $Q_2$ whereas for the median the result is rounded off to one decimal place. Carrying two decimal places for the quartile points insures sufficient accuracy to obtain a value of $Q$ correct to one decimal place when $Q$ is rounded off.

Calculating $Q_3$, the third quartile point for the sample data above, gives:

$$12 \times 0.75 = 9 \text{ (75 percent of } N)$$

$$Q_3 = 59.5 + \frac{9-7}{3} \cdot (1)$$

$$Q_3 = 59.5 + .67 = 60.17$$

For $Q_3$, the interval 59.5 to 60.5 holds the point below which 9 cases (75 percent of $N$) fall. Seven cases are below 59.5 and 3 cases lie in the interval so $(9 - 7)/3$ is the fraction of the interval width (1 point) that must be added to 59.5 to obtain $Q_3$.

For $Q_1$, the first quartile point, the calculations are:

$$12 \times 0.25 = 3 \text{ (25 percent of } N)$$

$$Q_1 = 24.5 + \frac{3-2}{2}$$

$$Q_1 = 24.5 + 0.50 = 25.00$$

Three cases must fall below $Q_1$ and the interval 24.5 to 25.5 must hold this point since below 24.5 there are two cases and below 25.5 there are four cases. Since two cases lie below 24.5 only $3 - 2$, or 1 case, is needed from this interval. Since there are two cases in this interval $\frac{1}{2}$ of the interval width, or $(\frac{1}{2}) \times 1$, must be added to 24.5, the lower limit of the interval, to obtain $Q_1$.

Substituting $Q_3$ and $Q_1$ in Formula 4–1 gives:

$$Q = \frac{60.17 - 25.00}{2} = \frac{35.17}{2}$$

$$Q = 17.585 = 17.6$$

In the calculations for $Q$, at least two decimal places are carried until $Q$ is obtained but in reporting the final answer, $Q$ is usually rounded off to one decimal place following the same rounding rules used for measures of central tendency.

## Interpreting Q

Many students have difficulty grasping the meaning of $Q$ because it is so abstract. When $Q$ is determined to be 17.6, the student asks, "Seventeen point six what?" The answer, "Points!", is often not very satisfying. To provide meaning to this figure, it is really necessary to consider several distributions, all with different values of $Q$. Then, this distribution, with $Q = 17.6$, can be compared with others and adjudged to be more or less variable, depending on the values of $Q$ for the other distributions. Distributions can be rank ordered as to the amount of variability of their scores by putting them in order of size with respect to $Q$. $Q$ permits, therefore, a comparison of one distribution with another but it does not permit any intuitively obvious statements about one distribution by itself.

There are a few statements that can be made which help some, although they are inclined to be somewhat cumbersome. For example, the interval $2Q$ will cover the middle 50 percent of the scores if the interval is properly located on the score scale. Of course, the proper location calls for placing the lower end of the interval at $Q_1$ and the upper end of the interval at $Q_3$. The statement then must be true because $Q_3$ has 75 percent of the cases below it and $Q_1$ has 25 percent below it so the middle 50 percent fall between $Q_3$ and $Q_1$ which is an interval of $2Q$ points.

If the distribution is symmetric, i.e. not skewed, then it can be stated that the interval Mdn $\pm Q$ (Median plus and minus $Q$) encompasses the middle 50 percent of the cases. A symmetric, or unskewed distribution, has no long tail on either side. If the distribution frequency polygon is cut in half by a vertical line, the left-hand side would be a mirror image of the right-hand side. This rule is not particularly useful because most distributions are not exactly symmetrical so the Mdn is not exactly half way between $Q_3$ and $Q_1$.

In the example above, $Q_1 = 25.00$, $Q_2 = 35.00$, and $Q_3 = 60.17$. The distance from $Q_1$ to $Q_2$ is 10 points while the distance from $Q_2$ to $Q_3$ is 25.17 points, over twice as great an interval. This shows that there is a longer tail on the right-hand side of the frequency polygon, or of the distribution, and hence the distribution is *positively* skewed. Thus, if $Q_3 - Q_2$ is greater than $Q_2 - Q_1$, the distribution is positively skewed, with a longer tail to the right. If $Q_2 - Q_1$ is greater than $Q_3 - Q_2$, the distribution is negatively skewed, with a longer tail to the left. The greater the difference between $Q_3 - Q_2$ and $Q_2 - Q_1$, the greater the skew in the distribution. If the distribution is symmetric, there will be no skew and $Q_3 - Q_2 = Q_2 - Q_1$, although finding $Q_3 - Q_2 = Q_2 - Q_1$ does not guarantee that the distribution will be symmetric. Thus $Q_3 - Q_2 = Q_2 - Q_1$ is a necessary but not sufficient condition to assure symmetry in the distribution. Examples of skewed distributions are shown in Figures 4–1 and 4–2.

FIGURE 4–1
A Positively Skewed Distribution

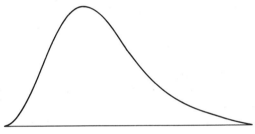

FIGURE 4–2
A Negatively Skewed Distribution

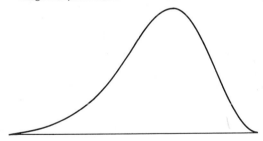

## When to Use Q

Q has about the same advantages as a measure of variability that the Mdn has as a measure of central tendency. That is, the semi-interquartile range is free of influence by the extreme scores. The scores below $Q_1$ and above $Q_3$ do not affect the size of Q at all, so if such extremes are exaggerated, inaccurate, or not precisely known, Q can still be computed and reported as a statistic relatively unaffected by these extraneous influences. The generally accepted rule is that where the median has been selected as the most appropriate measure of central tendency for describing a given body of data, $Q$ is usually selected as the most appropriate measure of variability.

## Problem 8
### DETERMINE THE SEMI-INTERQUARTILE RANGE (Q) FROM GROUPED DATA

The principal difference between computing $Q$ by the method described in Problem 7 and computing $Q$ from grouped data is that with Problem 7 the data are merely numbers arranged in order of increasing size whereas in Problem 8 the data are arranged in a frequency distribution. In both cases, $Q_1$ and $Q_3$ must be computed and substituted in Formula 4–1 to obtain $Q$, i.e. $Q = (Q_3 - Q_1)/2$. Whereas in Problem 7, $Q_3$ and $Q_1$ were computed by methods analogous to those used to compute the median $(Q_2)$ from ungrouped data (Problem 5), in Problem 8, $Q_3$ and $Q_1$ are computed by methods analogous to those used to compute the median $(Q_2)$ from grouped data (Problem 5). Regardless of the quartile point being located and the form in which the data are arranged, the task is to find a point on the score scale with a specified percentage of the cases below it. In each case, this is accomplished using the same basic method.

### Computing Q from Grouped Data

The method of finding the semi-interquartile range from grouped data will be illustrated by showing the computations to obtain $Q$ for the data in Table 3–1. First, computing $Q_3$, it is necessary to find the point below which 75 percent of the cases fall; this is $209 \times 0.75 = 156.75$. Thus, 156.75 cases must fall below $Q_3$ in the distribution of scores. Reference to the cumulative frequency column in Table 3–1 shows that 156 cases fall below 89.5, the upper limit of the interval 88.5 to 89.5. This is less than the 156.75 needed. Below 90.5, the upper limit of the next interval, there are 171 cases, which is more than 156.75. It follows that $Q_3$ must lie in the interval 89.5 to 90.5. Therefore:

$$Q_3 = 89.5 + \frac{156.75 - 156}{15} \cdot (1)$$

That is, $Q_3$ is the lower limit of the interval containing $Q_3$ plus the class interval multiplied by the ratio of the number of cases needed from this interval to the interval frequency. Continuing:

$$Q_3 = 89.5 + \frac{0.75}{15} = 89.5 + .05 = 89.55$$

$Q_1$, on the other hand, has only 25 percent of the scores below it, i.e. $209 \times .25 = 52.25$ cases. The cumulative frequency column in Table 3–1 shows 38 cases below 83.5 and 54 cases below 83.5 so:

$$Q_1 = 82.5 + \frac{52.25 - 38.00}{16} \cdot (1)$$

$$= 82.5 + \frac{14.25}{16} = 82.5 + .89 = 83.39$$

Substituting $Q_3$ and $Q_1$ into Formula 4–1 gives:

$$Q = \frac{Q_3 - Q_1}{2} = \frac{89.55 - 83.39}{2} = \frac{6.16}{2}$$

$$= 3.08 = 3.1$$

To obtain some indication concerning the degree of skew in the distribution, $Q_2$ may be computed as follows:

$$Q_2 = 86.5 + \frac{10.45 - 95}{25} \cdot (1) = 86.5 + \frac{9.5}{25} = 86.5 + 0.38 = 86.88$$

Then:

$$Q_3 - Q_2 = 89.55 - 86.88 = 2.67$$

$$Q_2 - Q_1 = 86.88 - 83.39 = 3.49.$$

Thus $Q_2 - Q_1$ is larger than $Q_3 - Q_2$ showing a longer tail toward the left and hence the frequency distribution in Table 3–1 is negatively skewed. The difference is less than one point, however, so the amount of skew relative to the range is very slight.

*A Second Example.* The example shown above involved a frequency distribution with a class interval of $i = 1$. $Q$ will be computed now for the frequency distribution in Table 3–2, where $i = 5$. Computing $Q_3$ and $Q_1$:

Number of cases below $Q_3 = 160 \times 0.75 = 120$.

$$Q_3 = 44.5 + \frac{120 - 109}{25} \cdot (5)$$

$$= 44.5 + \frac{55}{25} = 44.5 + 2.20 = 46.70$$

Number of cases below $Q_1 = 160 \times 0.25 = 40$

$$Q_1 = 19.5 + \frac{40 - 33}{23} \cdot (5) = 19.5 + \frac{35}{23} = 19.5 + 1.52 = 21.02$$

Then:

$$Q = \frac{Q_3 - Q_1}{2} = \frac{46.70 - 21.02}{2} = \frac{25.68}{2} = 12.84$$

$$Q = 12.8$$

The distribution in Table 3–2 is much more variable than the distribution in Table 3–1 since $Q$ for the latter was only 3.1 while for the former it is 12.8. This is also apparent in comparing the ranges for the two distributions, 17.0 and 66.0, respectively.

## Problem 9
### COMPUTE THE STANDARD DEVIATION (S.D.) BY THE DEVIATION SCORE METHOD

The standard deviation is a measure of variability that is affected by every score in the distribution. A deviation score, i.e. $x = X - M$, is computed for every raw score, $X$, in the distribution. These deviation scores are squared to get rid of the negative signs, since all raw scores below the mean have negative deviation scores. Then, the squared deviation scores are averaged. The square root of this average squared deviation score is the standard deviation (S.D.) of the distribution. A score far away from the mean will have a large squared deviation score, elevating the S.D. A score close to the mean will have a small squared deviation score, tending to restrict the size of the S.D. This definition of the standard deviation is expressed symbolically by the formula below.

### Deviation-Score Formula for the Standard Deviation (S.D.)

$$S.D. = \sqrt{\frac{\Sigma x^2}{N}} \tag{4-1}$$

Where

$$S.D. = \text{the standard deviation}$$

$$x = \text{a deviation score, } X - M$$

$$N = \text{the number of scores in the distribution}$$

For simplicity, in Formula 4–1, the limits have been omitted from the summation sign and the subscript from the variable term, $x$. With these restored $\Sigma x^2$ would become $\sum_{i=1}^{N} x_i^2$, or with a different subscript, $\sum_{j=1}^{N} x_j^2$. In Formula 4–1 it is understood without being explicitly stated that the summation is from 1 to $N$ and that $x$ is a variable term.

### Computing the S.D. by the Deviation Score Formula

Formula 4–1 represents the simplest statement of the definition for the standard deviation and is therefore typically employed to represent it in algebraic formulations. Formula 4–1 is not a good formula to use in computing the standard deviation, however, since it requires the actual computation and squaring of deviation scores. Deviation scores are typically decimal fractions, since $M$ is usually reported to one decimal place, and hence are difficult to work with. Problems 10 and C will present more practical methods for computing the S.D., but the method of computing the size of the S.D. by Formula 4–1 will be shown here with a simple example to insure that its meaning is clear.

Consider the following raw scores ($X$ values):

$$11, 12, 10, 9, 9, 5, 14, 13, 7, \text{ and } 10.$$

Adding these scores gives $\Sigma X = 100$ and hence $M = 10$. Computing the deviation scores ($x$ values) corresponding to these 10 raw scores gives:

$$1, 2, 0, -1, -1, -5, 4, 3, -3, \text{ and } 0.$$

If these deviation scores are added up, the sum is zero. This will always be true if the computations have been done correctly. In fact, the mean may be defined as the point about which the sum of the deviation scores equals zero.

Formula 4–1 calls for the sum of the squared deviations, however, rather than for the sum of the deviations themselves, so the squared deviation scores for the example above are:

$$1, 4, 0, 1, 1, 25, 16, 9, 9, \text{ and } 0.$$

Summing these values gives a total of 66. Substituting this total and $N = 10$ in Formula 4–1 gives S.D. $= \sqrt{66/10} = \sqrt{6.6} = 2.57$ which becomes 2.6 when rounded off to 1 decimal place.

*Determining the Square Root Using Table A.*    Whenever an S.D. is calculated, a square root must be obtained. If no electronic calculator is available that takes square roots, the square root may be obtained by using tables or it may be calculated. Table A in Appendix A will provide the square root for any number from 1 to 999. These square roots can also be used to obtain the square roots of certain other numbers that are not given in the table. For example, the square root of 316 is 17.7764. This information can be used to obtain the square root of 31600 as follows:

$$31600 = 316 \times 100 = 316 \times 10^2;$$

$$\sqrt{316 \times 10^2} = 10\sqrt{316} = 10 \times 17.7764 = 177.764.$$

Or, suppose the square root of 3.16 is desired. Then,

$$3.16 = 316 \div 100 = 316 \div 10^2;$$

$$\sqrt{\frac{316}{10^2}} = \frac{1}{10}\sqrt{316} = \frac{1}{10}(17.7764) = 1.7764.$$

The rule can be stated as follows: To obtain the square root of a number 100 times as large as a number given in the table, *multiply* the tabled square root value by 10 (move the decimal point one place to the right); to obtain the square root of a number one hundredth as large as the number given in the table, *divide* the tabled square root value by 10 (move the decimal point one place to the left). Square roots of numbers 10,000 times as large or $1/10,000$ of the tabled numbers are obtained by moving the decimal points in the tabled square roots two places to the right or to the left. Thus, the square root of 561 is 23.6854; the square root of 5,610,000 is 2368.54; and the square root of .0561 is .236854. Note, however, that there is no simple relationship between the square root of a number and a number 10 times as large. The square root of 25 is 5 and the square root of 250 is 15.8114, which is not 5 × 10. Actually the square root of $250 = \sqrt{25 \times 10} = \sqrt{25} \times \sqrt{10} = 5 \times 3.1623 = 15.8115$. This differs from the tabled square root of 250 only by 1 in the last place, due to rounding error. To obtain square roots by shifting decimal points of tabled square roots, therefore, the number desired must differ from the tabled number by multiples of 100, not multiples of 10.

Usually, however, the number for which a square root is desired is not tabled and cannot be obtained exactly by multiplying or dividing a tabled number by multiples of 100. For example, suppose the square root of 4572 is desired. The square root of 4500 is obtained from the square root of 45 as 6.7082 × 10 = 67.082 and the square root of 4600 is obtained from the square root of 46 as 6.7823 × 10 = 67.823. The in-between value, however, cannot be obtained from the table. The in-between value can be obtained to one decimal place accuracy without undue trouble through trial and error using the bracketing numbers 4500, 4600, and their square roots. Looking at the position of 4572 between 4500 and 4600, and considering their rounded off square roots of 67.1 and 67.8, it would be reasonable to expect the square root of 4572 to be 67.5, 67.6, or 67.7, correct to one decimal place. Multiplying out these three possibilities gives 4556.25, 4569.76, and 4583.29. Since 4569.76, the square of 67.6, is closer to 4572 than the squares of 67.5 and 67.7, 67.6 is taken as the square root correct to one decimal place. The tabled square roots with adjusting decimal points on the square roots, and some trial and error adjusting of the roots, will serve adequately to obtain the square root in many cases. This trial and error process works well with an electronic calculator that does not have the capability to extract square roots directly. The square root also may be obtained by an arithmetic process described in the Mathematical Notes section below.

## Variance

The term "variance" appears frequently in statistical material, unfortunately with varying connotations. The simplest and most direct usage of the term is that in the following sentence: "The variance of this sample of scores is 4.0." In this usage, the "variance" of the distribution is just the square of the standard deviation. Thus, when S.D. = 2, the variance, S.D.$^2$, equals 4. Other usages of this term will be encountered in connection with later problems.

## Mathematical Notes

A method for extracting square roots will be reviewed here without proof. The square root of 1456.25 will be calculated as an example (See Table 4-1):

1. Starting with the decimal point separate the digits in this number by groups of 2, as shown in Table 4-1.
2. Place above the left hand pair of digits (14) the largest number for which the square is less than or equal to 14. Three is the largest such number in this example, because 4 squared is 16, or larger than 14. Square 3 and subtract from 14 and also bring down the next two digits, 56.
3. Double the number above the line at this point (3) and bring it down on the same line as the 556, but allow space for a second digit to the right of the 6 and put the two digit number in parentheses. The number to go in the blank space, which is underlined, has yet to be determined.
4. The same number will go in the underlined blank space to the right of the 6 as goes above the second group of two digits. It must be the largest number such that when multiplied times the number in parentheses, the result will still be less than or equal to 556. Nine times 69 would give 621 which is too large. Seven times 67 would give 469 which is more than 67 less than 556, so the correct number is 8, as shown.
5. Subtract 8 times 68, or 544, from 556 and bring down the next two digits, 25, to give a remainder of 1225.
6. Double the 38 above the line to get 76 and put the 76 in parentheses on the same line as the remainder 1225, but leave an underlined blank space to the right of the 76 for the next digit of the square root to be determined.
7. Determine the number to go above the next pair of digits, 25; this number will also go in the blank space to the right of the 76 in parentheses at the left. The correct number is 1 here because

TABLE 4–1
Extracting the Square Root Arithmetically

|  | Extraction | | | | | | Check |
|---|---|---|---|---|---|---|---|
|  | 3 | 8 . 1 | 6 | 0 | 8 |  | 38.1608 |
|  | √14 | 56 . 25 | 00 | 00 | 00 |  | 38.1608 |
|  | 9 | | | | | | 3052864 |
| (6**8**) | 5 | 56 | | | | | 22896480 |
|  | 5 | 44 | | | | | 381608 |
| (76**1**) | | 12 25 | | | | | 3052864 |
|  | | 7 61 | | | | | 1144824 |
| (762**6**) | | 4 | 64 00 | | | | 1456.24665664 |
|  | | 4 | 57 56 | | | | |
| (7632**0**) | | | 6 44 00 | | | | |
|  | | | 00 | | | | |
| (76320**8**) | | | 6 44 00 00 | | | | |
|  | | | 6 10 56 64 | | | | |

|  | Extraction | | | | | | Check |
|---|---|---|---|---|---|---|---|
|  | 2 | 0 | 6 . 5 | 1 | 5 |  | 206.515 |
|  | √4 | 26 | 48 . 51 | 00 | 00 |  | 206.515 |
|  | 4 | | | | | | 1032575 |
| (4**0**) | 0 | 26 | | | | | 206515 |
|  | | 00 | | | | | 1032575 |
| (40**6**) | | 26 | 48 | | | | 1239090 |
|  | | 24 | 36 | | | | 4130300 |
| (412**5**) | | | 2 12 51 | | | | 42648.445225 |
|  | | | 2 06 25 | | | | |
| (4130**1**) | | | 6 26 00 | | | | |
|  | | | 4 13 01 | | | | |
| (41302**5**) | | | 2 12 99 00 | | | | |
|  | | | 2 06 51 25 | | | | |

1 times 761 is the largest number to go into 1225. Two times 762 would give too large a number. Subtract 761 from 1225 and bring down the next two digits, zeroes in this case, to get the remainder 46400.

8. Double the 381 to get 762 and put it in parentheses on the same line as the remainder, leaving a blank underlined space for the next digit of the square root.

9. If a 6 is placed above the next pair of digits, 00, and in the blank space to the right of 762, the product will give 45756 which is subtracted from 46400 to give the next remainder 64400 when the next two digits (00) are brought down. Seven times 7627 would be larger than 46400 and 6 times 7626 would be too small.

10. Double 3816 to get 7632 and put it in parentheses on the remainder

line with an underlined blank space to the right for the next digit of the square root. The correct next digit is 0 because even 1 times 76321 is larger than the remainder 64400. A zero, therefore, is placed above the next pair of digits to the right of 6 as the next digit of the square root and also in the underlined blank space to the right of the 7632. Zero is multiplied by 76320 to get zero which is subtracted from 64400 to get the next remainder when the next two digits are brought down.

11.   Double 38160 to get 76320 and place this number in parentheses on the same line as the remainder having an underlined blank space to the right. This will be filled by the digit 8 which is also placed above the last pair of digits as the last digit of the square root since 8 times 763208 is near to but less than the remainder. Nine times 763208 would be too large and 7 times 763207 would leave a remainder larger than 763207.

12.   The square root, 38.1608 is multiplied out to check its accuracy, giving 1456.24665664 which rounds off to 1456.25, the number for which the square root was sought. This process could have been continued to obtain the square root to more decimal places.

When the number is partitioned into digit pairs on either side of the decimal point, the extreme left hand "pair" will sometimes consist of one digit instead of two. An example of a worked-out square root fitting this case is shown in Table 4–2. This square root is computed only to three decimal places.

## Problem 10
### COMPUTE THE STANDARD DEVIATION (S.D.) BY THE RAW SCORE METHOD

Formula 4–1 is a conceptually simple formula but not a computationally simple one for calculating the standard deviation. Formula 4–1 involves deviation scores $(x)$ which are obtained by subtracting the mean $(M)$ from the raw scores $(X)$. As pointed out in the previous problem, it is computationally laborious to obtain and then square deviation scores. It is more convenient to compute the standard deviation using a formula that involves only raw scores.

### Raw Score Formula for the Standard Deviation (S.D.)

$$\text{S.D.} = \frac{1}{N} \sqrt{N\Sigma X^2 - (\Sigma X)^2} \qquad (4\text{–}2)$$

Where

$$X = \text{a raw score}$$
$$N = \text{the number of scores}$$
$$\text{S.D.} = \text{the standard deviation}$$

Formula 4–2 is mathematically equivalent to Formula 4–1, but all the $X$ values are upper case, i.e., $X$ not $x$, representing raw scores, not deviation scores. Thus it is merely necessary to obtain the sum of the raw scores, $\Sigma X$, and the sum of the squares of the raw scores, $\Sigma X^2$. Note that $\Sigma X^2$ means the raw scores are squared and *then* summed; $(\Sigma X)^2$, i.e. the sum of $X$ the quantity squared, indicates that the $X$ scores are summed first and then the entire quantity is squared. As in Formula 4–1 the limits on the summation signs in Formula 4–2 are implied rather than being explicitly stated. The limits on the summation are from 1 to $N$ and $X$ is a variable term rather than a constant.

## Computing the S.D. by the Raw Score Formula

To show that the same value for the S.D. is obtained whether it is computed by the deviation score formula or the raw score formula, the same data used for demonstration purposes in Problem 9 will also be used here. The raw scores, to repeat, were:

11, 12, 10, 9, 9, 5, 14, 13, 7, and 10.

The sum of these scores is $\Sigma X = 100$. The squared scores, $X^2$, are:

121, 144, 100, 81, 81, 25, 196, 169, 49, and 100,

respectively. Summing these squared raw scores gives $\Sigma X^2 = 1066$. The number of cases, $N$, is 10. Substituting these values in Formula 4–2 gives:

$$\text{S.D.} = \frac{1}{10} \sqrt{10(1066) - (100)^2}$$

$$= \frac{1}{10} \sqrt{10660 - 10{,}000}$$

$$= \frac{1}{10} \sqrt{660}$$

$$= \sqrt{\frac{660}{100}} = \sqrt{6.6} = 2.57$$

This is the same result that was obtained in applying the methods of Problem 9 to the same data.

With the small amount of data involved here and an integral value of the mean, computation of deviation scores was simple for this problem resulting in little difference between the methods of Problem 9 and 10 as far as the amount of work involved. For most examples, however, the methods of Problem 10 involve substantially less work. This is particularly true if a desk calculator or computer is available to square the raw scores and accumulate the sum of $X$ and the sum of $X^2$. Then, computing the S.D. is merely a matter of plugging these totals and $N$ into Formula 4–2. There are some desk computers now available that will compute the S.D. using Formula 4–2 once the raw scores have been entered. The operator need merely push a button.

## Summary of Steps for Calculating the Standard Deviation (S.D.) by the Raw Score Method

1. Sum the raw scores to get $\Sigma X$.
2. Sum the square raw scores to get $\Sigma X^2$.
3. Square $\Sigma X$, the sum of the raw scores, to get $(\Sigma X)^2$.
4. Multiply the sum of the squared raw scores by $N$, the number of cases, to get $N\Sigma X^2$.
5. Subtract 3 from 4 to get $N\Sigma X^2 - (\Sigma X)^2$.
6. Take the square root of 5 to get $\sqrt{N\Sigma X^2 - (\Sigma X)^2}$.
7. Divide 6 by $N$ to get the S.D.

### Mathematical Notes

This section may be skipped by the reader who has no immediate interest in the origin of Formula 4–2. It will be shown here that Formula 4–2 can be derived algebraically from Formula 4–1.

$$\text{S.D.} = \sqrt{\frac{\Sigma x^2}{N}} \tag{4–1}$$

but $x = X - M$, hence

$$\text{S.D.} = \sqrt{\frac{\Sigma(X - M)^2}{N}}$$

$$\text{S.D.} = \sqrt{\frac{\Sigma(X^2 - 2MX + M^2)}{N}}$$

$$= \sqrt{\frac{(\Sigma X^2 - \Sigma(2MX) + \Sigma M^2)}{N}}$$

Constants may be placed in front of the summation sign in the sum of a product of constant and variable terms; also, the summation of a constant term is $N$ times the constant. The expression above, therefore, may be rewritten as:

$$\text{S.D.} = \sqrt{\frac{(\Sigma X^2 - 2M\Sigma X + NM^2)}{N}}$$

Multiplying numerator and denominator under the radical by $N$ and removing $1/N^2$ from under the radical gives:

$$\text{S.D.} = \frac{1}{N} \sqrt{N\Sigma X^2 - 2(NM)\Sigma X + (NM)^2}$$

but $\text{NM} = N(\Sigma X/N) = \Sigma X$, so

$$\text{S.D.} = \frac{1}{N} \sqrt{N\Sigma X^2 - 2(\Sigma X)(\Sigma X) + (\Sigma X)^2}$$

$$= \frac{1}{N} \sqrt{N\Sigma X^2 - 2(\Sigma X)^2 + (\Sigma X)^2}$$

$$\text{S.D.} = \frac{1}{N} \sqrt{N\Sigma X^2 - (\Sigma X)^2}$$

## Problem C
## COMPUTE THE STANDARD DEVIATION (S.D.) BY THE CODE METHOD

Since small electronic calculators have become so widely available, there is less need for the labor saving code methods. The code method for computing the standard deviation has been placed, therefore, in the letter series of problems. Letter series problems may be omitted without interfering with the study of the number series problems.

The methods used in Problems 9 and 10 for calculating the standard deviation are not very satisfactory when no computing machinery is available and the sample contains more than a few scores. The method to be described here is a labor saving method which can be applied when the data have been arranged in a frequency distribution. Computing equipment can even be used to expedite the computations with this method, although the method of Problem 10 would ordinarily be used in preference to this one where such equipment is available. Probably the main exception to this rule would be the case where the frequency distribution is needed for some other purpose anyway. In that case, it would be easier to obtain the S.D. by the code method from the frequency distribution rather than from the raw scores using the raw score formula. It should be remembered, however, that grouping the data in class intervals results in some loss of information (unless $i = 1$) so that the S.D. obtained by the code method from a frequency distribution is an approximation. The S.D. computed from raw scores by the raw score method, on the other hand, is exact and not an estimate.

The errors of grouping will tend to make the S.D. computed by the code method slightly too large. The reason for this is that all the scores in an interval are considered as being concentrated at the midpoint of the interval. In most cases, however, since scores tend to pile up in the middle of the distribution and thin out at the extremes, the mean of the scores in any given interval will tend to be closer to the overall mean of the distribution than will the midpoint of the interval. This will give slightly larger than justifiable negative deviation scores below the mean and slightly larger than justifiable positive deviation scores above the mean. Since these deviation scores are squared, the negative signs disappear and all squared deviations tend to be slightly too large and positive, spuriously increasing the size of the S.D. In computing the mean by the code method these errors tend to balance each other out hence no systematic distortion in the computed mean value occurs. The extent of the error in the S.D. computed by the code method is not practically important in most situations. If the investigator has reason to believe it is important to obtain a precisely accurate S.D., he should compute the S.D. by the raw score formula from raw scores instead of using the code method and the frequency distribution.

## Code Method Formula for the Standard Deviation (S.D.)

$$S.D. = i\sqrt{\frac{\Sigma fx'^2}{N} - \left(\frac{\Sigma fx'}{N}\right)^2}$$   (4–3)

Where   $= \frac{i}{N}\sqrt{N\Sigma fx'^2 - \left(\Sigma fx'\right)^2}$

S.D. = the standard deviation of a sample of scores

$i$ = the class interval size

$x'$ = a deviation from the guessed mean in terms of class interval units

$N$ = the number of scores in the sample

The terms in this formula are analogous to those used in Problem B to compute the mean by the code method. The main difference is the appearance in this formula of $x'^2$, i.e. a square of $x'$, the deviation score computed about the guessed mean $(M_G)$ in numbers of class intervals. Note that the $f$ is not squared; only the $x'$ is squared.

## Computation of the S.D. by the Code Method

The computation of the S.D. by the code method will be illustrated by a worked-out example, shown in Table 4–2. The same frequency distribution was used in Table 3–3 to illustrate the computation of the mean by the code method. Re-examination of Table 3–3 in comparison with Table 4–2 will reveal that an additional column, headed by $fx'^2$, is required in computing the S.D. The other columns in the Table 4–2 are identical to those required for computing the mean by the code

TABLE 4–2
Computation of the S.D. by the Code Method

| $X$ | $f$ | $x'$ | $fx'$ | $fx'^2$ |
|---|---|---|---|---|
| 110–119 | 3 | 4 | 12 | 48 |
| 100–109 | 8 | 3 | 24 | 72 |
| 90– 99 | 12 | 2 | 24 | 48 |
| 80– 89 | 20 | 1 | 20 | 20 |
| 70– 79 | 36 | 0 | 0 | 0 |
| 60– 69 | 33 | −1 | −33 | 33 |
| 50– 59 | 24 | −2 | −48 | 96 |
| 40– 49 | 16 | −3 | −48 | 144 |
| 30– 39 | 7 | −4 | −28 | 112 |
| 20– 29 | 2 | −5 | −10 | 50 |
| $\Sigma$ | 161 | | −87 | 623 |

$$\text{S.D.} = i\sqrt{\frac{\Sigma fx'^2}{N} - \left(\frac{\Sigma fx'}{N}\right)^2} \tag{4–3}$$

$$\text{S.D.} = 10\sqrt{\frac{623}{161} - \left(\frac{-87}{161}\right)^2} = 10\sqrt{3.870 - (0.540)^2}$$

$$\text{S.D.} = 10\sqrt{3.870 - 0.292} = 10\sqrt{3.578}$$

$$\text{S.D.} = 10(1.892) = 18.92 = 18.9$$

method, as shown in Table 3–3. The last row shows the same sum of the frequencies, i.e., $N = 161$, and the same algebraic sum of the $fx'$ column values, −87. The sum of the values in the $fx'^2$ column, i.e. $\Sigma fx'^2$, is equal to 623. Note that all these $fx'^2$ values are positive.

The $fx'^2$ values are obtained by multiplying in each row the values in the $x'$ and $fx'$ columns, since $f$ times $fx'$ gives $fx'^2$. Thus, in row 1, 4 times 12 equals 48. In row 2, 3 times 24 equals 72; and finally, for the last row, −5 times −10 equals 50; the negative signs cancel each other, giving a positive product.

Given the values of $i$, $N$, $\Sigma fx'$, and $\Sigma fx'^2$, it is merely necessary to insert them in the formula for computing the S.D. by the code method. Note that the sum of the $fx'$ column is divided by $N$ and this result is squared in the right-hand term under the radical in Formula 4–3. The superscript, or power, 2, on the right-hand term under the radical in Formula 4–3 applies to the entire term within the parentheses, not just to the $x'$. In the left-hand term under the radical, however, only the $x'$ is squared and not the entire term.

This method is based upon the idea of obtaining the sum of squared deviation scores from a computationally more convenient reference point, i.e., $M_G$, than the actual mean which is usually a decimal fraction. The guessed mean, $M_G$, is a value that deviates from the midpoints of the intervals only in whole number units, i.e. $-6, -3, 0, 1, 5$, etc. The sum of the squared deviations can be obtained for these "coded" scores, in units of $i$ points, and then divided by $N$ to obtain an average squared coded score deviation. This average coded value can be corrected, using the right-hand term under the radical in Formula 4–3, to take into account the fact that deviations were taken from $M_G$ instead of $M$. If it were to happen that $M_G$ and $M$ were identical, then $\Sigma fx'$ would equal zero and this correction would be zero. After taking the square root of the corrected average squared coded score deviation, the result is rescaled from class interval units back to raw score units to get the value of the S.D.

*A Second Example.* To give another example of the calculation of the standard deviation by the code method, computations are shown in Table 4–3 for the same frequency distribution used to illustrate the calculation of the mean by the code method in Table 3–4.

TABLE 4–3
Computation of the S.D. by the Code Method

| $X$ | $f$ | $x'$ | $fx'$ | $fx'^2$ |
|---|---|---|---|---|
| 18–19 | 1 | 4 | 4 | 16 |
| 16–17 | 6 | 3 | 18 | 54 |
| 14–15 | 10 | 2 | 20 | 40 |
| 12–13 | 15 | 1 | 15 | 15 |
| 10–11 | 18 | 0 | 0 | 0 |
| 8– 9 | 12 | −1 | −12 | 12 |
| 6– 7 | 9 | −2 | −18 | 36 |
| 4– 5 | 6 | −3 | −18 | 54 |
| 2– 3 | 3 | −4 | −12 | 48 |
| 0– 1 | 2 | −5 | −10 | 50 |
| $\Sigma$ | 82 | | −13 | 325 |

$$\text{S.D.} = i \sqrt{\frac{\Sigma fx'^2}{N} - \left(\frac{\Sigma fx'}{N}\right)^2} \qquad (4\text{–}3)$$

$$\text{S.D.} = 2 \sqrt{\frac{325}{82} - \left(\frac{-13}{82}\right)^2}$$

$$\text{S.D.} = 2 \sqrt{3.963 - (0.159)^2} = 2 \sqrt{3.963 - 0.025}$$

$$\text{S.D.} = 2 \sqrt{3.938} = 2(1.984) = 3.968$$

$$\text{S.D.} = 4.0$$

## Mathematical Notes

This section may be skipped by the reader who has no immediate interest in the origin of Formula 4–3. It will be shown here that Formula

4–3 can be derived algebraically from Formula 4–1. Since all scores in a frequency distribution are treated as though they fall at the midpoint of the interval in which they are contained, any raw score, $X$, may be represented as follows:

$$X = M_G + ix'$$

Then, a deviation score, $x$, is given by:

$$x = X - M = [M_G + ix'] - \left[ M_G + \frac{\Sigma fx'}{N} \cdot i \right]$$

substituting the formula for the mean by the code method on the right side instead of $M$. The expression above simplifies, after cancelling the $M_G$ values, to:

$$x = ix' - \frac{\Sigma fx'}{N} \cdot i = i\left( x' - \frac{\Sigma fx'}{N} \right)$$

Substituting this expression for $x$ in Formula 4–1,

$$\text{S.D.} = \sqrt{\frac{\Sigma x^2}{N}}$$

gives:

$$\text{S.D.} = \sqrt{\frac{\Sigma \left[ i\left( x' - \frac{\Sigma fx'}{N} \right) \right]^2}{N}}$$

$$\text{S.D.} = \sqrt{\frac{i^2 \Sigma \left[ x'^2 - 2x'\left( \frac{\Sigma fx'}{N} \right) + \left( \frac{\Sigma fx'}{N} \right)^2 \right]}{N}}$$

$$\text{S.D.} = i\sqrt{\frac{\Sigma x'^2 - 2\left( \frac{\Sigma fx'}{N} \right)(\Sigma x') + N\left( \frac{\Sigma fx'}{N} \right)^2}{N}}$$

since the constant term $\Sigma fx'/N$ can be moved out in front of the summation sign in the middle term above and the sum of a constant is $N$ times the constant in the right hand term above. But,

$$\sum x'^2 = \sum_{i=1}^{N} x'_i{}^2 = \sum_{i=1}^{n} f_i x'_i{}^2 = \sum fx'^2$$

and

$$\sum x' = \sum_{i=1}^{N} x'_i = \sum_{i=1}^{n} f_i x'_i = \sum fx'$$

where $n$ = the number of class intervals. These relationships hold because in adding up the $x'$ and $x'^2$ values one number is added for each of $N$ persons. It does not matter if the numbers are grouped into collections of identical values by class interval, multiplying the common value by the frequency for the class interval, or whether the values are all added separately. The last expression above for the S.D. can be rewritten, therefore, as:

$$S.D. = i \sqrt{\frac{\Sigma fx'^2 - 2\left(\frac{\Sigma fx'}{N}\right)(\Sigma fx') + N\left(\frac{\Sigma fx'}{N}\right)^2}{N}}$$

$$S.D. = i \sqrt{\frac{\Sigma fx'^2}{N} - 2\left(\frac{\Sigma fx'}{N}\right)\left(\frac{\Sigma fx'}{N}\right) + \left(\frac{\Sigma fx'}{N}\right)^2}$$

$$S.D. = i \sqrt{\frac{\Sigma fx'^2}{N} - \left(\frac{\Sigma fx'}{N}\right)^2} \qquad (4\text{--}3)$$

*Effect on* M *and S.D. of Adding or Subtracting a Constant.* Let C be any constant, positive or negative. If C is added to every score, the mean will be affected as follows:

$$M = \frac{\Sigma X}{N}$$

Adding a constant, C, to every score gives

$$M = \frac{\Sigma(X + C)}{N} = \frac{\Sigma X}{N} + \frac{\Sigma C}{N}$$

$$M = M + \frac{\Sigma C}{N}$$

but the sum of a constant is N times the constant, so

$$M = M + \frac{NC}{N} = M + C$$

Adding a constant to every score, therefore, will give a set of scores with a mean equal to the old mean plus the constant.

Adding a constant to every score has no effect on the standard deviation, however, as shown by the following:

$$S.D. = \sqrt{\frac{\Sigma x^2}{N}} = \sqrt{\frac{\Sigma(X - M)^2}{N}}$$

Adding a constant to every score gives

$$S.D. = \sqrt{\frac{\Sigma[X + C - (M + C)]^2}{N}}$$

$$S.D. = \sqrt{\frac{\Sigma(X + C - M - C)^2}{N}} = \sqrt{\frac{\Sigma(X - M)^2}{N}} = \sqrt{\frac{\Sigma x^2}{N}}$$

which is the same as the S.D. of the scores before the constant was added.
These relationships can be utilized to ease the labor in computing

the $M$ and S.D. If all the scores in a distribution are between 101 and 130, for example, it is possible to subtract 100 from every score before computing $M$ and S.D. The S.D. will not be changed by this scaling process. It will only be necessary to add 100 points to the mean of the scaled scores to get the mean of the original unscaled scores.

*Effect on* M *and* S.D. *of Multiplying by a Constant.* If every score in a distribution is multiplied by a constant, $C$, the effect on the mean will be as follows:

$$M = \frac{\Sigma X}{N}$$

Multiplying each score by a constant, $C$, gives:

$$\text{Scaled score } M = \frac{\Sigma CX}{N} = \frac{C\Sigma X}{N} = CM$$

Thus, the effect of multiplying each score by $C$ is to multiply the mean by $C$, too. This also implies that if each score is divided by $C$, the mean will be divided by $C$.

For the standard deviation,

$$\text{S.D.} = \sqrt{\frac{\Sigma x^2}{N}} = \sqrt{\frac{\Sigma(X - M)^2}{N}}$$

multiplying each score by a constant, $C$, gives

Scaled score S.D.

$$= \sqrt{\frac{\Sigma(CX - CM)^2}{N}} = \sqrt{\frac{\Sigma C^2(X - M)^2}{N}} = \sqrt{\frac{C^2\Sigma(X - M)^2}{N}}$$

$$\text{Scaled score S.D.} = C\sqrt{\frac{\Sigma(X - M)^2}{N}} = C\sqrt{\frac{\Sigma x^2}{N}} = C(\text{S.D.})$$

Thus, multiplying every score by a constant, $C$, will multiply the standard deviation by the same constant.

This relationship will occasionally permit further economies in computing the $M$ and S.D. If all the scores in a distribution are multiples of 10 and the smallest is 430, subtract 430 from every score and divide the difference scores by 10, i.e., $X_s = (X - 430)/10$. Having computed the mean and S.D. of these scaled scores, the $M$ and S.D. of the unscaled scores are obtained as follows:

$$\text{Unscaled } M = 10(\text{Scaled Mean}) + 430$$

$$\text{Unscaled S.D.} = 10(\text{Scaled S.D.})$$

## Exercises

4–1.  Find the semi-interquartile range, $Q$, for the following scores: 3, 10, 10, 20, 25, 25, 25, 31, 31, 31, 40, 50.

4–2. Find the semi-interquartile range, $Q$, for the following scores: 100, 110, 110, 119, 130, 135, 135, 140, 140, 145, 145, 150, 160.

4–3. Find $Q$ for the following scores: 2, 4, 4, 4, 4, 8, 8, 8, 8, 12, 12, 12, 13, 14.

4–4. Find $Q$ for the following scores: 30, 35, 40, 42, 42, 42, 42, 42, 50, 55, 60, 65, 65, 70, 71.

4–5. Compute $Q$ for the frequency distribution given as the answer to Exercise 2–1 in the back of the book. Compute $Q_1$ working up from the bottom and $Q_3$ working down from the top. Is the distribution skewed? How? Why?

4–6. Compute $Q$ for the frequency distribution given as the answer to Exercise 2–3 in the back of the book. Compute $Q_1$ working up from the bottom and $Q_3$ working down from the top. Is the distribution skewed? How? Why?

4–7. Compute $Q$ for the frequency distribution in Table 3–1.

4–8. Compute $Q$ for the frequency distribution in Table 3–2.

4–9. Find the S.D. of the following scores using both the deviation score method and the raw score method: 2, 5, 6, 6, 8, 8, 10, 10, 11, 14. Find $M$. Add two points to each score and recompute the $M$ and the S.D. Multiply each score by three points and recompute the $M$ and the S.D. What is the effect of adding or multiplying by a constant on the $M$ and the S.D.?

4–10. Find the S.D. of the following scores using both the deviation score and the raw score method: 1, 1, 3, 3, 3, 4, 5, 5, 6, 6, 6, 7. Divide each score by 2 and recompute the S.D. What is the effect on the S.D. of dividing by a constant?

4–11. Using the arithmetic method, extract the square root of 1562.43 correct to three decimal places. (See mathematical Notes for Problem 9)

4–12. Using the arithmetic method, extract the square root of 0.4567 correct to four decimal places. (See Mathematical Notes for Problem 9)

4–13. Compute the S.D. by the code method for the frequency distribution shown in Table 3–3. (Problem C)

4–14. Compute the S.D. by the code method for the frequency distribution shown in Table 3–4. (Problem C)

4–15. Using the code method, find the $M$ and the S.D. of the following frequency distribution: (Problem C)
55–59, 3; 50–54, 8, 45–49, 14; 40–44, 19; 35–39, 25;
30–34, 17; 25–29, 15; 20–24, 10; 15–19, 6; 10–14, 2.
Place the $M_G$ in the interval 35–39.

4–16. Using the code method, find the $M$ and the S.D. of the following frequency distribution: (Problems B and C)
40–59, 2; 60–79, 2; 80–99, 4; 100–119, 5; 120–139, 7;
140–159, 12; 160–179, 20; 180–199, 15; 200–219, 6;
220–239, 3; 240–259, 1.
Place $M_G$ in the interval 160–179.

# chapter 5

# Scaled Scores and the Normal Curve

Measurements taken in the social sciences typically yield numbers, called "raw scores," on a rank-order scale. For example, a raw score on a psychological test may be calculated as the number of test items correctly answered by the subject. This number is on a rank-order scale of measurement because the scores derived in this way merely rank the subjects in order of their ability rather than specifying a precise quantity of ability. Thus, a person who passes 10 items does not necessarily have only half as much ability as someone who passes 20 items, although he presumably has less ability. The test scores only rank people within the limits of error characteristic of the test as applied to this group of individuals at this given administration.

Raw scores have a number of drawbacks that have prompted social scientists to find ways of converting them to other types of scores with more desirable properties. One of the difficulties with using raw scores is that they are not interpretable without additional information. For example, if a subject is known to have a raw score of 70 on a particular test, the information cannot be utilized without also knowing what scores other subjects had on the test to provide a standard of comparison. Only with such a standard of comparison available is it possible to know if a raw score of 70 is high, medium, or low. Raw scores can be converted to various kinds of scaled scores that have built into them a standard of comparison so that they can be interpreted directly without additional information about the scores of other subjects. This chapter will be devoted to describing several kinds of scaled scores and to the normal curve which is used in deriving some of them.

## Problem 11
### GIVEN A RAW SCORE, FIND ITS CENTILE RANK

One of the most common types of raw score transformations is to convert them to centile ranks. There are 99 centile ranks, i.e., the numbers from 1 to 99. They represent percentages of subjects falling below the given subject. For example, if a raw score of 70 is found to correspond to a centile rank of 44, this means that 44 percent of the subjects in the reference group had raw scores lower than 70. The centile ranks corresponding to raw scores are always with reference to a particular group of subjects. A raw score of 70 might be at the 44th centile rank with respect to a group of high school seniors but only at the 22nd centile rank with respect to a group of college students. The advantage, of course, of knowing a subject's centile rank, instead of just the raw score, is that the centile rank score immediately specifies the percentage of subjects whose raw scores are lower than the raw score which this subject earned. The usefulness of the centile rank is considerably diminished, however, if the reference group with respect to which it has been computed is not known. Having an I.Q. test score falling at the 75th centile rank for elementary school children would clearly be less impressive than if it were at the 75th centile rank for Phi Beta Kappa college graduates.

### Converting a Raw Score to a Centile Rank

To convert a raw score to a centile rank, first find the number of scores in the reference group that fall below this given raw score. Then, divide by the number of scores in the reference group to obtain the proportion of scores in the reference group below this one. This proportion is multiplied by 100 to convert it to a percentage which is rounded to the nearest whole number value from 1 to 99. Note that centile ranks of 0 and 100 are not used. Any percentage less than one is rounded up to 1 and any percentage above 99 is rounded down to 99.

In making these calculations, however, it is important to remember that a given score covers a range of a full point. A score of 70 goes from 69.5 to 70.5. In computing the number of scores below 70, half of the subject's own score is below 70.0 and half is above 70.0. So, the number of scores below 70 will include all other subjects below 70 plus half of the subject's own score which is at 70. If there are other subjects whose scores are also at 70, half of these also will be below 70.0 and half above 70.0. The rule then is take the number of scores below the given score plus half of those at or equal to the given score to determine the number below that score.

*An Example.* Consider the following raw scores representing number correct on a high school history examination for a class of 15 students:

10, 15, 20, 20, 25, 25, 26, 27, 28, 28, 29, 32, 33, 35, and 42. What is the centile rank for a score of 28?

First count the number of scores below 28. There are 8 scores between 9.5 and 27.5, i.e., below the score 28 which runs from 27.5 to 28.5. Now add half the scores at 28, i.e., one half of 2, or 1. This gives $8 + 1 = 9$ scores below 28.0, the midpoint of the range for the score 28. Dividing 9 by 15 gives 0.600 as the proportion of scores below 28.0. This proportion is multiplied by 100 to convert it to a percent, i.e., $0.600 \times 100 = 60.0$, which is rounded off to 60. The centile rank corresponding to a raw score of 28, then, is 60 in this reference group.

*An Example Using a Frequency Distribution.* Suppose the same history examination were given to a larger class in a different school, yielding the frequency distribution shown in Table 5–1. What would be the centile rank for a raw score of 28 with respect to this reference group?

Since the score of 28 is contained within a class interval in Table 5–1, it will be necessary to use an interpolative process in calculating

TABLE 5–1
History Examination Scores

| X | f | cf |
|---|---|---|
| 50–54 | 1 | 45 |
| 45–49 | 2 | 44 |
| 40–44 | 5 | 42 |
| 35–39 | 9 | 37 |
| 30–34 | 10 | 28 |
| 25–29 | 8 | 18 |
| 20–24 | 6 | 10 |
| 15–19 | 2 | 4 |
| 10–14 | 1 | 2 |
| 5–9 | 1 | 1 |
| N | 45 | |

the number of scores below 28.0. Below the interval 24.5 to 29.5, which contains the score of 28, there are 10 cases, as shown in the cumulative frequency column in Table 5–1. To these 10 scores must be added the portion of the 8 scores in the interval 25–29 (exact limits 24.5 to 29.5) which falls below 28.0. This is computed as follows:

$$10 + \frac{(28.0 - 24.5)}{5} \cdot (8) =$$

$$10 + (0.7)(8) =$$

$$10 + 5.6 = 15.6$$

The interval from the lower limit up to the score point 28.0 is divided by the class interval width ($i = 5$) to give the proportion of the interval below 28.0, i.e., 0.7. Since the cases are assumed to be evenly spread throughout the interval, this is also the proportion of cases below 28.0 in the interval. Multiplying 0.7 by 8, the total number of cases in the interval, gives 5.6 as the number of cases in the interval below 28.0. Adding 5.6 to 10 gives a total of 15.6 cases below 28.0 in the entire distribution. Dividing 15.6 by 45 gives 0.3467 as the proportion of cases below 28. Multiplying by 100 and rounding off to the nearest whole number gives 35 as the centile rank corresponding to a raw score of 28 on the History examination. Thus with respect to one reference group, the centile rank for a given raw score was 60 and with respect to another reference group, the same raw score gave a centile rank of 35. Clearly, then, interpreting a centile rank requires knowledge of the reference group with respect to which it was computed. Publishers of tests typically provide tables of centile ranks corresponding to raw scores on their tests for different normative groups, e.g., high school students, college students, general population males, and female work force. When a particular individual's test results are to be evaluated, his centile rank scores are determined with respect to the available comparison "norm" group that is most appropriate for him. His centile rank scores will then reveal directly how he compares with this group.

The student will encounter other names for the centile rank score. Often this score will be referred to merely as the "centile," or the "centile score." Thus, a particular individual's raw score, or the individual himself, will be described as being "at the seventy-fifth centile," or whatever centile rank is correct. This might be abbreviated as $P75$ or $P_{75}$. The "P" here comes from the fact that the centile score or centile rank score is also called a "percentile rank score," "percentile score," or just "percentile." The term "percentile" is gradually being replaced by the term "centile."

## Problem 12
## GIVEN A CENTILE RANK, FIND THE CORRESPONDING RAW SCORE

Instead of finding the centile rank corresponding to a particular raw score, it is often more pertinent to find the raw score that corresponds to a particular centile rank. Examples where this is the case have already been encountered, i.e., with the median, $P_{50}$, and the two quartile points $Q_1$ and $Q_3$, or $P_{25}$ and $P_{75}$. These are all examples of finding the raw score corresponding to a given centile rank. Since this is the case, the methods used to solve those problems, i.e., problems 4, 5, 7, and 8, can be used to solve Problem 12. In obtaining the raw score for some centile rank other than $P_{25}$, $P_{50}$, or $P_{75}$, say for example, $P_{68}$, the only difference is that instead of finding the raw score point below which 25, 50, or 75 percent of the cases falls, the raw score point below which 68 percent of the cases falls must be found. Thus, N will be multiplied by 0.68

instead of 0.25, 0.50, or 0.75 to find the number of scores below the desired point on the raw score scale.

*An Example.* Although these methods are already familiar to the reader by this time, an example will be given here for review purposes and to remove any possible doubts about how they are applied in this slightly different situation. Consider the distribution of history scores in Table 5–1, already discussed in connection with Problem 11. As an illustration, the raw score corresponding to a centile rank of 68, i.e., the sixty eighth centile, will be computed.

First find 68 percent of $N$, i.e., $45 \times 0.68 = 30.6$. Below the point corresponding to $P_{68}$, then, there are 30.6 cases. Using the methods of Problem 6, this point is found as follows (See Table 5–1):

$$P_{68} = 34.5 + \frac{(30.6 - 28)}{9} \cdot (5)$$

$$P_{68} = 34.5 + \frac{2.6}{9} \cdot (5)$$

$$P_{68} = 34.5 + \frac{13.0}{9}$$

$$P_{68} = 34.5 + 1.44$$

$$P_{68} = 35.94 = 35.9$$

The answer is always rounded off to one decimal place if the raw scores are integers or whole numbers.

*Norms.* Norms for published measuring instruments typically include tables which list the possible raw scores and the corresponding centile ranks for various norm groups. These tables can be constructed, with appropriate rounding off procedures, using either the method of Problem 11 or that of Problem 12. In some instances, the table will be constructed by reading from a graph which has been prepared using only a certain proportion of the centile points or raw scores. For example, the raw score equivalents for every fifth centile rank, e.g., 5, 10, 15, etc. could be computed and plotted on a chart. Then a smooth curve is drawn through these points and extrapolated at either end to the first and 99th centile ranks. The centile ranks corresponding to all possible raw scores can be read off this graph to construct the desired table. Raw scores above and below the end points are merely listed as being either at the 99th or the first centile, respectively.

## Characteristics of the Centile Score Scale

Conversion of raw scores to centile scores is a non-linear transformation that does not preserve the spacing characteristics of the raw score scale. Consider Formula 5–1:

$$P = b \cdot R + a \tag{5-1}$$

Where:

$P$ = centile rank score

$b$ = a constant, i.e., some number

$R$ = raw score

$a$ = a constant, i.e., some number

If it were possible to find values for $a$ and $b$ in Formula 5–1 that would convert raw scores to the corresponding centile scores, then conversion to centile scores would be a linear transformation and a graph of the centile scores plotted against the raw scores would give a straight line. This is not the case, however, since it is impossible to find values for $a$ and $b$ in Formula 5–1 to effect this transformation.

The centile score transformation distorts the spacing of individuals in comparison with what it was on the raw score scale. It tends to spread people out too much in the middle of the score scale and bunch them together too much at the ends of the scale. This is because an equal step on the centile scale represents one percent of the subjects. Thus, a score at the 51st centile has 51 percent of the scores below it and a score at the 50th centile has 50 percent of the scores below it. Since there tend to be a great many more people who have scores in the middle of the distribution, one percent covers a smaller interval on the raw score scale in the middle of the distribution than at the ends. Thus to improve from the 50th to the 53rd centile might require getting only one point more on a test but to go from the 96th to the 99th centile might require an increment of 10 points. The magnitude of this distortion is revealed by the fact that the following centile scores represent very roughly equidistant points on the scale: 1, 3, 16, 50, 84, 97, 99. That is, the difference between the first and third centiles is roughly equivalent to the difference between the 50th and the 84th centile. In using the centile scale therefore, it must be remembered that the differences in the middle of the scale do not mean as much as at the end of the scale. The real amount of performance difference between people at the 40th and 60th centile, for example, is not really very great as a rule even though there is a 20 point separation. A separation half this large at the extremes of the distribution would be of greater significance.

## Usefulness of Centile Scores

Why are raw scores converted to centile scores? The main reason is that raw scores have no standard interpretable meaning whereas centile scores do. If a subject passes 80 items on a test, giving him a raw score of 80, and 60 items on another test, it is impossible to say on the basis of this information alone how well he did in each test or even whether he did better on the first test than he did on the second one.

On the other hand, if it were known that the subject obtained a centile score of 60 on the first test and 10 on the second test, it would be immediately known that he did a bit above average in test one and rather poorly on test two, relative to the particular reference group on which the centile equivalents were obtained. Centile scores give directly the percentages of cases in the normative group that failed to do as well on the test as the subject under consideration.

These features of centile scores make them popular for normative purposes. When a test author develops a new test instrument, he will often prepare centile norms for the test with respect to various reference groups. Thus, he might have norms for high school males, high school females, college males, college females, and male industrial workers. Each set of norms consists of the possible raw scores on the test with their corresponding centile equivalents. An individual's raw score is used to enter the norm table most relevant for him and his centile score is taken from the table. This centile score will show where he stands relative to this particular reference group on this particular test. Such information can be used for such practical goals as vocational and educational guidance as well as personnel selection.

## Problem 13
## CONVERT RAW SCORES TO STANDARD SCORES USING A LINEAR TRANSFORMATION

A common transformation of raw scores that preserves the spacing proportions among the scores is the conversion of raw scores to standard scores using the following linear transformation:

$$Z = \frac{X - M}{\text{S.D.}} \qquad (5\text{--}2)$$

Where:

$$Z = \text{standard score}$$

$$X = \text{a raw score}$$

$$M = \text{the mean}$$

$$\text{S.D.} = \text{the standard deviation}$$

In terms of Formula 5–1, Formula 5–2 would be rewritten as:

$$Z = \frac{1}{\text{S.D.}}(X) + \frac{-M}{\text{S.D.}} \qquad (5\text{--}3)$$

Where:

$$b = \frac{1}{\text{S.D.}} \quad \text{and} \quad a = \frac{-M}{\text{S.D.}}$$

Thus, constants can be found that will convert raw scores to standard scores by means of a linear transformation.[1] If the standard scores obtained by Formula 5–2 are plotted against the corresponding raw scores, the resulting graph will be a straight line. This means that if two raw scores are close together, their standard scores will be likewise relatively close together. If two raw scores are far apart, their standard scores will also be relatively as well separated on the standard score scale. This means that the distortions in the scale produced by the raw score to centile rank transformation are absent in the linear transformation of raw score to standard score.

Since the relative spacing of raw scores is preserved in the linear transformation of raw scores to standard scores, it follows that whatever shape was characteristic of the raw score distribution will also be found in the transformed scores. Thus, if the raw scores are badly skewed in the positive direction, the linearly transformed standard scores will also be badly skewed in the positive direction. Transformation of raw scores to standard scores by Formula 5–2, therefore, will not "normalize" the resulting $Z$ scores. These linearly transformed $Z$ scores will be normally distributed only if the original $X$ scores were normally distributed.

Standard scores have a mean score of 0 and a standard deviation of 1. So if an individual's standard score is $-1$, it is known immediately that he falls at one standard deviation below the mean of the reference group with respect to which the score transformation is being obtained. If his standard score is near zero, it is immediately known that his raw score falls near the mean of the distribution for the reference group. Linearly transformed standard scores preserve the spacing among scores and reveal something about the individual's standing relative to the reference group but they do not give the percentage of cases falling below the score that is so conveniently available with the centile score.

*An Example.* Application of Formula 5–2 merely requires knowing the mean and standard deviation of the reference group as well as the raw score to be transformed. In a distribution with $M = 25$ and S.D. = 5, a raw score of 21 yields a standard score by Formula 5–2 as follows:

$$Z = \frac{21 - 25}{5} = \frac{-4}{5} = -0.8 \tag{5–4}$$

Standard scores for the vast majority of the cases will lie between $-3.0$ and $+3.0$. They are typically reported to one decimal place.

---

[1] The transformation in Formula 5–3 is "linear" because the variable $X$ appears in the equation only to the first degree, i.e., not $X^2$, $X^3$, and so on. If linearly transformed scores are plotted against the original scores, the points will all fall on a straight line; hence the term "linear transformation."

## Usefulness of Standard Scores

It has been pointed out already that centile scores are distorted in the sense that subjects in the middle of the distribution have centile scores that vary over too big a range while subjects at the extremes have scores that are bunched together too much. This type of distortion is eliminated in all standard score scales. Subjects with centile scale scores from 50 to 84 are spread over the standard scores of 0.0 to 1.0; subjects with centile scores from 84 to 97 are spread over the standard scores of 1.0 to 2.0, and subjects with centile scores from 97 to 99 are spread over the standard scores of 2.0 to $+\infty$, although standard scores beyond 3.0 are rare. Centile scores from 1 to 50 give negative standard scores in a mirror-image fashion to the positive standard scores for centile scores from 50 to 99. When test norms are put into standard score form instead of centile score form, it is to eliminate the distortions found in the centile score scale while still retaining the advantage of being able to determine from the score itself where in the distribution the person's score falls. This cannot be done with raw scores.

## Mathematical Notes

The reader may have wondered why the mean and standard deviation of standard scores are 0 and 1, respectively. To find the mean of $Z$-scores, substitute Formula 5–2 for $Z$ into the formula for the mean as follows:

$$M = \frac{\Sigma X}{N} \tag{5-5}$$

Substituting $Z$ for $X$ gives:

$$M_Z = \frac{\sum \left( \dfrac{X - M}{\text{S.D.}} \right)}{N}$$

$$M_Z = \frac{\sum \left( \dfrac{X - M}{\text{S.D.}} \right)}{N} = \frac{1}{\text{S.D.}} \sum \left( \frac{X - M}{N} \right) = \frac{1}{\text{S.D.}} \left[ \frac{\Sigma X}{N} - \frac{\Sigma M}{N} \right]$$

$$M_Z = \frac{1}{\text{S.D.}} \left( \frac{\Sigma X}{N} - \frac{N \cdot M}{N} \right)$$

$$M_Z = \frac{1}{\text{S.D.}} (M - M) = 0$$

Since the mean of the $Z$-scores is zero, they are deviation scores and to find the standard deviation of $Z$-scores the deviation score formula for the standard deviation can be used:

$$\text{S.D.} = \sqrt{\frac{\Sigma x^2}{N}} \tag{5-6}$$

Substituting $Z = \dfrac{X - M}{\text{S.D.}}$ for $x$ gives:

$$\text{S.D.}_Z = \sqrt{\frac{\sum \left(\dfrac{X - M}{\text{S.D.}}\right)^2}{N}}$$

$$\text{S.D.}_Z = \sqrt{\frac{\sum \left(\dfrac{x^2}{\text{S.D.}^2}\right)}{N}}$$

$$\text{S.D.}_Z = \sqrt{\frac{\sum \dfrac{x^2}{N}}{(\text{S.D.})^2}}$$

$$\text{S.D.}_Z = \sqrt{\frac{(\text{S.D.})^2}{(\text{S.D.})^2}} = 1$$

## Problem 14
## CONVERT STANDARD SCORES TO SCALED SCORES WITH A DIFFERENT $M$ AND S.D.

Standard scores are rather inconvenient for use in clinics, schools, and industry because about half the scores are negative and all scores are decimal fractions rather than whole numbers. To get rid of negative numbers and fractions, standard scores are very commonly further transformed to a more convenient scale by means of a linear transformation.

### Converting Standard Scores to Scaled Scores with a Different $M$ and S.D.

The linear transformation that will convert standard scores to a more convenient scale is given by Formula 5–7:

$$S = (\text{S.D.})(Z) + M \tag{5–7}$$

Where:

S.D. = desired standard deviation in the scaled scores

$Z$ = the standard score to be transformed to a more convenient scale

$M$ = desired mean of the scaled scores

Thus, it is merely necessary to select a new mean and standard deviation that will avoid the problems caused by the zero mean and unit standard deviation of the standard score scale. The standard score is multiplied by the new standard deviation and the result added to the new mean score to get the scaled score.

*Examples.* A commonly used example of this type of scaled score is the McCall T-Score scale, used with such tests as the *Minnesota Multiphasic Personality Inventory*[1] and the *Comrey Personality Scales.*[2] This score scale uses a mean of 50 and a standard deviation of 10. To convert standard scores to this scale, each standard score is multiplied by 10 to get rid of the decimal fraction and then added to 50 to get rid of the negative scores. The T-score scale is shown in Figure 5–1 just below the standard score scale. A T-score of 20 corresponds to a standard score of −3, a T-score of 30 corresponds to a standard score of −2, a T-score of 60 corresponds to a standard score of +1 and so on.

Just below the T-score scale in Figure 5–1 is another scale that has

FIGURE 5–1
The Normal Curve

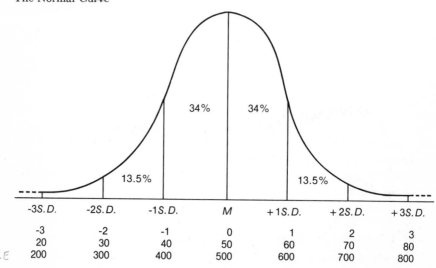

| | | | | | | |
|---|---|---|---|---|---|---|
| -3S.D. | -2S.D. | -1S.D. | M | +1S.D. | +2S.D. | +3S.D. |

| Z | -3 | -2 | -1 | 0 | 1 | 2 | 3 |
|---|---|---|---|---|---|---|---|
| T | 20 | 30 | 40 | 50 | 60 | 70 | 80 |
| GRE | 200 | 300 | 400 | 500 | 600 | 700 | 800 |

been used for the Graduate Record Examination scores. In that score scale the mean is 500 and the standard deviation is 100. For this score scale, standard scores must be determined to two decimal places since they are multiplied by 100 before adding to the new mean, 500. This score scale gives more points on the scale, differentiating more finely between individuals. This reduces the number of tied scores. Whether such precision is justified by the accuracy of the measuring instrument is another question. In most cases, T-scores provide as much discrimination between subjects as the measuring instrument accuracy can justify.

---

[1] S. R. Hathaway and J. C. McKinley, *Minnesota Multiphasic Personality Inventory: Manual for Administration and Scoring.* (New York: Psychological Corporation, 1967).

[2] Andrew L. Comrey, *Manual for the Comrey Personality Scales.* (San Diego: Educational and Industrial Testing Service, 1970).

Another popular score scale is the Stanine Scale. This scale has a mean of 5 and a standard deviation of 2.0. The conversion to this score scale takes place somewhat differently, however, than for the T-score scale. A table of Stanine-Standard Score equivalents is used to make the conversion rather than a formula translation like Formula 5–7. This table of equivalents is shown in Table 5–2. The Stanine Score Scale

TABLE 5–2
Stanine-Standard Score Equivalents

| Standard Score Equivalents | Stanine Scores |
|---|---|
| Above +1.75 | 9 |
| +1.25 to +1.75 | 8 |
| +0.75 to +1.25 | 7 |
| +0.25 to +0.75 | 6 |
| −0.25 to +0.25 | 5 |
| −0.75 to −0.25 | 4 |
| −1.25 to −0.75 | 3 |
| −1.75 to −1.25 | 2 |
| Below −1.75 | 1 |

only has nine points on it, which requires lumping scores over a half standard deviation interval into the same stanine score. At the extremes, e.g., for Stanine Scores of 1 and 9, a much larger range of standard scores is lumped into one stanine. The nine point Stanine Score Scale offers adequate differentiation of subjects for most practical personnel decisions and in fact may represent as much precision as most tests actually warrant. The main reason Stanine Scores are popular, however, is that such scores will fit on one column of an IBM card making it possible to pack more data on a card. Two digit scores would require double the space on the data cards.

## Problem 15
## USE THE NORMAL CURVE TABLE TO FIND
## NUMBERS OF CASES EXPECTED IN DIFFERENT PARTS
## OF THE DISTRIBUTION

Although many variables studied in the behavioral sciences have population frequency distributions that follow the bell-shaped normal curve when plotted for visual display (see Figure 5–1), sample distributions may vary substantially from this form. Consider the distribution of height for all adult females in the United States. The frequency polygon based on the frequency distribution for this variable would have a shape similar to that for the distribution in Figure 5–1. Such a distribution is called a normal curve. It has a precise mathematical equation (see Mathematical Notes section) that describes the shape exactly. Of course the curve of heights for all women in the United States might depart from this theoretical normal curve shape to a slight degree, but for all practical purposes it can be stated that the actual distribution for this variable is normal in form.

If a "random" sample is drawn from this very large population, the frequency distribution for the sample can be expected to differ from that of the population as a result of random sampling fluctuations. In drawing a "random" sample, each person in the population must have an equally likely chance of being selected for inclusion in the sample. If all adult women in the United States were listed alphabetically by name and every 100,000th person selected from the list, this would yield a random sample of more than 500 cases.

Even in a random sample where each person in the population has just as much chance of being selected as any other person, there still may be a disproportionate number of individuals in a given category selected by chance. Too many tall women might be chosen accidentally, resulting in a departure of the sample curve from the population curve.

The social science research worker rarely works with population distributions since it would be extremely costly and probably impossible to measure everybody in a large population on any variable. The usual situation is that a research worker has data only on a sample. Quite often, his sampling procedures have not been exactly random in character. He may have information leading him to think the population is normally distributed with respect to the variable under investigation, but he is not sure his sample has been drawn randomly from that population. Even if his sample is drawn randomly from a normally distributed population he can expect the sample distribution to depart from normality to some extent. The smaller the sample, the greater the departure from normality he may expect. If his sample was not drawn randomly, there is an even greater chance that he will have certain segments of the population over represented and other segments under represented, leading to a distortion in the sample distribution.

Whatever the situation, the investigator will often want to be able to compare the number of cases in his sample distribution in a particular part of the distribution with what should be there if the sample distribution were in fact normal in form. He may wish to know how many scores should be above a given raw score, how many below a given raw score, or how many there should be between two given raw scores if the sample distribution were normal. The obtained values may be compared with the actual number of scores in these regions to provide some indication of the degree to which the sample distribution approximates the normal distribution in form.

## The Normal Curve

The bell-shaped normal curve (or normal distribution) is shown in Figure 5–1. Dividing vertical lines are shown in Figure 5–1 at one, two, and three standard deviation distances above and below the mean. The mean, marked $M$ in Figure 5–1, is right at the center of the curve, where the height of the curve above the base line is greatest. In the region between the mean $(M)$ and one standard deviation (S.D.) above the mean, there are approximately 34 percent of all the scores in the normal distribution. In the region between the mean and one standard deviation below the mean there are also approximately 34 percent of the scores. The regions below the mean contain the same percentages as the regions above the mean since the curve is symmetric. The left half is a mirror image of the right half. From the one-standard-deviation point to the two-standard-deviation point there are about 13.5 percent of the cases. This same percentage of cases falls in the corresponding region below the mean.

These figures show that in the interval from minus one standard deviation to plus one standard deviation in the normal curve there are approximately 34 + 34 or 68 percent of the cases; $M \pm 1$ S.D. covers about 68 percent of the cases. In the interval from minus two S.D. to plus two S.D., there are approximately 13.5 + 34 + 34 + 13.5 or 95 percent of the cases; thus the $M \pm 2$ S.D. covers about 95 percent of the cases. This leaves about five percent of the cases outside the region $M \pm 2$ S.D., 2½ percent above plus-two S.D. and 2½ percent below minus-two S.D.

When raw scores are converted to standard scores, the transformed scores will have a mean of 0 and a standard deviation of 1.0. The standard score scale is shown in Figure 5–1 just below the row of symbols identifying the break points in the normal curve. Thus, a person whose raw score is at the mean of the distribution will receive a standard score of 0. If this raw score is one standard deviation above the mean, his standard score will be 1.0; if his raw score is 1.2 standard deviations below the mean, his standard score will be −1.2, and so on. Raw scores are also transformed to distributions with means and standard deviations different from 0 and 1, respectively (see Problem 14). The bottom two scales in Figure 5–1 are examples. The bottom one has a mean of 500 and a standard deviation of 100. The next one up has a mean of 50 and a standard deviation of 10 (The McCall T-Score distribution). A score of 3 on the standard score scale would be equivalent to 80 in the scale next below it and 800 in the bottom scale. All these points are at three standard deviations above the mean. It is just the units of the scale that are changing.

*The Normal Curve Table.* The proportions of total area under the normal curve for positive standard scores ($Z$ values) are given in Table

B, Appendix A. The table lists values of $Z$ by increments of 0.01 between $Z = 0.00$ and $Z = 3.99$. Each entry for a given $Z$ score is the proportion of the total area under the curve that falls below that particular $Z$ value. Note that at $Z = 0.00$, the tabled value is 0.5000, or one half the total area below $Z = 0.00$. This is because $Z = 0.00$ is at the mean of the distribution and the normal distribution is symmetrical about the mean, half above the mean and half below.

It has already been stated that the mean plus and minus one S.D. covers approximately the middle 68 percent of the cases. This statement can be verified by reference to the normal curve table. Entering Table B with $Z = 1.00$ gives an area of 0.8413, or 84.13 percent below plus one S.D., i.e., one S.D. above the mean. Subtracting 0.50 from 0.8413, i.e., the half below the mean, gives 0.3413 as the proportion of cases between the mean and plus one S.D. Since the normal curve is symmetric, there is an equivalent proportion of the total area between the mean and minus one S.D. Adding these two proportions together gives $0.3413 + 0.3413 = 0.6826$, or approximately 68 percent of the total area. The cases in a distribution are arranged as the area under the curve, so approximately 68 percent of the cases will fall in the interval $M \pm 1$ S.D. provided that the distribution is normal in form.

It was also stated above that the interval $M \pm 2$ S.D. contains about 95 percent of the cases in a normal distribution. Again this can be verified by entering Table B with $Z = 2.00$ which gives a value of 0.9772 for the area under the curve below $Z = 2.00$. Subtracting 0.5000, the area below the mean, from 0.9772 gives 0.4772 as the area between the mean and plus two standard deviations. There is an equal area between the mean and minus two S.D.s, so adding $0.4772 + 0.4772$ gives 0.9544 as the proportion of the total area in the interval $M \pm 2$ S.D. This rounds to a proportion of 0.95, which corresponds to about 95 percent of the cases falling between minus-two S.D.s and plus-two S.D.s in a normal distribution.

Table values are given for $Z$ by increments of 0.01 up to $Z = 3.99$. The last tabled value of $Z$ is 3.99 where the area is 1.0000. This value is the rounded off value, correct to four decimal places. The normal curve actually runs from minus infinity to plus infinity but the proportion of area beyond plus or minus 3.99 S.D.s is so small that it is closer to 0.0000 than it is to 0.0001.

## Using the Normal Curve Tables to Find Frequencies

Use of the normal curve tables can be illustrated by examples. Given a distribution of cases for a sample with $N = 300$ which has a mean of 60 and a standard deviation of 15, find the number of scores that would fall below a raw score of 55 in this distribution if it were distributed normally.

The first step is to convert the raw score of 55 to a standard score:

$$Z = \frac{X - M}{\text{S.D.}} = \frac{55 - 60}{15} = \frac{-5}{15} = -0.33$$

There is no negative $Z$ score in Table B, the Table of the Normal Curve, but since the normal curve is symmetric, the table may be entered with $Z = +0.33$ to get an area proportion equal to 0.6293 falling below the $Z$ score of $+0.33$. The area above $Z = +0.33$ may be obtained as the difference between 0.6293 and the total area, so $1 - 0.6293 = 0.3707$ as the proportion of area above $Z = +0.33$. This proportion is the same as the proportion of area below $Z = -0.33$, since the normal curve is symmetric. Thus the proportion of cases below $Z = -0.33$ is 0.3707. The total number of cases in the sample is 300, so the frequency, i.e., the number of cases below 55, is $0.3707 \times 300 = 111.2100$ or 111.2 cases, rounding to one decimal place. In a normal distribution of 300 cases, a mean of 60 and an S.D. of 15, there would be 111.2 scores below 55. This figure of 111.2 can be compared with the actual number of cases below 55 in this sample distribution. In making this comparison, be sure to include half the scores of 55 as being below 55.0 and half as being above 55.0.

*A Second Example.* Another type of problem that may be encountered is to find the number of cases *above* a given raw score point in a normal distribution rather than *below* that raw score point. As an example, assume a sample distribution with a mean of 110 and an S.D. of 20. There are 500 cases in the sample. The problem is to find the number of scores that should be above 85 in the sample distribution if it were normal in shape.

As before, convert the raw score to a standard score:

$$Z = \frac{X - M}{\text{S.D.}} = \frac{85 - 110}{20} = \frac{-25}{20} = -1.25$$

Looking up $Z = +1.25$ in Table B, since negative values are not tabled, gives 0.8944 as the proportion of area below $Z = +1.25$. Since the normal curve is symmetric in shape, however, the proportion of area below $Z = +1.25$ is the same as the proportion of area above $Z = -1.25$. Thus, the proportion of cases above 85 is 0.8944. Multiplying 0.8944 by 500 gives 447.2 cases above 85.

*A Third Example.* As a final illustration, the value of $Z$ will be determined such that $M \pm (Z \times \text{S.D.})$ will include the middle 99 percent of the cases in a normal distribution. Solving this problem requires using Table B in the reverse way, not entering with a Z value to get a proportion but entering with a proportion to find a Z value.

If the middle 99 percent is desired, this means there will be one half of one percent above the Z value and one half of one percent below the negative Z value at the other end of the distribution. These two halves add up to the missing one percent. A Z score is sought, therefore, that has 0.9950 as the proportion of cases below it. Inspection of Table B shows that $Z = 2.57$ has 0.9949 as the proportion below it and $Z = 2.58$ has 0.9951 as the proportion below it. Since 0.9950 is half way between the two tabled proportions, an arbitrary preference for the even value will be made. So, in the normal curve, $M \pm 2.58$ S.D. will encompass the middle 99 percent of the cases. In the distribution for the second example above where the mean was 110 and the S.D. was 20, this would give:

$$110 \pm (2.58 \times 20) = 110 \pm 51.6$$

$$= 58.4 \text{ to } 161.6$$

In that sample of 500 cases, then, if the distribution were exactly normal, there would be $0.99 \times 500 = 495$ cases between 58.4 and 161.6. Only 5 cases of the 500 would lie outside this interval.

### Finding Normal Curve Frequencies Corresponding to Obtained Frequencies in a Sample Distribution

The methods of the third example above may be applied to determine what the normal curve frequencies would be for a distribution with a specified $M$, S.D., and $N$. These normal curve frequencies may be compared with the actual obtained frequencies to determine how well the sample distribution approximates the normal distribution in form. The steps will be outlined below for finding these normal curve frequencies for the distribution in Table 5–1. This will be called the "Area Method" for getting normal curve frequencies.

1. For each class interval, write down the lower limit of the interval to the right of the obtained frequency value, $f_o$. These values are shown in Table 5–3. These values are in the column headed by $X_{11}$. The value for the bottom interval is not recorded because it will not be used.

2. Convert these lower limit scores to standard scores by the linear transformation $Z = (X - M)/\text{S.D.}$, i.e., $(X - 31.3)/9.2$ in this problem. For the interval 50–54 this will be $(49.5 - 31.3)/9.2 = 18.2/9.2 = 1.98$. This value is placed in the column headed by Z in Table 5–4. Each other Z value may be obtained in the same way. As a check on the accuracy of these Z values, the difference between adjacent values may be checked. They should be $5/9.2 = 0.54$ points apart since they are separated by one class interval which is 5 raw score points or $5/9.2$

TABLE 5-3
Computing Normal Curve Frequencies by the Area Method

| $X$ | $f_0$ | $X_{11}$ | $Z$ | Area $-\infty$ to $Z$ | Interval Area | $f_e$ |
|---|---|---|---|---|---|---|
| 50–54 | 1 | 49.5 | 1.98 | 0.9761 | 0.0239 | 1.1 |
| 45–49 | 2 | 44.5 | 1.44 | 0.9251 | 0.0510 | 2.3 |
| 40–44 | 5 | 39.5 | 0.89 | 0.8133 | 0.1118 | 5.0 |
| 35–39 | 9 | 34.5 | 0.35 | 0.6368 | 0.1765 | 7.9 |
| 30–34 | 10 | 29.5 | −0.19 | 0.4247 | 0.2121 | 9.5 |
| 25–29 | 8 | 24.5 | −0.74 | 0.2296 | 0.1951 | 8.8 |
| 20–24 | 6 | 19.5 | −1.28 | 0.1003 | 0.1293 | 5.8 |
| 15–19 | 2 | 14.5 | −1.82 | 0.0344 | 0.0659 | 3.0 |
| 10–14 | 1 | 9.5 | −2.37 | 0.0089 | 0.0255 | 1.1 |
| 5– 9 | 1 | | | | 0.0089 | 0.4 |
| $N$ | 45 | | | | | 44.9 |

standard score points. Rounding errors may cause this to be 0.53 or 0.55.

3. From the normal curve table, Table B, find the areas from $-\infty$ to $Z$ for each $Z$ score and place these areas in the next column (See Table 5–3), headed "Area $-\infty$ to Z."

4. Find the area within each class interval and place in the column headed by "Interval Area." For the top interval, subtract 0.9761 from 1.00 because all the area above the lower limit of the top interval will be included in the top interval. This gives $1.0000 - 0.9761 = 0.0239$ as the area for the top interval. For the bottom interval, use the area below the lower limit of the next higher interval since everything below the upper limit of the bottom interval is placed in the bottom interval. Thus 0.0089 is the area for the bottom interval.

For the other intervals, the interval area is the difference between the areas from $-\infty$ to $Z$ for adjacent intervals. Thus for the interval 45–49, the area is $0.9761 - 0.9251 = 0.0510$. Since the area below 49.5 is 0.9761 and the area below 44.5 is 0.9251, the area between must be the difference between them, i.e., 0.0510. The remaining areas are computed in a similar manner.

5. The interval areas are multiplied by $N = 45$ to convert areas to frequencies. These normal curve frequencies, often referred to as "theoretical" frequencies, or "expected frequencies," are shown in the last column of Table 5–3, headed by "$f_e$." The failure to have the normal curve frequencies add up to $N$ with the area method is due to rounding error, presuming no computational mistake has been made. Typically one or two of the values is adjusted by 0.1 as necessary to make the total come out to equal $N$ exactly.

In a later chapter on chi square, $\chi^2$, a statistical test will be described for determining if the differences between the obtained and theoretical frequencies are large enough to reject the hypothesis that the distribution of scores in the population from which the sample was drawn is in fact normal. In the present case, as shown in Table 5-3, no normal curve frequency value is different from its corresponding obtained frequency, $f_o$, by more than 1.1 cases, so this distribution is very close to being normal in form.

### Mathematical Notes

The simplest normal curve formula is that for the standard normal curve:

$$y = \frac{1}{\sqrt{2\pi}} e^{-(Z^2/2)} \tag{5-8}$$

Where:

$y$ = the ordinate, i.e., the height of the curve above a point, $Z$, on the base line

$\pi$ = pi, approximately 3.14159

$e$ = base of the Napierian logarithm system, approximately 2.71828

$Z$ = point on the standard score scale to be selected, i.e., on the base line

Formula 5-8 can be used to find $y$, the height of the standard normal curve (the ordinate), above any selected point, $Z$, on the standard score scale.

To find the height, $y$, above any given point on the base line, $Z$, first square $Z$ and divide by 2. Then $e$ is raised to minus this power. Thus if $Z$ were $+2$, then $e$ would be raised to the $-(4/2)$ or $-2$ power. This would be the same as $1/e^2$ since raising a number to a negative power is the same as 1 over the number to the positive power. The result of $1/e^2$ would be $1/(2.718)^2 = 1/7.388 = 0.1354$. This value must be multiplied by $1/\sqrt{2\pi}$ or $1/\sqrt{2(3.1416)} = 1/\sqrt{6.2832} = 1/2.505 = 0.3989$; thus $0.1354 \times 0.3989$ gives $0.05401$ as the height of the standard normal curve above the standard score value of plus two.

This is a small value, as would be expected since the curve is very low (see Figure 5-1) at $Z = +2.0$. As a standard of comparison, the height of the curve above the point $Z = 0$ will be computed since this is the point where the curve is highest.

Squaring $Z = 0$ still gives zero, so $e$ is raised to the minus 0 over 2 power, i.e., $e^{-(0/2)}$. This is $e^0$; but any number to the zero power is equal to 1.0. So 1.0 is multiplied by $1/\sqrt{2\pi}$ which was already computed to be 0.3989. Thus, the height of the standard normal curve above the zero point (the mean) on the base line is 0.3989. It should be noted in computing the value of $y$ for a given value of $Z$, the value of $Z$ is always squared. This means that $y$ will come out to be the same for a $Z$ value whether or not it has a minus sign on it. The height of the curve (the ordinate) above $-3$ is the same as the height of the curve above $+3$.

Computing the value of $y$ for other values of $Z$ is not so easy since $e$ must be raised to a power that is a decimal fraction. This is accomplished most easily through the use of logarithms. The process is tedious at best. Fortunately, tables are available giving the value of $y$ for selected values of $Z$ within the usual normal limits. Most values of $Z$ lie between $-3$ and $+3$. The normal curve actually runs, however, from minus infinity to plus infinity. If $Z$ is a very large number, positive or negative, then $e$ is raised to a large negative power, giving 1 divided by a large number. This will make the value of $y$, the ordinate, very small when $Z$ is very large. Beyond $+4$, the value of the ordinate is so small that it is ordinarily not even listed in the normal curve tables. The values of the ordinates, or $y$ values, that correspond to the various standard scores are listed in the normal curve table, Table B in Appendix A. This table will be described at greater length in a later section of this chapter.

*Areas Under the Normal Curve.* The ordinate values, or $y$ values, described in the previous section can be used to obtain approximations to the areas under the normal curve within given intervals. Suppose, for example, the area under the standard normal curve from $-0.25$ S.D. to $+0.25$ S.D. is desired, i.e., the mean plus and minus a quarter standard deviation. This represents an interval of $(+0.25)$ to $(-0.25)$ or 0.50 on the standard score scale, i.e., one half point along the base line in Figure 5-1. In the last section, it was determined that the height of the standard normal curve above the zero point (the mean) on the base line is 0.3989. If a rectangle of height 0.3989 and base equal to 0.50 standard score points is superimposed on the standard normal curve as shown in Figure 5-2, it will be seen that the area of the rectangle is very close to the area under the curve between $Z = -0.25$ to $Z = +0.25$. The area of the rectangle which equals the base times the height, i.e., $0.50 \times 0.3989 = 0.1994$, is just slightly greater than the area under the curve because the ends of the top of the rectangle stick up above the curve a small amount. The correct area under the curve in this interval is 0.1974 which is two thousandths less than the rectangular approximation to it.

The approximation to the area under the curve by taking the area of a rectangle becomes better and better as the base of the rectangle

FIGURE 5–2
Areas under the Normal Curve

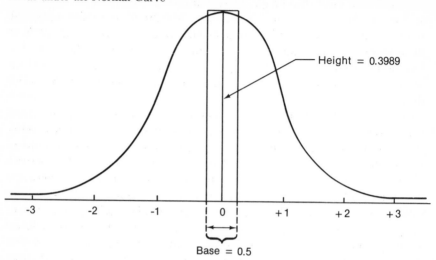

becomes smaller and smaller. The area obtained is an approximation to the proportion of the cases in the standard normal curve that falls within that interval. That is, when the area under the standard normal curve is found for a particular interval along the base line, that area represents the proportion of the standard scores (Zs) that would fall in that interval. This applies only for normalized Z scores (see Problem D), not for linearly transformed Z scores (Problem 13).

Since the areas under the curve for given intervals represent proportions of cases falling in those intervals, it follows that the total area under the curve between $-\infty$ and $+\infty$ must add up to 1.0. A proportion of 1.0 is necessary to encompass all the cases. By dividing the area under the curve into very narrow strips, using very small intervals along the base line, it is possible to estimate the areas with sufficient accuracy for practical purposes using areas of rectangles. The rectangles have the same base size as the narrow strips and heights equal to the ordinates at the center of the strips.

It is impossible to go clean to minus and plus infinity of course but if the areas between $-4.0$ and $+4.0$ are computed, the remaining areas in the tails are quite small. The area between $-4.0$ and $+4.0$ can be subtracted from 1.0 to obtain the area in the two tails, one half of it below $-4.0$ and one half of it above $+4.0$. The area above 4.0 is less than one part in 10,000, i.e., 0.0001.

By adding up the areas in these rectangles with the very small bases, the areas under the standard normal curve between $-\infty$ and any given Z score can be approximated with sufficient accuracy for all practical purposes. Fortunately, these areas have been computed and are available in Table B in Appendix A.

The methods just described can be used to obtain approximations to the normal curve frequencies corresponding to obtained frequencies in a given sample distribution. The midpoint of each class interval is first converted to a Z score, by the formula $Z = (X - M)/\text{S.D.}$ and the $y$ values, or ordinates, determined from Table B. These are the heights of rectangles approximating the areas under the curve for the class intervals. The base of each rectangle is given by $i/\text{S.D.}$ in standard score units. The base, $i/\text{S.D.}$, is multiplied by the height, $y$, to obtain the interval area. This area value, a proportion between 0 and 1, is multiplied by $N$ to obtain the normal curve frequency estimate for the interval. These frequencies usually add up to less than 1.0 because there is some area in the normal curve beyond the ends of the extreme intervals. This difference between the sum of the normal frequencies and $N$ is usually divided equally and added to the end interval normal curve frequencies so the sum of all normal curve frequencies will equal $N$, the total number of cases and the sum of the actual obtained frequencies. Normal curve frequencies computed by this method, called the "ordinate method," for the data in Table 5–1 are shown in Table 5–4 together with the corresponding Area Method normal curve frequencies for the same data.

TABLE 5–4
Normal Curve Frequencies Compared with
Actual Frequencies*

| X Score Intervals | Actual Frequencies | Ordinate Method Frequencies | Area Method Frequencies |
|---|---|---|---|
| 50–54 | 1 | 0.8 | 1.1 |
| 45–49 | 2 | 2.5 | 2.3 |
| 40–44 | 5 | 4.9 | 5.0 |
| 35–39 | 9 | 8.0 | 7.9 |
| 30–34 | 10 | 9.7 | 9.5 |
| 25–29 | 8 | 8.7 | 8.8 |
| 20–24 | 6 | 5.9 | 5.8 |
| 15–19 | 2 | 2.9 | 3.0 |
| 10–14 | 1 | 1.1 | 1.1 |
| 5– 9 | 1 | 0.3 | 0.4 |
| N | 45 | 44.8 | 44.9 |

* The sum of the normal curve frequencies computed by the ordinate method do not equal 45 here, because the areas above 54.5 and below 4.5 have been ignored. To include these areas, add 0.1 to the top interval frequency and 0.1 to the bottom interval frequency, making them 0.9 and 0.4, respectively. Then, the frequencies by the ordinate method will add up to $N$ exactly. One of the frequencies for the area method may be adjusted arbitrarily by adding 0.1 so that these frequencies will also add up exactly to equal 45.0.

**Problem D**
## CONVERT RAW SCORES TO NORMALIZED STANDARD SCORES[1]

It has already been pointed out that conversion of raw scores to standard scores by means of a linear transformation, e.g., Formula 5–2, does not alter the general shape of the distribution of scores. If the distribution were badly skewed in the negative direction before the conversion, for example, it would still have that character after the transformation. If there is reason to believe that the distribution of scores should be normal in form, i.e., bell-shaped, it may be desirable to carry out a transformation of raw scores to standard scores that are normalized. That is, the distribution of the transformed scores will be normal or bell-shaped regardless of the shape of the raw score distribution. This type of transformation may be preferred where there is reason to believe that irregularities in the raw score distribution reflect inadequacies in the measuring instrument rather than fundamental properties of the underlying trait which the instrument was designed to assess. A transformation that will convert skewed raw scores to normalized standard scores is a nonlinear transformation. It is not possible in such a case to find values of $b$ and $a$ in Formula 5–1 that will convert the raw scores to their corresponding normalized standard scores. This nonlinear transformation will be effected using the normal curve.

### Computing Procedures for Converting Raw Scores to Normalized Standard Scores

The areas from $-\infty$ to $Z$ in the normal curve table (see Table B, Appendix A) give the proportions of cases under the normal curve that correspond to the $Z$ scores themselves. By looking up a given standard score or $Z$-score in the table and taking the corresponding area from $-\infty$ to $Z$, the proportion of cases in the standard normal curve below that $Z$ score is immediately given. If this value is multiplied by 100 and rounded off to the nearest whole number from 1 to 99, the centile rank for that $Z$ score is obtained. Thus with very little effort, the centile rank can be obtained for every tabled value of $Z$ in the normal curve table. This process can also be reversed. That is, given a certain centile rank, it is possible to use the areas from $-\infty$ to $Z$ to find the $Z$ score corresponding to that centile rank. This will require some interpolation, however, since the areas from $-\infty$ to $Z$ do not fall on whole numbers in the normal curve table shown in Table B (See Appendix A).

The basic principle in converting raw scores to normalized standard scores is to find that $Z$ score that has the same centile rank as the

---

[1] It will be recalled that problems in the letter series may be omitted, if desired, without interfering with study of problems in the number series.

raw score. This will be the normalized $Z$ score corresponding to that raw score. To accomplish this, find the proportion of cases in the raw score distribution falling below the raw score that is to be converted to a normalized $Z$ score. The proportion is usually computed to three decimal places. Four places may be carried if greater precision is needed. Once the proportion of cases below the raw score is obtained this value is used to enter the normal curve table (Table B, Appendix A). This value is treated as an area from $-\infty$ to $Z$ and by interpolation the value of $Z$ is found that would correspond to this proportion. If an approximation is sufficient, the $Z$ value for that area closest to the proportion may be taken as the normalized standard score corresponding to the raw score that was converted. The interpolation process by which a more accurate $Z$ value is obtained will be explained in the following example.

*An Example.* As an illustration of the procedure, a raw score of 38 in the frequency distribution shown in Table 5–1 will be converted to a normalized standard score, $Z$. The first step is to find the number of cases in this distribution falling below 38.0. The computations for this are as follows:

$$\text{Number of cases below } 38.0 = 28 + \left(\frac{39.5 - 38.0}{5}\right) \times 9$$

$$\text{Number of cases below } 38.0 = 28 + (0.3)(9)$$

$$\text{Number of cases below } 38.0 = 28 + 2.7 = 30.7$$

The number 28 above is the number of cases below the lower limit of the interval containing the score of 38. To this is added the proportion of the 9 cases in the interval that fall between 34.5, the lower limit of the interval, and 38.0. The number 30.7 is divided by 45, the number of scores in the distribution, to obtain the proportion of cases in the distribution below 38.0. This value is 0.682, to three decimal places.

The proportion of cases below 38.0 in Table 5–1, i.e., 0.682, is used to enter the normal curve table (see Table B, Appendix A) to find the corresponding normalized standard score, $Z$. This will require interpolation between tabled values as follows:

| Standard Score, $Z$ | Area from $-\infty$ to $Z$ | | |
|---|---|---|---|
| 0.47 | 0.6808 | | |
| | 0.6820 | 0.0012 | 0.0036 |
| 0.48 | 0.6844 | | |

The proportion 0.6808 is tabled and corresponds to a $Z$ score of 0.47; the proportion 0.6844 corresponds to a $Z$ score of 0.48. These two adjacent values were picked from the table because the proportion of cases below the score of 38 is 0.682 which lies between 0.6808 and 0.6844. This means that the Z score corresponding to the proportion 0.6820 must lie between 0.47 and 0.48. The unknown Z value is determined by finding that Z which lies relatively at the same point between 0.47 and 0.48 as 0.6820 does between 0.6808 and 0.6844. The process by which this value is found is called "linear interpolation." The distance between 0.6808 and 0.6820 is 0.0012. The distance between 0.6808 and 0.6844 is 0.0036. This means 0.6820 is 0.0012/0.0036 = one third of the way between 0.6808 and 0.6844. Therefore, Z is one third of the way between 0.47 and 0.48. The distance between 0.47 and 0.48 is 0.01. Multiplying this by one third gives 0.01/3 or 0.0033 which rounds to 0.003 for three decimal places. Adding this to 0.47 gives 0.473 as the normalized standard score, Z, corresponding to the raw score of 38 in the distribution shown in Table 5–1.

If this value is rounded to two decimal places, it becomes 0.47. If two decimal place accuracy is sufficient for the Z score, as it usually is, it is sufficient merely to find the tabled value in the area from $-\infty$ to Z column which is nearest to the look-up value and use the corresponding Z score without interpolation. Three decimal place accuracy usually would be called for only if the Z score is to be further transformed to some other score scale.

## Mathematical Notes

Formula 5–8 is the formula for the standard normal curve; that is, the normal curve for which the mean equals zero, the standard deviation equals one, and the area under the curve equals one. Actually the normal curve is not a single curve but rather a family of curves. As each value of the area under the curve, the mean, and the standard deviation changes, the particular normal curve that results changes. All of these are bell-shaped curves. They vary in height and width as the values of these constants, or *parameters,* change.

Consider a hypothetical population distribution that is normal in form. Let these parameters of this population be designated as follows:

$\sigma = $ the standard deviation of the population
(not the standard deviation of a sample
from the population, i.e., the S.D.)

$\mu = $ the mean of the population
(not $M$, the standard deviation of a sample
from the population)

$N = $ the number of cases in the population
(In a theoretical population this value
would normally be infinitely large.
Practically, populations are finite.)

The $\mu$ and $\sigma$ above are the small Greek letters *mu* and *sigma*, which are commonly used to represent the mean and standard deviation, respectively, of a population. It is convenient to make a distinction between population parameters and sample statistics. For this reason, the mean and standard deviation of a sample are labeled in this book as M and S.D., respectively, while population values are labeled as $\mu$ and $\sigma$.

Using these symbols for the population parameters, the general equation for the normal distribution family of curves is given by Formula 5–9 as follows:

$$Y = \frac{N}{\sigma \sqrt{2\pi}} e^{-\left[\frac{(X-\mu)^2}{2\sigma^2}\right]} \qquad (5\text{--}9)$$

Where:

$Y$ = height of the curve

$N$ = number of cases in the distribution and total area under the curve

$\pi$ = pi, approximately 3.14159

$e$ = base of the Napierian logarithm system, approximately 2.71828

$X$ = a raw score in the distribution

$\mu$ = mu, the mean of the distribution

$\sigma$ = sigma, the standard deviation of the distribution

When $N = 1$, $\mu = 0$, and $\sigma = 1$ in Formula 5–9, the equation reduces to that for the standard normal curve shown in Formula 5–8. This requires exchanging the symbol $Z$ for the symbol $X$ in the resulting equation, but in fact if the $X$ score distribution consists of $X$ scores that have a mean of zero and a standard deviation of one, and the distribution is normal, the $X$ scores are in fact normalized standard scores, or $Z$ scores.

### Exercises

5–1. Find the centile rank for the raw score of 89 in the frequency distribution shown in Table 3–1.

5–2. Find the centile rank for the raw score of 44 in the frequency distribution shown in Table 3–2.

5–3. Find the centile rank for the raw score of 42 in the frequency distribution shown in Table 3–3.

5–4. Find the centile rank for the raw score of 8 in the frequency distribution shown in Table 3–4.

5-5.   Find the raw score that corresponds to the centile rank of 68 in the frequency distribution shown in Table 4–2. Report the score to one decimal place.

5-6.   Find the raw score that corresponds to the centile rank of 16 in the frequency distribution shown in Table 4–3. Round off the answer to one decimal place.

5-7.   Given a distribution with a mean of 26.1 and a S.D. of 5.7, convert a raw score of 21 to a standard score, using a linear transformation. Compute $Z$ to two decimal places.

5-8.   Given a distribution with a mean of 99.9 and a standard deviation of 15.1, find the standard score corresponding to a raw score of 121, using a linear transformation. Compute $Z$ to two decimal places.

5-9.   Given the distribution in Table 4–2, find the standard score corresponding to a raw score of 52, using a linear transformation. Compute the score to two decimal places. (Use $M = 69.1$, S.D. $= 18.9$)

5-10.  Given the distribution in Table 4–3, find the standard score corresponding to a raw score of 9, using a linear transformation. Compute the score to two decimal places. (Use $M = 10.2$, S.D. $= 4.0$)

5-11.  Given a distribution with $M = 65$, S.D. $= 20$, and $N = 200$, find the number of cases that would fall below a score of 73 if the distribution were normal.

5-12.  Given a distribution with $M = 48$, S.D. $= 10$, and $N = 300$, find the number of cases that would fall below a score of 44 if the distribution were normal.

5-13.  Given a sample distribution with $M = 80$, S.D. $= 15$, and $N = 150$, find the number of cases that should be above 98 if the distribution were normal.

5-14.  Given a sample distribution with $M = 100$ and S.D. $= 16$, find the percentage of cases that should be above 132 if the distribution were normal.

5-15.  The McCall T-score distribution has a mean of 50 and an S.D. of 10. Convert a standard score of $-1.2$ to this scale.

5-16.  Convert a standard score of 2.5 to a McCall T-score.

5-17.  Convert a standard score of $-3.12$ to a scale with $M = 500$ and S.D. $= 100$.

5-18.  Convert a standard score of $+1.68$ to a scale with $M = 500$ and S.D. $= 100$.

5-19.  Using the Area Method, find the normal curve frequencies for the distribution in Table 4–2 ($M = 69.1$, S.D. $= 18.9$). Make the $f_e$ values add up to $N = 161$ by slight adjustments to the computed values.

5-20.  Using the Ordinate Method, find the normal curve frequencies for the distribution in Table 4–2 ($M = 69.1$, S.D. $= 18.9$). Make the $f_e$ values add up to $N = 161$ by slight adjustments to the computed values. (See Mathematical Notes for Problem 15.)

5–21. Given the distribution in Table 4–2, convert a raw score of 52 to a normalized standard score. Report the score to two decimal places.

5–22. Given the distribution in Table 4–3, convert a raw score of 9 to a normalized standard score. Report the score to two decimal places.

5–23. Given the distribution in Table 4–2, convert a raw score of 86 to a normalized standard score. Report the score to two decimal places.

5–24. Given the distribution in Table 4–3, convert a raw score of 18 into a normalized standard score. Report the score to two decimal places.

# chapter 6

# Regression Analysis and Correlation

In any scientific field, a very important activity is the study of the relationships between major variables of importance in that domain. In the more highly developed sciences, it is sometimes possible to express these relationships in terms of rather exact mathematical equations. For example, the distance that a body will fall in a vacuum in a given time period may be expressed by the following equation:

$$S = \tfrac{1}{2} g t^2 \qquad (6\text{-}1)$$

Where:

$S$ = distance fallen in feet

$g$ = a gravitational constant = 32 ft./sec.$^2$

$t$ = elapsed time in seconds

The relationship expressed in Formula 6–1 may be represented graphically as shown in Figure 6–1. By selecting a time ($t$) along the base line (abscissa or $X$ axis) and drawing a vertical line from that point up to where it intersects the curve and then drawing a horizontal line over to the ordinate ($Y$ axis), the number of feet that a body will fall in a vacuum in $t$ seconds can be read off. Actual measured values of t and $S$ would be expected to give points falling directly on this line, within the limits of measuring error and the approximate nature of Formula 6–1. The relationship expressed by Formula 6–1 and graphically represented in Figure 6–1 is an exact mathematical relationship relating the variables $S$ and $t$ ($g$ is a constant). The measured value of $S$ is an exact function of $t$, at least within very close limits under proper measurement conditions. $S$ does not depend on other variables under the specified conditions where this equation holds.

100

FIGURE 6–1
A Mathematical Relationship

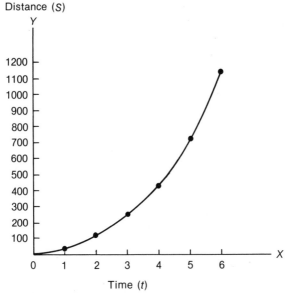

Distance (*S*)

Time (*t*)

In the social sciences on the other hand, relationships among variables are rarely expressible as exact mathematical relationships. If two variables are related at all in the social sciences, the nature of the relationship is apt to be *statistical* rather than *mathematical*. This is illustrated in Figure 6–2 which represents graphically the relationship between Intelligence Quotients (I.Q.s) and scores on a reading comprehension test for a fictitious sample of high school students. Each point in Figure 6–2 represents the pair of scores for one student. The student's I.Q. score is located on the $X$ axis (abscissa) and his Reading Comprehension Score is located on the $Y$ axis (ordinate). The intersection of lines through these points parallel to the coordinate axes locates the point on the graph for that student.

Inspection of the points readily reveals that it is impossible to draw any smooth, regular curve that will go through or even near all these points. The best that can be done here is to draw a straight line that comes as close as possible to as many of the points as possible. It is the fact that the points in the graph scatter about the line, on either side of it at some distance, that makes the relationship between these variables statistical rather than mathematical in character.

Thus, a person who has an I.Q. of 95 may have a Reading Comprehension Score somewhere between 30 and 50, according to the appearance of this graph. It will not necessarily be 40, as would be expected

FIGURE 6–2
A Statistical Relationship
Reading Comprehension

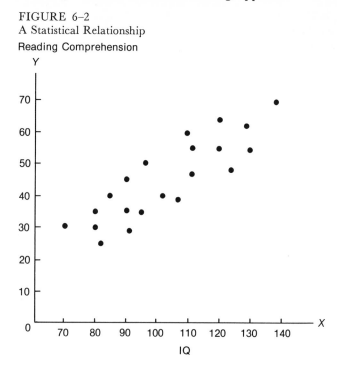

if all points fell right on a line running through the middle of these points. Figure 6–3 shows the same points with such a best-fitting line running through them. The points scatter about the line because the relationship between I.Q. scores and Reading Comprehension Scores is an imperfect one. One student with a modest I.Q. might have had intensive training and experience in reading difficult material. Another person of moderate I.Q. might have had less than the usual exposure to reading. Another person of very high I.Q. might resent reading because his parents had pushed him too hard, developing resistance on his part that decreased his performance. These and other factors may be at work to keep the relationship between these two variables from approaching a mathematical type of relationship in precision.

The relationship shown in Figures 6–2 and 6–3 is still substantial in that people with high I.Q. scores generally have high Reading Comprehension Scores. It is just not possible to determine exactly what the Reading Comprehension Score would be, knowing only the student's I.Q. Most of the relationships between variables encountered in the social sciences will not be as strong as the one shown in Figure 6–2 and 6–3.

Another difference that may be noted between the relationships shown in Figures 6–1 and 6–3 is that Figure 6–1 shows a *curvilinear* relationship

FIGURE 6–3
A Regression Line

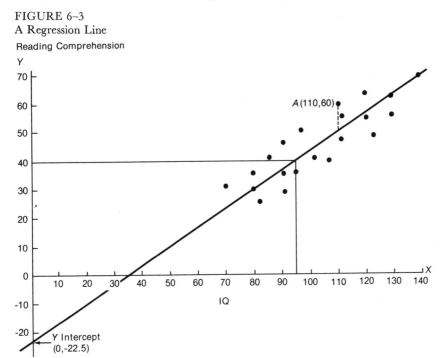

and Figure 6–3 shows a *linear* relationship. The best-fitting line through the points in Figure 6–1 is curved, whereas the best-fitting line through the points in Figure 6–3 is a straight line. Of course it would be possible to snake a very curved line through the points in Figure 6–3 in such a way as to touch all of them, but this would not be admissible. Only regular curves are permitted and no such curve would be appropriate for the points in Figure 6–3. Most of the relationships dealt with in the social sciences involve linear relationships. That is, a straight line fits the points relating the two variables better than any other regular smooth curve. For this reason, only methods of expressing the degree of linear relationship will be considered in this book. Methods for expressing the degree of curvilinear relationship are seldom used in the social sciences in comparison with linear methods.

### Expressing Linear Relationships Quantitatively

The general equation for a straight line, from high school algebra, is shown in Formula 6–2:

$$Y = bX + a \qquad (6\text{–}2)$$

Where:

$$Y = \text{a raw score in the } Y \text{ variable}$$

$$X = \text{a raw score in the } X \text{ variable}$$

$$b = \text{the slope of the line of best fit}$$

$$a = \text{the } Y \text{ intercept}$$

By reference to the plot of Reading Comprehension Scores against I.Q. Scores in Figure 6–3 it is apparent that the approximate line of best fit, as drawn by eye, passes through the $Y$ axis at about the point $(0, -22.5)$, where 0 is the $X$ axis coordinate of the point and $-22.5$ is the $Y$ axis coordinate. The value $-22.5$ is the $Y$ *intercept* in this plot. This value is substituted for the constant $a$ in Formula 6–2 in obtaining the equation of the line shown in Figure 6–3.

The value $b$ in Formula 6–2 represents the *slope* of the line. In Figure 6–3 the slope of the line may be determined by dropping a perpendicular from any point on the line to the $X$ axis and taking the ratio of this height to the distance from the base of this perpendicular to the point where the line crosses the $X$ axis (the $X$ intercept). For example, a triangle is formed by the line of best fit, the base line, and the perpendicular dropped from the line of best fit to the point $(95, 0)$ on the $X$ axis. The slope of the line is the height of this triangle divided by the base. This gives approximately 0.66.

Any other perpendicular dropped from the line of best fit to the $X$ axis will also form a triangle for which the height to base ratio will be approximately 0.66, the slope of the line of best fit. The equation for the line of best fit in Figure 6–3 then is given by:

$$Y = 0.66X - 22.5 \qquad\qquad (6\text{–}3)$$

Substituting $X = 95$ in Formula 6–3 gives

$$Y = 0.66(95) - 22.5 = 62.5 - 22.5 = 40$$

The point $(95, 40)$ indeed falls on the line of best fit. Any number substituted for $X$ in Formula 6–3 will produce a value of $Y$ such that the point $(X, Y)$ will fall on the line of best fit.

*The Regression Line for Deviation Scores.* The equation of the line of best fit, which will also be called a "regression line," becomes simplified when scores are expressed in terms of deviation scores instead of raw scores:

$$y' = bx \qquad\qquad (6\text{–}4)$$

Where:

$$y' = \text{a predicted deviation score in } y$$

$$b = \text{slope of the regression line}$$

$$x = \text{deviation score in } x$$

Formula 6–2 simplifies to Formula 6–4 for deviation scores because the mean of deviation scores is zero for both variables and the regression line, or line of best fit, passes through the origin in every case, i.e. the point $(0, 0)$. This is illustrated in Figure 6–4. The regression line passes

FIGURE 6–4
A Deviation Score Regression Plot

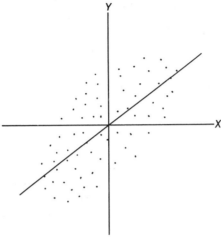

through the origin because the point on the $Y$ axis most appropriately associated with the mean point on the $X$ axis would be the mean of the $Y$ scores, i.e., 0. In other words, if it is known that a person's I.Q. is at the mean, giving him a deviation score of zero, the best prediction of his score on Reading Comprehension would be the mean of all the Reading Comprehension Scores, a reading comprehension deviation score of zero. If the regression line, or line of best fit, passes through the origin, it follows that the $Y$ intercept will be zero. Hence, Formula 6–2 reduces to Formula 6–4 for plots involving deviation scores.

It would be possible to prepare a graph of the points showing the relationship between any two variables of interest, draw in a best-fitting straight line by eye and then determine from the placement of the line what the slope and $Y$ intercept values are. This would make it possible to obtain approximate equations, such as that in Formula 6–3, relating raw scores in $Y$ to raw scores in $X$. By using deviation scores to make the plots instead of raw scores, only the slope needs to be determined and substituted for $b$ in Formula 6–4.

Determining the location of regression lines by visual approximation, however, is seldom carried out in practice. It will be demonstrated in the mathematical notes section below that a least squares determination of the value of $b$ in Formula 6–4 leads to the following equation for the regression line for predicting deviation scores in $Y$ given the deviation scores in $X$:

$$y' = r \times \frac{S.D._y}{S.D._x} \times x \qquad (6\text{-}5)$$

Where:

$y'$ = a predicted deviation score in $Y$

$x$ = a deviation score in $X$

$r$ = the correlation coefficient relating $Y$ and $X$ scores (See next section for definition)

$S.D._y$ = the standard deviation of the $Y$ scores

$S.D._x$ = the standard deviation of the $X$ scores

A prime is placed on the $y$ at the left in Formula 6–5 to designate that it is a *predicted* score rather than an actual score. If the relationship between $y$ and $x$ were mathematically exact, rather than statistical, there would be one and only one value of $y$ associated with a given value of $x$. With a statistical relationship, however, where the points scatter about the line of best fit rather than falling directly on it, there may be several values of $y$ corresponding to a given value of $x$. That is, for all the people who have a score in $x$ of 10, the $y$ values may vary considerably, e.g., $y = -1, 2, 4, 5, 8$, etc. The value of $y$, predicted from $y'$, will ordinarily be near the mean of the actual $y$ values for that particular value of $x$. The values of $(x, y')$ are all on the regression line. The values of $(x, y)$ scatter over the entire plot.

*The Correlation Coefficient.*    The correlation coefficient, $r$, is defined by the following formula:

$$r = \frac{\Sigma xy}{N(S.D._x)(S.D._y)} \qquad (6\text{-}6)$$

Where:

$r$ = the correlation coefficient, which varies between $-1.0$ and $+1.0$

$x$ = a deviation score in $X$

$y$ = a deviation score in $Y$

$N$ = the number of cases

$S.D._x$ = the standard deviation of the $X$ scores

$S.D._y$ = the standard deviation of the $Y$ scores

The proof of this formula is given in the mathematical notes section

below. It states that the correlation coefficient may be computed as
follows:

1. For each person in the sample, multiply his deviation score in $X$
   by his deviation score in $Y$.
2. Add these products for all $N$ cases to get the numerator of Formula
   6–6.
3. Multiply $N$, the number of cases, by the product of the two standard
   deviations for variables $X$ and $Y$ (standard deviations for raw scores
   and deviation scores are identical) to obtain the denominator of
   Formula 6–6.
4. Divide the numerator from step 2 above by the denominator from
   step 3 to obtain $r$.

Once the correlation coefficient has been determined, it is multiplied
by the ratio of the standard deviation of the $Y$ scores to the standard
deviation of the $X$ scores, as called for in Formula 6–5, to produce the
equation relating predicted deviation scores in $Y$, i.e. $y'$ scores, to actual
deviation scores in $X$.

The correlation coefficient gives a numerical index of the degree of
relationship or association between two variables. For example, if
$r = +1.0$, there is a perfect positive relationship. This would be a mathematical type relationship in which all the observed points fall right on
the regression line rather than scattering on either side of the line. The
highest score in $Y$ would be held by the same person who has the highest
score in $X$. The lowest score in $X$ would belong to the person who also
had the lowest score in $Y$.

If $r = -1.0$, the relationship would still be a perfect mathematical
relationship, but negative. That is, the highest score in $Y$ would be associated with the lowest score in $X$; the lowest score in $Y$ would belong
to the person who had the highest score in $X$. Scatter plots for several
types of correlation situations are shown in Figure 6–5. The more closely
the points are clustered about the line of best fit, the higher the correlation. The larger the standard deviation in $X$, relative to $Y$, the more
the scatter plot is stretched in the horizontal direction. The larger the
standard deviation in $Y$, relative to $X$, the more the scatter plot is
stretched in the vertical direction. When the correlation is zero, there
is no association between the two variables. The graphs in Figure 6–5
are for deviation scores since the origin is at the center of every plot.

When the standard deviations of the two variables are equal, Formula
6–5 reduces to $y' = r \times x$. Since standard scores all have $M = 0$ and
S.D. $= 1$, Formula 6–5 for standard scores would reduce to:

$$Z'_y = r \cdot Z_x \qquad (6\text{–}7)$$

FIGURE 6–5
Regression Plots for Different Correlations

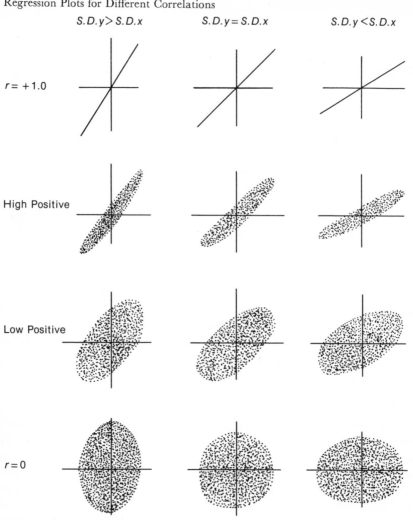

Where:

$$Z'_y = \text{a predicted standard score in } Y$$

$$Z_x = \text{a standard score in } X$$

$$r = \text{the correlation between } Y \text{ and } X$$

Formula 6–7 shows most clearly how the correlation coefficient expresses the relationship between $Y$ and $X$. For example, if the standard score in $X$ is increased one point, the increase in the predicted standard score

FIGURE 6–5 (*Continued*)

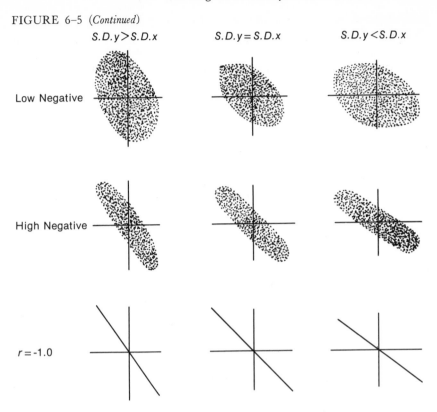

of $Y$ is $r$ points. If the correlation coefficient is 1.0, i.e. perfect, then an increase of one point in the standard score in $X$ leads to a one point increase in the predicted standard score in $Y$. It is also clear that for standard scores, or even scores with the same standard deviations, the slope of the line of best fit equals the correlation coefficient. This slope will be 1.0 for a perfect correlation and the regression line will be at 45° to the $X$ axis. As the correlation decreases from 1.0 toward zero, the angle which the regression line makes with the $X$ axis declines from 45° to zero degrees and becomes coincident with the $X$-axis when $r = 0$. With $r = -1$, the regression line is at −45° and moves toward zero degrees as $r$ moves from −1.0 toward zero.

*A Second Regression Line.* By convention, in this book $Y$ will be used to designate the variable to be predicted, i.e., the dependent variable, and $X$ will be used to designate the variable from which the predictions are made, i.e., the independent variable. For example, grade point average might be predicted from a knowledge of scholastic aptitude test score. G.P.A. would be the dependent variable to be predicted, $Y$, and the aptitude scores would be the independent variable, $X$. It would make little sense to predict aptitude scores from G.P.A. values. It is always possible to specify that the predicted score will be placed on the ordinate, or the $Y$ axis, making Formula 6–5 appropriate. If it is desired to obtain an equation, however, predicting $x$ scores from a knowledge of $y$ scores, the corresponding equation would be:

$$x' = r \cdot \frac{\text{S.D.}_x}{\text{S.D.}_y} \cdot y \qquad (6\text{–}8)$$

Where the terms are defined as with Formula 6–5. There are, therefore, two distinct regression lines. They coincide and become one line only when $r = 1.0$ or $r = -1.0$. These two lines are illustrated in Figure 6–6. It should be emphasized, however, that prediction is almost always

FIGURE 6–6
Two Deviations Score Regression Lines

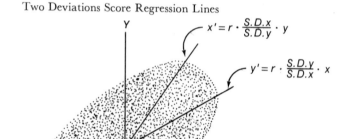

in one direction only, making it possible to use only Formula 6–5 rather than requiring the determination of two regression lines using both Formulas 6–5 and 6–8.

## Interpretation of the Correlation Coefficient

The coefficient of correlation provides a numerical index of the extent to which two variables covary, but it does not shed any light on the reasons for this covariation. Suppose, for example, that a study revealed a substantial negative correlation between age at time of marriage and number of years that the person has remained married. This correlation would indicate that people who get married at a younger age do not tend to stay married as long as those who get married at a later age.

A social reformer observing this correlation might be tempted to mount a program to prevent early marriages under the assumption that this would have the effect of reducing the divorce rate. This would be an example of confusing correlation with causation. It might be true that getting married at a younger age tends to cause divorce but the correlation does not establish this conclusion as fact. It might instead be true that the same traits which lead a person to marry early, e.g., impulsiveness and/or irresponsibility, also lead the person to get divorced. If this were true, postponing marriage until a later age would not necessarily affect the rate of divorce.

Higher incidence of crime is associated with lower income residential areas in urban communities. This represents a correlation between two variables. From the mere existence of this correlation, however, it cannot be inferred that the crime rate could be lowered by raising the income level in a given residential area. This might or might not be true, but to assume it to be true on the basis of the correlation alone would be to make the logical error of mistaking correlation for causation.

*Size of the Correlation Coefficient.* The magnitude of the correlation coefficient must be interpreted with caution, because it is not on an absolute scale like height or weight. For example, a correlation of 0.25 is not half as large as a correlation of 0.50, and a correlation of 0.50 is not half as large as a correlation of 1.00. The relationship is more complicated than that. A better indication of the magnitude of the correlation relative to other correlation coefficients is obtained by taking $r^2$, the coefficient of determination. A correlation of 0.707 would give $r^2 = .50$, and an $r$ of 0.50 would give $r^2 = 0.25$. Thus, an $r$ of 0.707 is in a sense approximately twice as great as a correlation of 0.50, since their coefficients of determination are on the order of two to one.[1]

## Mathematical Notes

The derivation of Formula 6–4 is based on the principle of least squares. Consider point A (110, 60) in Figure 6–3. The actual value of $Y$ for $X = 110$ is 60, whereas the value of $Y$ on the regression line for $X = 60$ is $Y = 50$. Thus, there is a discrepancy between the actual $Y$ and the predicted $Y'$ of $60 - 50$, or $+10$. The principle of least squares states that the regression line should be placed in such a way as to minimize the sum of squares of these discrepancies over all cases. The sum of squares of these discrepancies to be minimized may be expressed mathematically as follows for deviation scores:

$$\Sigma(y - y')^2 = \text{Sum of squared discrepancies between predicted and} \qquad (6\text{–}9)$$
$$\text{obtained deviation scores.}$$

[1] J. P. Guilford and B. Fruchter, *Fundamental Statistics in Psychology and Education,* 5th ed. (New York: McGraw-Hill Book Company, 1973).

By Formula 6–4, $y' = bx$. Substituting this in Formula 6–9 gives:

$$\Sigma(y - bx)^2 = \text{Sum of squared discrepancies} \qquad (6\text{–}10)$$

The expression for the sum of squared discrepancies between obtained and predicted deviation scores in $Y$, as shown in Formula 6–10, is to be minimized by the appropriate choice of $b$. That is, only the slope of the regression line determines the location of the line when scores are expressed as deviation scores since the line always goes through the origin, giving a zero $Y$ intercept. Thus, the task is to pick a value of $b$ that will minimize the expression in Formula 6–10.

Finding a value of $b$ that will minimize the sum of squared discrepancies between predicted and obtained deviation scores in $Y$ is a calculus problem. Students with no training in calculus cannot be expected to understand fully the rest of the proof. The expression in 6–10 is differentiated with respect to $b$ and set equal to zero to give:

$$2\Sigma(y - bx)(-x) = 0$$

$$\Sigma(-xy + bx^2) = 0$$

$$-\Sigma xy + b\Sigma x^2 = 0$$

$$b = \frac{\Sigma xy}{\Sigma x^2}$$

$$b = \frac{\Sigma xy}{N \cdot \dfrac{\Sigma x^2}{N}}$$

$$b = \frac{\Sigma xy}{N \cdot (\text{S.D.}_x)^2} \cdot \frac{\text{S.D.}_y}{\text{S.D.}_y}$$

$$b = \left(\frac{\Sigma xy}{N \cdot \text{S.D.}_x \text{S.D.}_y}\right) \cdot \frac{\text{S.D.}_y}{\text{S.D.}_x} \qquad (6\text{–}12)$$

$$b = r \cdot \frac{\text{S.D.}_y}{\text{S.D.}_x} \qquad (6\text{–}13)$$

The value of $b$ in Formula 6–4 is given by Formula 6–13 above. Making this substitution in Formula 6–4 gives Formula 6–5. The correlation coefficient is by definition that part of the formula for $b$ which is in parentheses in Formula 6–12.

Derivation of Formula 6–13 can be accomplished working with raw scores but it is more difficult. Since the raw score regression line does not pass through the origin, the line is determined by two constants, the slope, $b$, and the $Y$ intercept, $a$. It is necessary to differentiate the expression for the sum of the squared discrepancies partially with respect to $b$ and then $a$, set both expressions equal to zero, and then solve them simultaneously for the values of $b$ and $a$.

## Problem 16
### COMPUTE THE CORRELATION COEFFICIENT FROM RAW SCORES USING THE RAW SCORE FORMULA

Formula 6–6 gave a formula for the correlation coefficient in terms of deviation scores. That formula is very seldom used in actual calculation of the correlation coefficient because it requires the use of deviation scores. Deviation scores are obtained by subtracting the mean from raw scores $(x = X - M_x,\ y = Y - M_y)$. In addition to being a computational chore in itself, this process yields decimal fractions which are difficult to cross multiply to obtain the $xy$ cross product terms in Formula 6–6.

### Raw Score Formula for the Correlation Coefficient

A more convenient formula than Formula 6–6 for computation of the correlation coefficient directly from raw scores, rather than from deviation scores, is given by:

$$r = \frac{N\Sigma XY - (\Sigma X)(\Sigma Y)}{\sqrt{N\Sigma X^2 - (\Sigma X)^2}\ \sqrt{N\Sigma Y^2 - (\Sigma Y)^2}} \tag{6-14}$$

Where

$r =$ the correlation coefficient

$X =$ a raw score in the $X$ variable

$Y =$ a raw score in the $Y$ variable

The derivation of Formula 6–14 from Formula 6–6 is given in the mathematical notes section below. Formula 6–14 requires $\Sigma X$, $\Sigma Y$, $\Sigma X^2$, $\Sigma Y^2$, and $\Sigma XY$ as well as $N$. All but $\Sigma XY$ are terms that would be used in computing the standard deviations of these two variables by the raw score method. In fact, multiplying the left hand radical in the denominator of Formula 6–14 by $1/N$ gives the raw score formula for the standard deviation in $X$. Multiplying the radical on the right in the denominator of Formula 6–14 by $1/N$ gives the raw score formula for the standard deviation of the $Y$ variable. The term $\Sigma XY$ requires that for each person his raw scores in $X$ and $Y$ are multiplied together. These products are then added up for all $N$ cases in the sample to get $\Sigma XY$.

### Computation of the Correlation Coefficient by the Raw Score Formula

Computation of the correlation coefficient, sometimes called the *Pearson r* or the *product moment r,* by Formula 6–14 will be illustrated by an example which is shown worked out in Table 6–1. Let $X$ represent

TABLE 6-1
Computation of the Correlation Coefficient Using the
Raw Score Formula

| Subject Number | X | Y | $X^2$ | $Y^2$ | XY |
|---|---|---|---|---|---|
| 1 | 5 | 10 | 25 | 100 | 50 |
| 2 | 7 | 13 | 49 | 169 | 91 |
| 3 | 10 | 2 | 100 | 4 | 20 |
| 4 | 10 | 8 | 100 | 64 | 80 |
| 5 | 11 | 6 | 121 | 36 | 66 |
| 6 | 15 | 4 | 225 | 16 | 60 |
| 7 | 9 | 7 | 81 | 49 | 63 |
| 8 | 9 | 7 | 81 | 49 | 63 |
| 9 | 8 | 9 | 64 | 81 | 72 |
| 10 | 6 | 9 | 36 | 81 | 54 |
| 11 | 9 | 8 | 81 | 64 | 72 |
| 12 | 12 | 5 | 144 | 25 | 60 |
| *Sum* | 111 | 88 | 1107 | 738 | 751 |

$$r = \frac{N\Sigma XY - (\Sigma X)(\Sigma Y)}{\sqrt{N\Sigma X^2 - (\Sigma X)^2}\sqrt{N\Sigma Y^2 - (\Sigma Y)^2}}$$

$$r = \frac{12(751) - (111)(88)}{\sqrt{12(1107) - (111)^2}\sqrt{12(738) - (88)^2}}$$

$$r = \frac{9012 - 9768}{\sqrt{13284 - 12321}\sqrt{8856 - 7744}}$$

$$r = \frac{-756}{\sqrt{963}\sqrt{1112}} = \frac{-756}{(31.03)(31.81)}$$

$$r = \frac{-756}{987.06} = -0.766 = -0.77$$

the reported average number of hours studied per week for students in
a mathematics class, and let $Y$ represent the number of problems missed
on the final examination for the course. Thus the first person in the
group of 12 students reported studying an average of five hours per week
and missed 10 problems on the final examination. The squared raw scores
in $X$ are placed in the column headed by $X^2$, the squared raw scores
in $Y$ are placed in the column headed by $Y^2$, and the cross product
terms are placed in the last column. Substituting $N = 12$ and the sums
of the columns in Table 6-1 in Formula 6-14 results in a correlation
of $-0.77$ between hours of study and number of errors on the final
exam, as shown by the computations at the bottom of Table 6-1. The
negative correlation means that greater numbers of hours spent studying

tend to be associated with smaller numbers of errors on the final examination.

This method of computing the correlation coefficient is particularly useful when computing machinery, e.g., a desk calculator or an electronic computer, is available. If it is necessary to do the computations by hand without such aids, multiplying and adding large raw scores can be laborious, especially for large numbers of cases. Where no computing aids are available, therefore, it will probably be more economical to calculate $r$ by the code method which will be the subject of Problem E.

## Summary of Steps for Computing the Correlation Coefficient from Raw Scores Using the Raw Score Formula

1.  Obtain the sum of raw scores for the $X$ score variable, $\Sigma X$.
2.  Obtain the sum of the squared raw scores for the $X$ score variable, $\Sigma X^2$.
3.  Obtain the sum of raw scores for the $Y$ score variable, $\Sigma Y$.
4.  Obtain the sum of the squared raw scores for the $Y$ score variable, $\Sigma Y^2$.
5.  Obtain the sum of the cross product terms, $\Sigma XY$, by multiplying each $X$ score by the corresponding Y score and adding up these values over all cases from 1 to $N$.
6.  Substitute the sums obtained in Steps 1 through 5 above, along with the value of $N$, into Formula 6–14 and solve for $r$, the correlation coefficient.

## Mathematical Notes

The derivation of Formula 6–14 proceeds from Formula 6–6 as follows:

$$r = \frac{\Sigma xy}{N \, \text{S.D.}_x \text{S.D.}_y} \tag{6-6}$$

Substituting the raw score formula for the standard deviation and the raw score equivalent to the deviation score in Formula 6–6 gives:

$$r = \frac{\Sigma(X - M_x)(Y - M_y)}{N \cdot \dfrac{1}{N}\sqrt{N\Sigma X^2 - (\Sigma X)^2} \cdot \dfrac{1}{N}\sqrt{N\Sigma Y^2 - (\Sigma Y)^2}}$$

Multiplying numerator and denominator by $N$ and multiplying out the two terms in parentheses gives:

$$r = \frac{N\Sigma(XY - M_x Y - M_y X + M_x M_y)}{\sqrt{N\Sigma X^2 - (\Sigma X)^2}\,\sqrt{N\Sigma Y^2 - (\Sigma Y)^2}}$$

Remembering that the sum of a constant term times a variable term moves the constant term outside the summation sign, and the sum of the product of two constants is just $N$ times the product, gives:

$$r = \frac{N\Sigma XY - N \cdot M_x\Sigma Y - N \cdot M_y\Sigma X + N^2 M_x M_y}{\sqrt{N\Sigma X^2 - (\Sigma X)^2} \sqrt{N\Sigma Y^2 - (\Sigma Y)^2}}$$

Since $M_x = \Sigma X/N$, $NM_x = \Sigma X$ and $NM_x = \Sigma Y$, giving

$$r = \frac{N\Sigma XY - \Sigma X\Sigma Y - \Sigma X\Sigma Y + \Sigma X\Sigma Y}{\sqrt{N\Sigma X^2 - (\Sigma X)^2} \sqrt{N\Sigma Y^2 - (\Sigma Y)^2}}$$

$$r = \frac{N\Sigma XY - \Sigma X\Sigma Y}{\sqrt{N\Sigma X^2 - (\Sigma X)^2} \sqrt{N\Sigma Y^2 - (\Sigma Y)^2}} \tag{6-1}$$

## Problem 17
## DEVELOP A DEVIATION SCORE REGRESSION EQUATION

Formula 6–5 represents the formula that will usually be employed to set up a regression equation for predicting deviation scores in one variable, $Y$, from a knowledge of deviation scores in the other, $X$. From Formula 6–5 it is clear that three items of information are needed, the correlation between variables $Y$ and $X$, i.e., $r$, and the two standard deviations, S.D.$_x$ and S.D.$_y$:

$$y' = r \cdot \frac{\text{S.D.}_y}{\text{S.D.}_x} \cdot x \tag{6-5}$$

To set up a regression equation in deviation score form for predicting deviation scores in $Y$ from deviation scores in $X$ it is merely necessary to substitute S.D.$_x$, S.D.$_y$, and $r$ into Formula 6–5.

### An Example of a Deviation Score Regression Equation

Suppose that a regression equation were being set up to predict deviation scores in intellectual aptitude (S.D. = 10.9) from scores in verbal knowledge (S.D. = 5.1). Assume the correlation between these two variables to be 0.80. Substituting S.D.$_y$ = 10.9, S.D.$_x$ = 5.1, and $r$ = 0.80 in Formula 6–5 gives:

$$y' = (0.80) \times \frac{10.9}{5.1} \times x \tag{6-15}$$

$$y' = 1.71x$$

Formula 6–15 can be used to obtain predicted values of $y$, i.e. predicted deviation scores in the $Y$ variable, associated with any particular $x$ value. Substituting $x = +1$ in Formula 6–19, for example, gives 1.71 as the predicted score in $y$. Substituting $x = -2$ in Formula 6–19 gives $-3.42$ as the corresponding $y'$ value.

It must be remembered that the $y'$ values obtained by substituting $x$ values in Equation 6–15 are such that the points $(x, y')$ fall directly on the regression line. Actually $y$ values for real subjects with a given $x$ value will vary in a more or less random fashion about the $y'$ value associated with that given $x$. The variability of the $y'$ values for all $x$ values is less than the variability of the $y$ values themselves, unless the correlation between $x$ and $y$ is perfect. Predictions of $y$ values are restricted within a rather narrow range about zero (the mean of the $y$ values) for low correlations and gradually spread out over a greater range as the correlation increases.

Although it is not generally needed, the deviation score regression equation for predicting $x'$ from $y$ can be obtained. Using the data in this example and Formula 6–8 gives:

$$x' = r \cdot \frac{\text{S.D.}_x}{\text{S.D.}_y} \cdot y \qquad (6\text{–}8)$$

$$x' = (0.80) \frac{5.1}{10.9} \cdot y$$

$$x' = 0.37y$$

Regression equations in deviation scores are simpler to deal with because deviation scores have a mean of zero and the regression lines go through the origin. In practice, it is more common to deal with regression equations that are stated in terms of raw scores since conversion of raw scores to deviation scores represents additional labor to be avoided if possible.

## Problem 18
### DEVELOP A STANDARD SCORE REGRESSION EQUATION AND A RAW REGRESSION EQUATION

Formula 6–7 showed that the regression equation for standard scores involves only the two variables and the correlation coefficient:

$$Z'_y = r \cdot Z_x \qquad (6\text{–}7)$$

For the data shown in Problem 17 where the correlation between variables was 0.80, Formula 6–7 would become $Z'_y = 0.80Z_x$. The corresponding regression equation for predicting standard scores in $X$ from standard scores in $Y$ would be $Z'_x = rZ_y$ or $Z'_x = 0.80Z_y$ in this particular example.

Conversion of raw scores to standard scores is even more laborious than converting them to deviation scores, however, so Formula 6–7 is not often used to develop a regression equation for practical use.

The most commonly used type of regression equation will involve the prediction of raw scores in $Y$ from raw scores in $X$. An equation for this purpose can be obtained very easily by substituting the formulas for the deviation scores, $x = X - M_x$ and $y = Y - M_y$, into the deviation score formula for the regression equation:

$$y' = r \cdot \frac{\text{S.D.}_y}{\text{S.D.}_x} \cdot x \qquad (6\text{–}6)$$

$$Y' - M_y = r \cdot \frac{\text{S.D.}_y}{\text{S.D.}_x} \cdot (X - M_x)$$

$$Y' = r \cdot \frac{\text{S.D.}_y}{\text{S.D.}_x} \cdot X + \left( M_y - r \frac{\text{S.D.}_y}{\text{S.D.}_x} \cdot M_x \right) \qquad (6\text{–}16)$$

Formula 6–16 is a raw score formula for the regression line for predicting raw scores in the $Y$ variable from a knowledge of raw scores in the $X$ variable. Although it is not usually needed, since the variable to be predicted can always be designated arbitrarily as the $Y$ variable, the regression equation for predicting raw scores in the $X$ variable from raw scores in the $Y$ variable is:

$$X' = r \cdot \frac{\text{S.D.}_x}{\text{S.D.}_y} \cdot Y + \left( M_x - r \cdot \frac{\text{S.D.}_x}{\text{S.D.}_y} \cdot M_y \right) \qquad (6\text{–}17)$$

*An Example.* To apply Formula 6–16, the necessary items of information are $S.D._x$, $S.D._y$, $M_x$, $M_y$, and $r$. For example, substituting the values

$$S.D._x = 5.1,\ S.D._y = 10.9,\ M_x = 15.1,\ M_y = 48.3,\ \text{and}\ r = 0.80$$

into Formula 6–20 gives:

$$Y' = (0.80) \cdot \frac{10.9}{5.1} \cdot X + \left( 48.3 - (0.80) \cdot \frac{10.9}{5.1} \cdot (15.1) \right)$$

$$Y' = 1.71X + 22.5 \qquad (6\text{–}18)$$

Substituting an $X$ score of 15.1, for example, into the raw score regression Formula 6–18 yields $(1.71)(15.1) + 22.5$ or $25.8 + 22.5 = 48.3$, the mean score for variable $Y$. This was to be expected since the $X$ value entered into the equation was the mean value of the variable $X$. The largest value of $X$ is 27. Entered into Formula 6–18, $X = 27$ gives a predicted $Y$ value of 68.7 which is still a class interval below the top $Y$ score. Since the correlation is not perfect, predicted $Y'$ scores have a smaller range than obtained $Y$ scores.

Equations such as Formula 6–18 can be used for predicting success in a variety of activities. If test scores are available, for example, on a test known to be highly correlated with after-training job productivity, a regression equation can be developed to predict production scores. Prospective employees whose predicted scores exceed a certain level can be selected for hiring with a substantial likelihood that they will perform successfully on the job.

## Problem E
### COMPUTE THE CORRELATION COEFFICIENT FROM A SCATTER DIAGRAM USING THE CODE METHOD[1]

When computing machinery is not available, the best way to obtain the product moment correlation coefficient is to compute it from a scatter diagram using the code method. Even if a desk calculator or similar piece of equipment is to be used as an aid in the calculations this method may still be preferred over the raw score formula method, provided that a look at the regression plot is desired. If there is any question whether the regression will be linear or not, for example, it is prudent to prepare a scatter diagram so that a visual inspection of the regression plot can be made. Unless the regression plot appears to be linear, i.e., a straight line is the curve of best fit to the points in the plot, the product moment correlation coefficient should not be computed. If the regression plot appears to be curvilinear, i.e., a curved line appears to be the line of best fit through the points, some other method of treating the relationship between the variables, e.g., the correlation ratio,[2] should be used instead of the product moment correlation coefficient.

If a scatter plot is to be made anyway, for purposes of examining the regression pattern, computation of the product moment correlation coefficient by the code method follows rather easily and may be preferred to use of the raw score formula whether or not computing aids are available. If many correlations are to be computed in series by electronic computer, however, the raw score formula method (Problem 16) will be more effective. In this case if plots are to be made, they can be done separately, also by computer. Or, linearity of regression can be tested computationally using a Chi Square test without making a plot (see Chapter 10).

[1] This problem, which is in the letter series, may be omitted without disturbing the continuity of study for problems in the number series.

[2] J. P. Guilford and B. Fruchter, *Fundamental Statistics in Psychology and Education,* 5th ed. (New York: McGraw-Hill Book Company, 1973).

## Preparation of a Scatter Plot

Problem 1 was concerned with the preparation of a frequency distribution for one variable, $X$ or $Y$. This represents a one-dimensional scatter plot. That is, it shows how scores are spread out over the class intervals in a single dimension. The scatter plot is a two-dimensional frequency distribution. It shows how many cases fall in a particular class interval for the $X$ variable but *also* in a specified class interval for the $Y$ variable. If there are 10 class intervals in the frequency distribution for the $X$ variable and 12 class intervals in the frequency distribution for the $Y$ variable, there are $10 \times 12 = 120$ cells in the joint frequency distribution, or scatter plot.

Each person in the sample has two scores, an $X$ score and a $Y$ score, e.g., an aptitude test score and a grade point average. Two scores are required to locate the person in one of the cells in the two-dimensional frequency distribution, or scatter plot. Figure 6–7 shows a scatter plot for two variables as well as subtotals to be used for obtaining the correlation coefficient by the code method. Let the scatter plot in Figure 6–7 represent the relationship between scores on a test of general intellectual aptitude ($Y$), and scores on a test of verbal knowledge ($X$).

The class intervals for the $Y$ variable, general intellectual aptitude scores, are shown in the first column of Figure 6–7. The class intervals for the $X$ variable (verbal knowledge test scores), however, are not listed vertically, as usual, but horizontally across the top of the scatter plot. The lowest class interval in $X$ is 4–5, and the top class interval is 26–27.

The only information in Figure 6–7 that is concerned with the scatter plot, per se, is the column and row of class intervals for the $Y$ and $X$ variables, respectively, and the tallies that appear in the center of many cells in the table. Each tally represents a single person. If a person had a verbal knowledge score ($X$) of 8 and a general intellectual aptitude score ($Y$) of 39, the tally for this person would fall in the box located in the row to the right of class interval 35–39 in $Y$ and in the column beneath the class interval 8–9 in $X$. There are three tallies in that cell. The numbers below and to the left and above and to the right of the tally marks are used in computing $r$ and do not in themselves constitute part of the scatter plot. This is also true of the rows and columns of numbers below and to the right of the scatter plot in Figure 6–7.

In making a scatter plot then, the first step is to choose $i$, the class interval size for variables $X$ and $Y$ and then lay off the $Y$ class intervals in the left-hand column and the $X$ class intervals across the top of the chart. Lines are drawn to form boxes for tallying. A tally mark is made for each subject, locating the box where that subject belongs at the intersection of the column his $X$ score belongs in and the row his $Y$ score belongs in. In Figure 6–7, a straight line appears to be the curve of

## FIGURE 6-7
### Computation of r by the Code Method

| x \ y | 4-5 | 6-7 | 8-9 | 10-11 | 12-13 | 14-15 | 16-17 | 18-19 | 20-21 | 22-23 | 24-25 | 26-27 | $f_y$ | $y'$ | $f_y'$ | $f_y'^2$ | $x'y'$ |
|---|---|---|---|---|---|---|---|---|---|---|---|---|---|---|---|---|---|
| 70-74 | | | | | | | | | | | | | 3 | 4 | 12 | 48 | 52 |
| 65-69 | | | | | | | | | | | | | 8 | 3 | 24 | 72 | 57 |
| 60-64 | | | | | | | | | | | | | 11 | 2 | 22 | 44 | 32 |
| 55-59 | | | | | | | | | | | | | 14 | 1 | 14 | 14 | 8 |
| 50-54 | | | | | | | | | | | | | 23 | 0 | 0 | 0 | 0 |
| 45-49 | | | | | | | | | | | | | 23 | -1 | -23 | 23 | 10 |
| 40-44 | | | | | | | | | | | | | 17 | -2 | -34 | 68 | 56 |
| 35-39 | | | | | | | | | | | | | 17 | -3 | -51 | 153 | 144 |
| 30-34 | | | | | | | | | | | | | 9 | -4 | -36 | 144 | 156 |
| 25-29 | | | | | | | | | | | | | 5 | -5 | -25 | 125 | 125 |
| $f_x$ | 5 | 6 | 11 | 9 | 15 | 18 | 23 | 20 | 11 | 6 | 4 | 2 | 130 | | -97 | 691 | 640 |
| $x'$ | -6 | -5 | -4 | -3 | -2 | -1 | 0 | 1 | 2 | 3 | 4 | 5 | | | | | |
| $f_x'$ | -30 | -30 | -44 | -27 | -30 | -18 | 0 | 20 | 22 | 18 | 16 | 10 | -93 | | | | |
| $f_x'^2$ | 180 | 150 | 176 | 81 | 60 | 18 | 0 | 20 | 44 | 54 | 64 | 50 | 897 | | | | |
| $x'y'$ | 126 | 105 | 144 | 57 | 48 | 17 | 0 | 7 | 16 | 33 | 52 | 35 | 640 | | | | |

best fit for the data so it is appropriate to proceed with calculations to obtain the product moment correlation coefficient.

## Code Method Formula for the Product Moment Correlation Coefficient

The formula for calculating the product moment coefficient of correlation by the code method is given by:

$$r = \frac{\dfrac{\Sigma x'y'}{N} - \dfrac{\Sigma fx'}{N}\dfrac{\Sigma fy'}{N}}{\sqrt{\dfrac{\Sigma fx'^2}{N} - \left(\dfrac{\Sigma fx'}{N}\right)^2}\sqrt{\dfrac{\Sigma fy'^2}{N} - \left(\dfrac{\Sigma fy'}{N}\right)^2}} \tag{6-19}$$

Where:

$r$ = the product moment (Pearson) correlation coefficient

$x'$ = deviation from the guessed mean in class interval units for the $X$ variable

$y'$ = deviation from the guessed mean in class interval units for the $Y$ variable

$f$ = frequency

Formula 6–19 will be derived from the deviation score formula for the correlation coefficient (Formula 6–6) in the mathematical notes section below. The similarity of the radical terms in the denominator to the code method formula for the standard deviation (Formula 4–3) should be noted. Multiplying the left hand radical in the denominator of Formula 6–19 by $i_x$, the class interval in the $X$ distribution, gives the code method formula for the standard deviation in $X$. Multiplying the radical on the right by $i_y$, the class interval in the $Y$ distribution, gives the code method formula for the standard deviation in $Y$.

### Steps in Computing $r$ by the Code Method

Computation of the product moment correlation coefficient by the code method from a scatter diagram is illustrated by the worked out example shown in Figure 6–7 and Table 6–2. The steps are as follows:

1.  Prepare a scatter diagram in accordance with the rules already described above.
2.  To the right of the scatter diagram, add columns for $f_y$, $y'$, $fy'$, $fy'^2$, and $x'y'$. All of these columns except $x'y'$ are the same ones used in calculating the standard deviation by the code method.
3.  Lay off the corresponding $f_x$, $x'$, $fx'$, $fx'^2$, and $x'y'$ rows for the $X$ variable below the scatter diagram.
4.  Fill in the columns $f_y$, $y'$, $fy'$, and $fy'^2$ in the same manner as if the standard deviation in variable $Y$ were being computed by the code method. Do the same thing for $f_x$, $x'$, $fx'$, and $fx'^2$ as if the

TABLE 6–2
Computation of $r$ by the Code Method from
Data in Figure 6–7

$$r = \frac{\dfrac{\Sigma x'y'}{N} - \dfrac{\Sigma fx'}{N}\dfrac{\Sigma fy'}{N}}{\sqrt{\dfrac{\Sigma fx'^2}{N} - \left(\dfrac{\Sigma fx'}{N}\right)^2}\sqrt{\dfrac{\Sigma fy'^2}{N} - \left(\dfrac{\Sigma fy'}{N}\right)^2}}$$

$$r = \frac{\dfrac{640}{130} - \left(\dfrac{-93}{130}\right)\left(\dfrac{-97}{130}\right)}{\sqrt{\dfrac{897}{130} - \left(\dfrac{-93}{130}\right)^2}\sqrt{\dfrac{691}{130} - \left(\dfrac{-97}{130}\right)^2}}$$

$$r = \frac{4.92 - (-0.715)(-0.746)}{\sqrt{6.9000 - 0.5112}\sqrt{5.3154 - 0.5564}}$$

$$r = \frac{4.92 - 0.53}{\sqrt{6.3888}\sqrt{4.7589}}$$

$$r = \frac{4.39}{\sqrt{30.4037}}$$

$$r = \frac{4.39}{5.51} = 0.798$$

$$r = 0.80$$

standard deviation in variable $X$ were being computed by the code method. Obtain the row and column sums, except for the $y'$ column and $x'$ row. The sum of the $f_y$ column and the $f_x$ row should be identical, i.e., $N$. Mark the column boundaries for the $x' = 0$ column with heavier lines. Do likewise for the $y' = 0$ row. This facilitates reading the chart and helps to prevent errors.

5. Determine the $x'y'$ value for each cell in the chart that contains any tallies and write this value in the lower left-hand corner of the cell, as shown in Figure 6–7. Some other location in the cell would serve as well as long as it is used consistently for all cells. The $x'y'$ value is just the product of the $x'$ in the column where the cell is located by the $y'$ in the row where the cell is located. These values depend upon the choice of placement for the guessed means in the $X$ and $Y$ variables.

6. Multiply the $x'y'$ values in each cell by the number of tallies in the cell and place the product in the upper right-hand corner of the cell, as shown in Figure 6–7. Each person in the cell has the same $x'y'$ value so the product of this cell $x'y'$ value by the frequency count in the cell is the sum of the $x'y'$ values for that cell. Note

that this sum may be positive or negative depending upon whether the $x'y'$ product is positive or negative. If either $x'$ or $y'$ alone is negative, it is sufficient to make $x'y'$ negative. Of course, if both $x'$ and $y'$ are negative, $x'y'$ will be positive.

7. Sum the $x'y'$ totals for the cells in each row by adding algebraically the cell totals for $x'y'$ in the upper right-hand corners of the cells. Place these row totals in the column headed by $x'y'$, as shown in Figure 6–7. Adding these values algebraically means that the negative cell totals must be subtracted from the positive cell totals. It sometimes happens that the total for an entire row will be negative.

8. Sum the $x'y'$ totals for the cells in each column by adding algebraically the cell totals for $x'y'$ in the upper right-hand corners of the cells. Place these column totals in the row labeled $x'y'$.

9. Sum the $x'y'$ column and the $x'y'$ row algebraically, i.e., taking into account the signs of the values. These two sums should be identical unless an error has been made.

10. Substitute the sums of $fx'$, $fy'$, $fx'^2$, $fy'^2$, and $x'y'$ in Formula 6–19. This process is illustrated in Table 6–2, using the sums derived from the example shown in Figure 6–7. The student should take particular care to avoid the common error of failing to note that $\Sigma fy'^2$ and $(\Sigma fy')^2$ are different. In the first case, it is only the $y'$ values that are squared. This total is derived from the sum of the column headed by $fy'^2$. The term $(\Sigma fy')^2$ is obtained by squaring the sum of the column headed by $fy'$.

*Approximate Nature of the Code Method.* Grouping data and computing a correlation coefficient by the code method is an approximation method. This is due to the fact that all scores are treated as though they were at the midpoint of the interval. The possible error of approximation increases as the size of the interval increases. Ordinarily, the code method approximation to the correlation coefficient is sufficiently good for all practical purposes. If an exact value is needed, however, the raw score formula should be used rather than the code method.

**Mathematical Notes**

Formula 6–19 for the correlation coefficient may be derived from the deviation score Formula 6–6. The code method formula for the standard deviation is given by:

$$S.D. = i \sqrt{\frac{\Sigma fx'^2}{N} - \left(\frac{\Sigma fx'}{N}\right)^2} \qquad (4\text{–}3)$$

In the mathematical derivation of Formula 4–3 it was shown that:

$$x = i\left(x' - \frac{fx'}{N}\right) \qquad (6\text{–}20)$$

The deviation score formula for the correlation coefficient is given by:

$$r = \frac{\Sigma xy}{N \, S.D._x S.D._y} \tag{6-6}$$

Substituting Formulas 4–3 and 6–16 in 6–6 gives:

$$r = \frac{\Sigma i_x \left( x' - \dfrac{\Sigma fx'}{N} \right) \cdot i_y \left( y' - \dfrac{\Sigma fy'}{N} \right)}{N i_x \sqrt{\dfrac{\Sigma fx'^2}{N} - \left( \dfrac{\Sigma fx'}{N} \right)^2} \cdot i_y \sqrt{\dfrac{\Sigma fy'^2}{N} - \left( \dfrac{\Sigma fy'}{N} \right)^2}} \tag{6-21}$$

The $i$ values in 6–21 are constants that can be moved outside the summation sign and cancelled with the $i$ values in the denominator. Doing this and multiplying out the terms in parentheses in the numerator gives:

$$r = \frac{\Sigma \left( x'y' - \dfrac{\Sigma fx'}{N} \cdot y' - \dfrac{\Sigma fy'}{N} \cdot x' + \dfrac{\Sigma fx'}{N} \dfrac{\Sigma fy'}{N} \right)}{N \sqrt{\dfrac{\Sigma fx'^2}{N} - \left( \dfrac{\Sigma fx'}{N} \right)^2} \sqrt{\dfrac{\Sigma fy'^2}{N} - \left( \dfrac{\Sigma fy'}{N} \right)^2}}$$

Dividing numerator and denominator by $N$ and summing through in the numerator gives:

$$r = \frac{\dfrac{\Sigma x'y'}{N} - \dfrac{\Sigma fx'}{N} \dfrac{\Sigma y'}{N} - \dfrac{\Sigma fy'}{N} \cdot \dfrac{\Sigma x'}{N} + \dfrac{1}{N} \sum \left( \dfrac{\Sigma fx'}{N} \dfrac{\Sigma fy'}{N} \right)}{\sqrt{\dfrac{\Sigma fx'^2}{N} - \left( \dfrac{\Sigma fx'}{N} \right)^2} \sqrt{\dfrac{\Sigma fy'^2}{N} - \left( \dfrac{\Sigma fy'}{N} \right)^2}} \tag{6-22}$$

In the derivation of the code method formula for the standard deviation it was shown that $\Sigma fx'$ and $\Sigma x'$ are equal. It is also true that the sum of a constant term is $N$ times that constant so the $1/N$ in the last term in the numerator is cancelled by multiplying by this $N$ when the constant product term is summed. This leaves the last three terms in the numerator identical except for sign, so 6–22 reduces to:

$$r = \frac{\dfrac{\Sigma x'y'}{N} - \dfrac{\Sigma fx'}{N} \dfrac{\Sigma fy'}{N}}{\sqrt{\dfrac{\Sigma fx'^2}{N} - \left( \dfrac{\Sigma fx'}{N} \right)^2} \sqrt{\dfrac{\Sigma fy'^2}{N} - \left( \dfrac{\Sigma fy'}{N} \right)^2}} \tag{6-19}$$

## Exercises

6–1. Given the following pairs of Stanine Scores on two tests for 12 subjects, compute the correlation coefficient between the test scores using the raw score method: (6, 7); (5, 4); (4, 2); (5, 5); (6, 5); (2, 4); (9, 8); (7, 9); (1, 4); (8, 7); (6, 5); and (5, 6).

6-2.  Given the following pairs of observations on two variables for 10 subjects, compute the correlation coefficient using the raw score method: (3, 50); (6, 40); (12, 28); (5, 38); (7, 20); (6, 35); (4, 45); (9, 25); (8, 30); and (7, 26).

6-3.  Using the raw score formula compute the correlation coefficient for the first 20 cases in the data for Exercise 6–13. A calculator is needed for this exercise.

6-4.  Using the raw score formula compute the correlation coefficient for the second 20 cases in the data for Exercise 6–13 (i.e., subjects 21–40). A calculator is needed for this exercise.

6-5.  Using the raw score formula, compute the correlation coefficient for all 60 cases in the data for Exercise 6–13. (Use the totals from Exercises 6–3 and 6–4 if they have been obtained already.) A calculator is needed for this exercise.

6-6.  Using the raw score formula, compute the correlation coefficient for all 40 cases in the data for exercise 6–14. A calculator is needed for this exercise.

6-7.  Given the following pairs of error scores on successive administrations of a given test, compute the correlation between the sets of scores using the raw score formula:

(9, 10); (7, 2); (1, 5); (12, 14); (3, 0);
(6, 8); (9, 7); (10, 11); (10, 5); (2, 1).

6-8.  Given the following age and error scores on an ability test for a small sample of boys, find the correlation using the raw score formula:

(8, 8); (6, 12); (12, 5); (15, 1); (10, 10);
(5, 15); (12, 5); (9, 10); (13, 3); (10, 9).

6-9.  In a certain industrial plant, the average number of items produced per day per worker is 50 with a standard deviation of 10. A manual dexterity test correlates 0.50 with this criterion of production. Develop a deviation score regression equation to predict performance from test scores. The test score mean is 80 and the S.D. is 15. For a person with a test score of 60, what would be his predicted deviation production score?

6-10.  Given two sets of test scores, develop two deviation score regression equations to predict either score from the other where $r_{12} = 0.6$, $M_1 = 20$, S.D.$_1 = 3$, $M_2 = 40$, S.D.$_2 = 5$. What raw score in variable 2 would result in a predicted raw score of 18 in variable 1?

6-11.  Express the regression equation in exercise 6–9 in both standard score form and raw score form. Use the formula to predict the raw score in production.

6-12.  Express the regression equations in exercise 6–10 in both standard score form and raw score form. Use these formulas to solve for the standard score and raw score, respectively, in variable 2 that would result in a predicted score equivalent to a raw score of 18 in variable 1.

6–13.   Given the following scores for two personality factors 0, Orderliness vs Lack of Compulsion, and C, Social Conformity vs Rebelliousness, for a sample of 60 male subjects, prepare a scatter diagram and compute the correlation coefficient by the code method. Use $i = 5$ and 82 as the guessed mean for both $X$ and Y. (Problem E)

| Subject | 0 | C | Subject | 0 | C | Subject | 0 | C |
|---|---|---|---|---|---|---|---|---|
| 1..... | 72 | 59 | 21..... | 93 | 87 | 41..... | 77 | 69 |
| 2..... | 81 | 87 | 22..... | 111 | 83 | 42..... | 56 | 53 |
| 3..... | 54 | 36 | 23..... | 70 | 53 | 43..... | 97 | 83 |
| 4..... | 91 | 61 | 24..... | 103 | 81 | 44..... | 71 | 50 |
| 5..... | 74 | 58 | 25..... | 77 | 82 | 45..... | 71 | 84 |
| 6..... | 88 | 86 | 26..... | 97 | 93 | 46..... | 88 | 97 |
| 7..... | 78 | 70 | 27..... | 91 | 58 | 47..... | 103 | 84 |
| 8..... | 90 | 39 | 28..... | 53 | 95 | 48..... | 81 | 92 |
| 9..... | 77 | 104 | 29..... | 81 | 67 | 49..... | 56 | 56 |
| 10..... | 88 | 84 | 30..... | 73 | 101 | 50..... | 53 | 76 |
| 11..... | 90 | 70 | 31..... | 93 | 72 | 51..... | 77 | 77 |
| 12..... | 69 | 55 | 32..... | 70 | 49 | 52..... | 104 | 73 |
| 13..... | 94 | 89 | 33..... | 82 | 66 | 53..... | 83 | 60 |
| 14..... | 100 | 94 | 34..... | 86 | 92 | 54..... | 81 | 48 |
| 15..... | 74 | 51 | 35..... | 66 | 57 | 55..... | 70 | 44 |
| 16..... | 76 | 54 | 36..... | 81 | 90 | 56..... | 71 | 66 |
| 17..... | 96 | 101 | 37..... | 101 | 95 | 57..... | 75 | 105 |
| 18..... | 50 | 46 | 38..... | 94 | 68 | 58..... | 76 | 62 |
| 19..... | 103 | 80 | 39..... | 108 | 86 | 59..... | 94 | 73 |
| 20..... | 93 | 91 | 40..... | 69 | 51 | 60..... | 94 | 85 |

6–14.   Consider the following scores to represent weights and number of minutes played in a randomly selected football game for all team players. Prepare a scatter diagram and compute the correlation coefficient by the code method. (Problem E)

| Player | Weight | Min-utes | Player | Weight | Min-utes | Player | Weight | Min-utes |
|---|---|---|---|---|---|---|---|---|
| 1.... | 180 | 15 | 16.... | 215 | 0 | 31.... | 205 | 19 |
| 2.... | 220 | 20 | 17.... | 170 | 0 | 32.... | 200 | 17 |
| 3.... | 200 | 29 | 18.... | 185 | 50 | 33.... | 195 | 15 |
| 4.... | 175 | 8 | 19.... | 155 | 2 | 34.... | 220 | 49 |
| 5.... | 195 | 0 | 20.... | 230 | 46 | 35.... | 185 | 11 |
| 6.... | 200 | 38 | 21.... | 225 | 38 | 36.... | 190 | 5 |
| 7.... | 240 | 6 | 22.... | 180 | 25 | 37.... | 200 | 18 |
| 8.... | 190 | 21 | 23.... | 175 | 10 | 38.... | 175 | 12 |
| 9.... | 205 | 33 | 24.... | 200 | 25 | 39.... | 190 | 20 |
| 10.... | 230 | 40 | 25.... | 195 | 31 | 40.... | 200 | 17 |
| 11.... | 160 | 8 | 26.... | 210 | 35 | | | |
| 12.... | 180 | 0 | 27.... | 190 | 21 | | | |
| 13.... | 185 | 13 | 28.... | 200 | 26 | | | |
| 14.... | 200 | 19 | 29.... | 195 | 16 | | | |
| 15.... | 210 | 28 | 30.... | 185 | 12 | | | |

chapter 7

# Testing Hypotheses with Frequency
# Data Using the Binomial Distribution

Up to this point in the book, the emphasis has been on "descriptive" statistics. That is, the various statistical features of a body of data are described, much as a biographical sketch will describe a person. The remainder of the book will be concerned with "inferential" statistics. Whereas descriptive statistics are concerned with painting a picture of what the data are like, inferential statistics are concerned with the problem of specifying how accurate that description is.

Suppose, for example, that it is important to determine what proportion of males in the United States are smokers. A large random sample could be drawn from the population of U.S. males and the proportion of them who are smokers determined by interview, questionnaire, or whatever. This proportion would be a descriptive statistic for the sample and would provide an estimate of the proportion of smokers in the total population of U.S. males. The question "How accurate is this estimate?" raises a problem in statistical inference. The sample proportion is not apt to be exactly equal to the population proportion and in some cases may depart from it to an appreciable degree. Through the methods of statistical inference it is possible to estimate the probability that the sample proportion will deviate from the population proportion by more than a predetermined amount. This helps to fix the location of the true population proportion within certain limits rather than just to estimate its value without giving any idea about how far off the estimate could be.

The present chapter will be concerned with statistical inferences that can be made by applying the binomial distribution to frequency data. The next chapter will treat problems of statistical inference with continuous variable data. The last two chapters, on Chi Square and Analysis of Variance, are also concerned with problems in statistical inference.

## Problem 19
## GIVEN A SITUATION WITH TWO POSSIBLE OUTCOMES, DETERMINE THE PROBABILITY OF A PARTICULAR COMBINATION OF OUTCOMES IN $n$ TRIALS

One of the major tools for making statistical inferences with frequency data is the binomial distribution. Problem 19 represents an application of this distribution which is not an example of statistical inference in itself. It is, however, a necessary preliminary to more complicated applications of the binomial which do represent examples of statistical inference.

### The Binomial Distribution

Consider the classical example of tossing a coin $n$ times where the coin may fall heads or tails on any one trial. If $p$ is the probability of a head coming up on any one trial and $q$ is the probability of a tail coming up on any one trial, where $p + q = 1$, then the probability of getting exactly $x$ heads in $n$ trials is given by:

$$P_x = \frac{n!}{x!(n-x)!} p^x q^{n-x} \qquad (7\text{–}1)$$

Where:

$P_x$ = probability of exactly $x$ heads showing up in $n$ coin tosses

$p$ = probability of a head on any given trial (usually $p = \frac{1}{2}$ in this situation)

$q$ = probability of a tail (usually $\frac{1}{2}$) where $q = 1 - p$

$n$ = number of coin tosses

$x$ = number of heads being tested for the probability value

If 10 coins are tossed, or one coin is tossed 10 times, Formula 7–1 can be used to determine the probability that the 10 tosses will result in any particular combination of heads and tails totalling 10, e.g., 4 heads and 6 tails. Substituting $x = 4$, $n = 10$, and $p = \frac{1}{2}$ (remember $q = 1 - p$) in Formula 7–1 gives:

$$P_4 = \frac{10!}{4!(10-4)!} \left(\frac{1}{2}\right)^4 \left(\frac{1}{2}\right)^6 \qquad (7\text{–}2)$$

Remembering that 10! (read as 10 factorial) means $10 \cdot 9 \cdot 8 \cdot 7 \cdot 6 \cdot 5 \cdot 4 \cdot 3 \cdot 2 \cdot 1$, and that $(\frac{1}{2})^p(\frac{1}{2})^q = (\frac{1}{2})^{p+q}$, 7-2 above becomes:

$$P_4 = \frac{10 \cdot 9 \cdot 8 \cdot 7 \cdot 6 \cdot 5 \cdot 4 \cdot 3 \cdot 2 \cdot 1}{(4 \cdot 3 \cdot 2 \cdot 1)(6 \cdot 5 \cdot 4 \cdot 3 \cdot 2 \cdot 1)} \left(\frac{1}{2}\right)^{10}$$

$$P_4 = \frac{10 \cdot 9 \cdot 8 \cdot 7}{4 \cdot 3 \cdot 2 \cdot 1} \cdot \frac{1}{1028}$$

$$P_4 = \frac{210}{1028}$$

If the probability of no heads is called for, then Formula 7-2 gives:

$$P_0 = \frac{10!}{0!(10-0)!} \left(\frac{1}{2}\right)^0 \left(\frac{1}{2}\right)^{10}$$

$$P_0 = \frac{1}{1028}$$

This is due to the fact that 0! is defined to be equal to 1 and any quantitity to the zero power is defined to be 1.

Applying Formula 7-2 successively for all the values of $x$ from 0 to 10 with $n = 10$ and $p = \frac{1}{2}$ gives the table of probabilities shown in Table 7-1. It will be noted that the sum of all the probability values adds up to

TABLE 7-1
Probabilities for $x$ Heads
in Ten Coin Tosses

| $x$ | Probability |
|---|---|
| 0 | 1:1024 |
| 1 | 10:1024 |
| 2 | 45:1024 |
| 3 | 120:1024 |
| 4 | 210:1024 |
| 5 | 252:1024 |
| 6 | 210:1024 |
| 7 | 120:1024 |
| 8 | 45:1024 |
| 9 | 10:1024 |
| 10 | 1:1024 |
| *Sum* | 1024:1024 |

$1024/1024 = 1$; since these outcomes exhaust all the possibilities, the sum of their probabilities *must* equal 1.0. It should also be noticed that the distribution in Table 7-1 is symmetric about $x = 5$, the central value, i.e., 4 and 6 have the same probability, 3 and 7 have the same probability, and so on. The distribution will be symmetric in this way when $p = q$.

Formula 7–1, then, can be used to determine the probability of a particular combination of outcomes (e.g., numbers of heads and tails) in $n$ trials, or equivalently, the probability of a particular value of $x$, e.g., number of heads in tossing $n$ coins. It is assumed that $p$, the probability of a head, for instance, remains constant from trial to trial and it assumes that each trial is independent of each other trial. That is, what happened on previous trials does not affect the probabilities for the current trial. A derivation for Formula 7–1 is given in the mathematical notes section below.

*The Binomial Expansion.* Formula 7–1 gave the formula for the general term of the binomial distribution. That formula can be used to generate each and every term for all the values of $x = 0, 1, 2, \ldots n$. It is sometimes more convenient, however, to use the following scheme for generating these terms:

$$(p + q)^n = p^n + \frac{n}{1} p^{n-1} q^1 + \frac{n(n-1)}{2 \cdot 1} p^{n-2} q^2 \tag{7–3}$$

$$+ \frac{n(n-1)(n-2)}{3 \cdot 2 \cdot 1} p^{n-3} q^3 + \frac{n(n-1)(n-2)(n-3)}{4 \cdot 3 \cdot 2 \cdot 1} p^{n-4} q^4$$

$$+ \cdots + \frac{n!}{n!} p^{n-n} q^n$$

The sequence of terms in Formula 7–3 provides a device for producing all the terms of the binomial distribution. The terms follow a regular progression from the first term. The first term is always $p^n$. The exponent of the first term is divided by 1 and multiplied by the second term. The exponent of $p$ is reduced by 1 in the second term and the exponent of $q$ is increased by 1. It was $q^0$ in the first term, which is just 1, hence does not show there. The third term adds an additional multiplier, $(n - 1)$, in the numerator and 2 in the denominator. The numerator of the multiplier with each successive term will have an additional element multiplied in which is one less than that multiplied in for previous terms. The denominator of the term multiplier keeps multiplying in an extra element with each term that is one larger than the one for the previous term. This process will terminate, i.e., the last term will be reached, when the term multiplier is $n!/n! = 1$. The $p$ exponent decreases by 1 each time until with the last term it becomes $p^0$ and drops out since $p^0 = 1$. The exponent of $q$ increases by one each term until it becomes $q^n$ for the last term.

Formula 7–3 will yield $n + 1$ terms in the binomial expansion. The first term gives the probability of an $n$, 0 split, e.g., all heads and no tails, the second term gives the probability of an $n - 1$, 1 split, e.g., $n - 1$ heads and 1 tail, and so on. The last term gives the probability of a 0, $n$ split, e.g., no heads, all tails. The sum of all the terms in a particular expansion equals 1.0.

The terms of Formula 7–3, therefore, give the results that would be obtained by successively substituting $x = n, n - 1, n - 2, \ldots 1, 0$ in 7–1, the formula for the general term of the binomial distribution. Letting $x = 0$, for example, gives:

$$P_0 = \frac{n!}{0!(n - 0)!} p^0 q^{n-0}$$

$$= \frac{n!}{n!} q^n$$

$$P_0 = q^n \qquad (7\text{--}4)$$

The value of $P_0$ is $q^n$, the last term of Formula 7–3.
Substituting $x = n$ in Formula 7–1 gives:

$$P_n = \frac{n!}{n!(n - n)!} p^n q^{n-n}$$

$$= \frac{n!}{n!(0!)} p^n q^0$$

$$P_n = p^n \qquad (7\text{--}5)$$

The value of $P_n$ in Formula 7–5 is the first term of Formula 7–3.

To get all the terms of the binomial expansion, or the first few terms of the series, Formula 7–3 is probably most convenient. To obtain a particular single term, it is usually more convenient to use Formula 7–1, particularly if it is not an early term in the series.

### Application of the Binomial Distribution to Social Science Research Problems

There are many situations in social science research where a given trial could result in one of two possible outcomes and where many trials can be run. The binomial distribution can be applied to such situations to estimate the probability of obtaining a particular outcome, assuming $p = \frac{1}{2}$, or some other specified value, $n$ is known, and the trials are independent.

Consider the case of the T-maze in experimental psychological research. A rat placed on the runway proceeds to the choice point in the maze and then goes either left or right. No other possibilities are admitted and the rat is "encouraged" to continue until a choice has been recorded. Assuming that rats are making random choices at the choice point, i.e., $p = \frac{1}{2}$, the probability of obtaining a specified number of left and right turns when $n$ rats are run in the maze can be determined using the binomial distribution.

*An Example.* Suppose 15 rats are run in a T-maze. It turns out that 10 rats go left and 5 go right. What is the probability of obtaining this particular result if the rats are making random choices, i.e., if the probability of a right turn (or a left turn) is $\frac{1}{2}$?

Since this is a single term of the binomial not near the beginning, it will be more convenient to use Formula 7–1. Substituting $x = 5$ and $n = 15$ in Formula 7–1 gives:

$$P_5 = \frac{15!}{5!(15-5)!} \left(\frac{1}{2}\right)^5 \left(\frac{1}{2}\right)^{15-5}$$

$$= \frac{15 \cdot 14 \cdot 13 \cdot 12 \cdot 11}{5 \cdot 4 \cdot 3 \cdot 2 \cdot 1} \left(\frac{1}{2}\right)^{15}$$

$$P_5 = \frac{3003}{32768} = 0.09$$

Since $p = q = \frac{1}{2}$ in this problem, it did not matter whether $x = 5$ or $x = 10$ is substituted in Formula 7–1. When $p$ is not equal to $q$, however, the value of $x$ must be related to the event for which $p$ is the probability on a given trial, not $q$. For example, consider the probability of getting a 1 on exactly 4 out of 6 tosses of a die. There are six faces on a die, only one with a single dot, so if we denote $p$ as the probability of getting a 1 on a single throw of the die, $p = \frac{1}{6}$ and $q = 1 - \frac{1}{6} = \frac{5}{6}$. Here, $\frac{5}{6}$ is the probability of *not* getting a 1, i.e., 2, 3, 4, 5, or 6. To obtain the probability of getting a 1 exactly four times out of six, substitute $x = 4$ and $p = \frac{1}{6}$ in Formula 7–1:

$$P_4 = \frac{6!}{4!(6-4)!} \left(\frac{1}{6}\right)^4 \left(\frac{5}{6}\right)^2$$

$$= \frac{6 \cdot 5}{2 \cdot 1} \cdot \frac{1}{36 \cdot 36} \cdot \frac{25}{36}$$

$$P_4 = 15 \cdot \frac{25}{46656} = \frac{375}{46656} = 0.008$$

Inserting $x = 2$ (i.e., $n - 4$) in Formula 7–1 would not give the correct answer, however, unless $p$ is designated as $\frac{5}{6}$ and $q$ as $\frac{1}{6}$. Putting $p = \frac{1}{6}$ and $x = 2$ in Formula 7–1 would give the probability of two 1s out of 6 instead of 4.

Most applications of the binomial distribution in social science statistical inference will require addition of the probabilities for several terms rather than the use of the probability of a single term. Problem 19, therefore, is concerned with the solution of a part of a statistical inference problem more often than it represents a solution to the entire problem. In Problems 20, 21, and 22 the techniques used in Problem 19 will be applied to the solution of actual social science research statistical inference problems.

## Mathematical Notes

In this section, a derivation for Formula 7–1 will be given. Let there be two possible outcomes of an event, A with probability $p$, and B, with probability $q$, where $p + q = 1$ but $p$ does not necessarily equal $q$. Consider a given set of $n$ trials resulting in $x$ As and $n - x$ Bs. The obtained results and their associated probabilities may be schematized as follows:

$$\text{A} \quad \text{B} \quad \text{A} \quad \text{A} \quad \text{B} \quad \cdots \quad \text{B} \quad \text{A} \quad \text{A} \quad \text{A}$$
$$p \cdot q \cdot p \cdot p \cdot q \quad \cdots \quad q \cdot p \cdot p \cdot p$$

where there are $x$ As interspersed with $n - x$ Bs. The product of the probabilities is commutative, so the $p$s and $q$s above can be rearranged to give:

$$(p \cdot p \cdot p \cdots p)(q \cdot q \cdot q \cdots q) \tag{7–7}$$

where there are $x$ $p$s in the first parentheses and $n - x$ $q$s in the second parentheses. This would then give $p^x q^{n-x}$ as the probability of getting the precise order of As and Bs schematized above in Formula 7–6.

Any other sequence of As and Bs that have $x$ As and $n - x$ Bs could be schematized and the $p$s and $q$s rearranged to give Formula 7–7. It follows that every ordering of $x$ As and $n - x$ Bs has the probability of $p^x q^{n-x}$. Since each of these is an independent event, the probability of any one of them occurring would be the sum of their probabilities since there are no duplicate orders in the list of possible orders. This would be equivalent to multiplying $p^x q^{n-x}$ by the number of such possible arrangements of $x$ As and $n - x$ Bs since each has the same probability of occurring.

If each A and each B were different in some minor way so that each movement of a given A (or B) to the place of another A (or B) would create a different order, the total number of arrangements of these As and Bs would be $n(n - 1)(n - 2) \cdots 1 = n!$ since $n$ different ones could be picked for the first spot, $n - 1$ for the second, and so on down until only 1 is left.

A great many of these $n!$ different orders would become indistinguishable from each other, however, if the As could not be distinguished from each other and the Bs could not be distinguished from each other. In fact each different placement of the As in a given dispersal arrangement among the Bs would give $x(x - 1)(x - 2) \cdots 1 = x!$ different arrangements when the distinguishing marks on the As are left on. Thus, $x!$ different orders collapse to 1 when the distinguishing marks on the As are removed. This is true for every one of the possible placements of the As. Hence, the total number of orders with distinguishing marks, $n!$ reduces to $n!/x!$ when the distinguishing marks are removed from the As. Similarly, every different placement of the Bs without distinguishing marks on them gives $(n - x)!$ different orders when the marks are distinguishable. Thus $(n - x)!$ orders reduce to 1 when the marks on the Bs are removed. Thus the total number of different orders of As and Bs when the marks are removed from both is $n!/x!(n - x)!$. When this number is multiplied by $p^x q^{n-x}$, the probability of each such order, the result is Formula 7-1:

$$P_x = \frac{n!}{x!(n - x)!} p^x q^{n-x} \qquad (7\text{--}1)$$

## Problem 20

### USING THE BINOMIAL DISTRIBUTION, MAKE A STATISTICAL TEST OF AN HYPOTHESIS, SELECTING THE APPROPRIATE ONE- OR TWO-TAILED TEST

The methods of Problem 19 provide a way of determining the probability of a particular distribution of two alternate outcomes under the proper conditions, e.g., the probability of getting 12 heads and 4 tails in 16 coin tosses. The methods of Problem 19 are not sufficient, however, to answer the question of whether or not this particular outcome is reasonable.

A coin should turn up heads 50 percent of the time in the long run, if it is a properly balanced coin. On any given sequence of 16 tosses, however, it is not to be expected that there will be exactly 8 heads and 8 tails. Even if the coin is properly balanced, the number of heads will sometimes be more, sometimes less than the number of tails in a given finite sequence of trials.

If a coin is not balanced, the number of heads might consistently exceed the number of tails. How unequal does the split have to be to raise a question about the balance or fairness of the coin? Would a 12–4 split be enough to render the coin suspect?

Suppose that on a given series of trials suspicions were aroused about a coin because it turned up more heads than tails, so that a decision was made to conduct a test to see if the coin is biased in favor of giving too many heads. Suppose further that a series of 16 tosses were decided upon as the test and the results came out 12 heads and 4 tails. Can it now be concluded that the coin is biased?

The solution to this problem will require the use of Formula 7–3 to calculate the first five terms of the binomial corresponding to $x = 16$, 15, 14, 13, and 12 heads. These terms will give, respectively, the probabilities for splits of 16–0, 15–1, 14–2, 13–3, and 12–4 for the division of the 16 trial results into heads and tails. These terms are as follows:

$$\left(\frac{1}{2}+\frac{1}{2}\right)^{16} = \left(\frac{1}{2}\right)^{16} + \frac{16}{1}\left(\frac{1}{2}\right)^{15}\left(\frac{1}{2}\right) + \frac{16\cdot15}{2\cdot1}\left(\frac{1}{2}\right)^{14}\left(\frac{1}{2}\right)^{2}$$

$$+ \frac{16\cdot15\cdot14}{3\cdot2\cdot1}\left(\frac{1}{2}\right)^{13}\left(\frac{1}{2}\right)^{3} + \frac{16\cdot15\cdot14\cdot13}{4\cdot3\cdot2\cdot1}\left(\frac{1}{2}\right)^{12}\left(\frac{1}{2}\right)^{4} + \cdots$$

Evaluation of these terms gives the following probabilities for 16, 15, 14, 13, and 12 heads, respectively, out of 16 trials:

$$\frac{1}{65,536} + \frac{16}{65,536} + \frac{120}{65,536} + \frac{560}{65,536} + \frac{1820}{65,536}$$

In trying to answer the question of whether a 12–4 split of heads and tails is extreme enough to cast doubt on the balance of the coin, it is not enough to consider only the probability of the 12–4 split by itself. It is necessary to add the probabilities for those splits that were even more extreme to the probability of the 12–4 split, i.e., all the terms above, giving $2517/65536$ or 0.038. The probability for the 12–4 split above is $1820/65536$ or 0.028.

The reason it is necessary to add the probabilities for the more extreme heads-tails splits to that for the 12–4 split is that any one of these events could have occurred instead of the 12–4 split and if so, the result would have been taken as evidence against the fairness of the coin. Since any of these events could have occurred, the probability of getting one or the other of them is the sum of their separate probabilities. This sum, 0.038, gives the probability that 12 *or more* heads will occur in 16 tosses of a balanced coin.

Thus, it can be stated that the probability of getting as many heads as occurred in this test (i.e., 12) or more in 16 trials with a fair coin would be 0.038. On the basis of this probability, can it be said that the coin is biased in favor of heads when 12 out of 16 tosses resulted in heads?

About four series out of 100 where a fair coin is tossed 16 times, the number of heads obtained will be 12 or more out of 16. If this is the case it is clearly impossible to state categorically that a coin is biased in favor of heads if on a test series it gives 12 heads in 16 trials since in four series out of 100 on the average this would happen even with a balanced coin. Yet, it is disturbing to have a coin give such a disproportionate number of heads when it was specifically being put to the test because it was suspected of yielding too many heads. Only about four times in 100 could this happen with a fair coin. Is it to be supposed that this particular test is one of the four out of 100 in which such a result could occur with a fair coin and not one of the 96 times it would not occur? Or is it to be concluded that a probability of 0.038 or about four times in 100 is too unlikely for it to have happened on this particular test?

## Levels of Confidence

When it is asked whether the probability of 0.038 for 12 or more heads in 16 trials is too small to believe that the coin is fair, a little thought makes it clear that any answer must be based on an arbitrary standard. This is because there is always a finite probability of getting any possible result, no matter how extreme, even with a fair coin. There is one chance in 65,536 of getting 16 heads in 16 tosses with a fair coin. There is about one chance in 43 million of getting 32 heads in 32 tosses. It would certainly be unreasonable, however, to believe the coin is fair if it gave 32 heads in 32 tosses even if it does happen on the average 1 time in 43 million attempts. At what point does it become unreasonable to believe the coin is fair even though it gives a disproportionate number of heads?

Scientific custom has established two arbitrary "confidence levels" that have been used more than any others to provide an answer to such questions. These are the five percent level of confidence and the one percent level of confidence. In practice, the probability, $p$, for a given result, such as 12 or more out of 16, is compared with 0.05 and 0.01, respectively, with the following conclusion: (a) if $p > 0.05$, the result is said to be "not significant"; (b) if $0.05 \geqslant p > 0.01$, the result is said to be significant at the five percent level of confidence; and (c) if $p \leqslant .01$, the result is said to be significant at the one percent level of confidence.

General practice tends to conflict to some extent with what is theoretically appropriate in this situation. Strictly speaking, the investigator should decide before making the test what level of confidence he will use. If $p$ for the resulting outcome is less than or equal to the level selected, he states that the result is "significant" at that level. If $p$ for the resulting outcome of the test is greater than the confidence level selected, the result is declared to be "not significant." The investigator should not wait to see the result to decide what level of confidence he will select. To do so would tempt him to set the confidence level at a less stringent value (e.g., 0.05 instead of 0.01) if he wishes to obtain a "significant" result and the outcome failed to satisfy a more stringent criterion. On the other hand, if the obtained result were to yield a $p < 0.01$, he would be tempted to declare significance at the 0.01 level to increase the apparent clarity of his experiment when in fact he would not have been willing to declare the result "not significant" if $p$ had been less than 0.05 but greater than 0.01. The need for presetting of confidence levels is widely ignored, however, despite the lack of theoretical justification for this practice.

There is nothing to prevent a scientist from selecting some other value than 0.05 or 0.01 for his predetermined standard to use in deciding whether his results are to be labelled as "significant" or "not significant." In fact, it is not uncommon for social scientists in some areas to use the 10 percent level of confidence (0.10) as a criterion for identifying potentially meaningful results. It is also not uncommon for scientists to report the significance level as equivalent to the highest level of significance the results could justify. For example, if the value of $p$ for their obtained results were 0.019, they would report the result as "significant at the 0.02 level" or equivalently, "significant at the two percent level." If $p$ were 0.0009, they might report the results to be "significant at the 0.001 level." The reader should be reminded that, although common, this practice is not strictly appropriate from the standpoint of the theory of statistics.

*Testing Hypotheses.*    In the coin tossing experiment, an hypothesis was being tested, namely, that the coin was balanced. This hypothesis would be stated as follows:

$$h_0 : p = \tfrac{1}{2}$$

Where:

$h_0 =$ the hypothesis to be tested

$p =$ the true probability of a head

That is, $h_0$, the hypothesis to be tested, states that $p$, the true probability of getting a head on any given trial, is $\tfrac{1}{2}$. If $h_0$ is true, the coin is indeed balanced. Of course, in practice the number of heads obtained with an unbiased coin cannot be guaranteed to be exactly one half short of tossing the coin an infinite number of times, an impossibility. The obtained number of heads approaches 50 percent closer and closer, however, as the number of trials increases when the coin is fair.

In the coin tossing experiment where 12 out of 16 tosses were heads for the coin suspected of yielding too many heads, the $p$ value for the obtained result was 0.038. Had the 0.05 level of confidence (or five percent level of confidence) been adopted in advance, the result would have been declared "significant" since $0.038 < 0.05$. When the result is declared to be significant at a specified level of confidence, then $h_0$ is said to be "rejected" at that same level of confidence.

If, on the other hand, the result of the experiment were to prove to be "not significant" because the $p$ value was greater than the confidence level chosen, it would not be possible to reject $h_0$. The experimenter would "fail to reject" $h_0$. It should most specifically be noted that he would *not* "accept" $h_0$ as a consequence of the fact that the results were "not significant." Regardless of the results, he can only "fail to reject" $h_0$ or "reject" $h_0$; he can *never* "accept" $h_0$. The reason for this is that to "accept" $h_0$ would be to assert that $p$ is *exactly* $\frac{1}{2}$. To determine this to be true would have required tossing the coin an infinite number of times and getting exactly 50 percent heads, a condition clearly contrary to fact. On the other hand, he can "fail to reject" $h_0$ because the results are not sufficiently different from what one would predict, under the assumption that $h_0$ is true, to warrant the conclusion that $h_0$ is false.

The status of $h_0$ is not unlike that of the defendant in a trial who is deemed to be "innocent" until proved "guilty." If the trial results in a verdict of "not guilty," this does not mean the defendant is "innocent." It merely means that guilt could not be proved beyond a reasonable doubt. When the investigator fails to reject $h_0$, it does not mean $h_0$ is true, merely that $h_0$ could not be shown to be false beyond a "reasonable" doubt, e.g., at the 0.05 level or the 0.01 level.

## One-Tailed and Two-Tailed Tests

In the coin tossing experiment described above 12 out of 16 tosses yielded heads and four yielded tails. The first five terms of the binomial expansion added up to $2517/_{65536}$ or 0.038 as the probability of getting 12 or more heads in 16 tosses. This result was termed "significant at the five percent level" because 0.038 is less than 0.05, indicating that the obtained result would occur less than 5 percent of the time in a large number of such experiments with an unbiased or balanced coin in which the true probability of getting a head on any one trial is $\frac{1}{2}$. Since the result was significant, $h_0$, the hypothesis that $p$, the true probability of getting a head on any one trial, equals $\frac{1}{2}$ was rejected at the five percent level of confidence.

The test just described constituted a one-tailed test of the hypothesis, $h_0$. It was a one-tailed test because the coin was suspected of giving too many heads and only that particular type of bias was being considered. Only the probability for 12 or more heads was computed. There was no consideration of the probabilities of events involving an excess number of tails. When $p = q = \frac{1}{2}$, the binomial distribution is symmetric so the computed probabilities for 16, 15, 14, 13, and 12 tails out of 16 trials would be exactly the same as those for the corresponding number of heads. To compute the probability of getting a division of heads and tails as extreme or more extreme than 12–4 in either direction would require adding the probabilities for 12 or more heads and the probabilities for 12 or more tails. Since the sum of the probabilities for 12 through 16 heads was approximately 0.038, twice this figure, 0.076, would be the sum for the terms representing 12 or more heads *and* 12 or more tails.

If the question were being posed merely as whether the coin is biased or not, without reference to the direction of bias, then it would be necessary to add the probabilities from *both* ends of the binomial expansion, taking into account all terms for outcomes as extreme or more extreme than the outcome actually obtained. It would not matter whether heads or tails came up more often; both sides would be considered anyway. The reason for this is that the coin would be presumed to be biased if it gave an extreme result in *either* direction, too many heads *or* too many tails. Since any of these events has a finite possibility of occurring with an unbiased coin, their probabilities must all be included in the total representing the probability of getting a result as extreme or more extreme than the obtained result under the assumption that $h_0$ is true.

When both sides of the binomial are considered in this way, the test of $h_0$ is called a "two-tailed" test. The two-tailed test is actually more commonly used than the one-tailed test. There must be special justification to use the one-tailed test. In this particular situation, if a two-tailed test is made, the value of $p$ for the obtained outcome is 0.076. Since this probability is greater than 0.05, the result is found to be "not significant" and hence the investigator must "fail to reject" $h_0$. The hypothesis $h_0$ remains tenable according to the five percent level of significance. Of course, if the 10 percent level had been preset as the criterion of significance, then $h_0$ would have been rejected at the 10 percent level of confidence since 0.076 is less than 0.10. Most scientists, however, are reluctant to use such a lenient test as the 10 percent level.

It should be emphasized here that using a one-tailed test would have resulted in significance at the five percent level of confidence while using a two-tailed test would not. Use of a one-tailed test, therefore, effectively permits an apparently less extreme result to achieve significance in certain cases. This would appear to be getting "something for nothing." This is not the case, in fact, because the use of a one-tailed test is justified only in certain cases, must be labeled as such, and places a limit on the interpretations that can be made.

In the test of the coin suspected of giving too many heads, the experiment was specifically set up ahead of time to test *only* that notion, thereby permitting the use of a one-tailed test. In such a case, $h_0$ could be rejected *only* if the outcome showed too many heads. It would not be appropriate to reject $h_0$ if it should have happened that too many tails came up, even if it were 16 tails and no heads. If the experimenter had any intention to reject $h_0$ if an extreme number of tails appeared as well as if an extreme number of heads appeared, it would be obligatory to make a two-tailed test since considering either type of outcome increases his chances of getting an extreme result, i.e., either too many heads *or* too many tails.

Normally, the appropriate thing to do is to make a two-tailed test rather than a one-tailed test. Certainly a coin is biased if it gives an extreme result in either direction. In any event, it is most explicitly improper to wait to see how the experiment comes out to decide whether a one-tailed or a two-tailed test is to be made and then to choose in such a way as to make the results "significant" with a one-tailed test when they would not have been significant by a two-tailed test. There are indeed situations where one-tailed tests are appropriate but these are for the most part situations in which only a result in one direction has any importance or meaning. In the case of the coin with too many heads such a situation would be the following: a stranger comes to town and cleans everybody out betting on heads using his own coin. The sheriff seizes the stranger, impounds his coin, and sets up an experiment to see if the stranger's coin is honest. Since the stranger did not win betting on tails, it would not matter if too many tails showed up in the test, only if too many heads showed up. In such a case, a one-tailed test would be entirely justified. At a given level of confidence, e.g., five percent level, it would also take a less extreme result to incriminate the stranger. With a five percent level one-tailed test he would be incriminated with a 12 heads out of 16 trials result. With a two-tailed test 13 or more heads out of 16 trials would be required to yield a significant result at the five percent level. The probability of 13 or more heads plus the probability of 13 or more tails in 16 trials would add up to 0.04 which is less than 0.05, leading to the rejection of $h_0$ at the five percent level of confidence with a two-tailed test.

## Application of the Binomial to a "Crucial" Experiment

After a careful study of two competing theories of animal behavior, an independent experimenter deduces consequences from the two that contradict each other. Theory A implies that after a specified type of training rats will jump from a jumping stand through a door with a black circle painted on a white background, preferring it to the other alternative, a door with a black cross painted on a white background. Theory B, on the other hand, implies that with the exact same type of training, the rats would prefer to jump from the jumping stand through the door with the cross on it. The investigator, therefore, plans an experiment that will provide a crucial test of the two theories. He will put 14 rats on the jumping stand, one at a time, and see which door they prefer, having each door half the time on the left and half the time on the right to control for position effects. If the rats choose the circle, Theory A is more correct. If they choose the cross, the results are more consistent with Theory B.

In this case, a two-tailed test is clearly demanded since an extreme result in either direction will lead to rejection of $h_0$. The five percent level of confidence will be selected as the criterion (instead of the one percent level) since the number of rats is not large and this is an exploratory investigation.

The investigator runs the experiment and finds out that 12 of the 14 rats jump through the door with a circle on it and 2 prefer the door with the cross. The hypothesis to be tested here, $h_0$, is that $p = \frac{1}{2}$ where $p$ is the true probability of jumping through the door with a circle on it (or a cross). The probabilities associated with outcomes as extreme as the obtained result or more extreme include the first three terms and the last three terms of the binomial expansion for $(\frac{1}{2} + \frac{1}{2})^{14}$. Since the distribution is symmetric with $p = q$, the sum of the first three terms may be computed and doubled to give the sum for the first three *and* the last three terms together. The first three terms may be obtained using Formula 7–3 as follows:

$$\left(\frac{1}{2} + \frac{1}{2}\right)^{14} = \left(\frac{1}{2}\right)^{14} + \frac{14}{1}\left(\frac{1}{2}\right)^{13}\left(\frac{1}{2}\right) + \frac{14 \cdot 13}{2 \cdot 1}\left(\frac{1}{2}\right)^{12}\left(\frac{1}{2}\right)^{2} + \cdots$$

$$\left(\frac{1}{2} + \frac{1}{2}\right)^{14} = \frac{1}{4048} + \frac{14}{4048} + \frac{91}{4048} + \cdots$$

The sum of these first three terms is $^{106}/_{4048}$. Doubling this value to include the last three terms gives $^{212}/_{4048}$ or 0.052. This value is not less than or equal to 0.05 (unless rounded to 0.05) and hence the result is not quite significant so $h_0$ cannot be rejected.

Although $h_0$ could not be rejected at the five percent level of significance, the outcome of the experiment would encourage the investigator to repeat the experiment using a larger number of rats since the result is almost significant. If the rats continue to prefer the door with the circle on it in the same proportion, an experiment with a larger number of rats would certainly yield a significant result. In no case, however, would it be appropriate for the investigator to lower his criterion of significance to the 10 percent level or do a one-tailed test so that the results of the present experiment would then turn out to be "significant."

## Summary of Steps in Using the Binomial Distribution to Make a Statistical Test of an Hypothesis

1. Determine $h_0$, the hypothesis to be tested. Usually, $h_0$ states that the true probability of a given outcome (e.g., a left or right turn in a maze) is $\frac{1}{2}$ where there are two possible outcomes.

2. Decide whether $h_0$ is to be tested by means of a two-tailed test or a one-tailed test. A one-tailed test can be used only if it has been decided in advance of the experiment that $h_0$ would be rejected

by an extreme result in one direction only, not by an extreme result in either direction.

3. Determine $n$, the number of trials, and $x$, the number of trials in which a "success" occurred. A "success" is defined as an outcome in the expected direction, i.e., in the direction that would result in a rejection of $h_0$ if enough successes occur in $n$ trials. In a two-tailed test, either of the two possible outcomes may be designated as a "success," but usually the outcome with the greater frequency is chosen.

4. Using either Formula 7–1 or Formula 7–3, determine the probability of $x$ successes under the hypothesis that $h_0$ is true. This is the probability of getting exactly $x$ successes.

5. Using the same formula, determine the probability of getting $x + 1$ successes, $x + 2$ successes, and so on, up to and including $n$ successes.

6. Add all the probabilities together from Steps 4 and 5 to get the probability of getting $x$ or more successes in $n$ trials.

7. If a one-tailed test is being conducted, let $p$ equal the probability obtained in Step 6. If a two-tailed test is being conducted, let $p$ be *twice* the probability obtained in Step 6 (assumes two equally likely outcomes, i.e., $p = q$ in Formulas 7–1 and 7–3).

8. If $p$ is less than 0.05, reject $h_0$ at the five percent level of confidence. If $p$ is less than 0.01, reject $h_0$ at the one percent level of confidence. If $p$ is greater than 0.05, do *not* reject $h_0$. To be technically correct, the decision should be made in advance as to whether the 0.05 or the 0.01 level will be used to decide whether $h_0$ is to be rejected or not. In this case, $h_0$ is merely rejected or not rejected at the given level of confidence that has been preset as the criterion. If a one-tailed test has been made, this fact should be indicated in giving the results of the test.

## Problem 21
### USING THE NORMAL CURVE APPROXIMATION TO THE BINOMIAL DISTRIBUTION, MAKE A STATISTICAL TEST OF AN HYPOTHESIS

Although it is beyond the scope of this book to do so, it is possible to prove that the binomial distribution approaches the normal distribution in form as the number of trials approaches infinity. The plausibility of this theorem can be grasped intuitively by inspection of Figure 7–1 which gives the expected frequencies in 1,048,576 series for different combinations of heads and tails when tossing an unbiased coin 20 times per series. These expected frequencies are the coefficients of the terms in the binomial expansion by Formula 7–3 of $(\frac{1}{2} + \frac{1}{2})^{20}$. Each of these terms has a denominator of 1,048,576. Thus, the probability in one 20-

FIGURE 7-1
Normal Curve Approximation to the Binomial for $p = \frac{1}{2}$ and $n = 20$

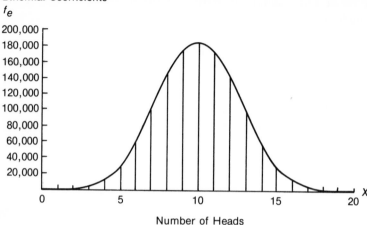

Binomial Coefficients

trial experiment of getting 20 heads would be 1/1,048,576. In 1,048,576 repetitions of this experiment (20 tosses in each experiment) it would be expected that one time there would be 20 heads out of 20. The sequence of terms in the binomial is represented as follows:

$$\left(\frac{1}{2} + \frac{1}{2}\right)^{20} = \left(\frac{1}{2}\right)^{20} + \frac{20}{1}\left(\frac{1}{2}\right)^{19}\left(\frac{1}{2}\right)^{1} + \frac{20 \cdot 19}{2 \cdot 1}\left(\frac{1}{2}\right)^{18}\left(\frac{1}{2}\right)^{2}$$
$$+ \frac{20 \cdot 19 \cdot 18}{3 \cdot 2 \cdot 1}\left(\frac{1}{2}\right)^{17}\left(\frac{1}{2}\right)^{3} + \cdots + \frac{20!}{20!}\left(\frac{1}{2}\right)^{0}\left(\frac{1}{2}\right)^{20}$$

There will be $n + 1$ or 21 of these terms the first 11 of which have the following coefficients: 1, 20, 190, 1140, 4845, 15,504, 38,760, 77,520, 125,970, 167,960, and 184,756. The last 10 will be the same as the first 10 in reverse order: 167,960, 125,970, . . . , 1. Each of these coefficients is multiplied by 1/1,048,576. The sum of these coefficients, of course, must equal 1,048,576 since the sum of the probabilities for all possible events must equal 1.

Figure 7-1 shows the plot of the binomial coefficients, or expected frequencies, $f_e$, in 1,048,576 sets of trials for the different possible numbers of heads, i.e., $x = 0, 1, 2, 3, . . . , 20$ when $p = \frac{1}{2}$. The vertical lines above the possible values of $x$ extend as high as the expected frequency in 1,048,576 repetitions of the experiment. These are the binomial coefficients when $n = 20$ and $p = \frac{1}{2}$. The end points of these vertical lines have been connected with a smooth curve. The closeness in appearance of this curve to the normal curve is obvious.

## Formulas for the Mean and Standard Deviation of the Binomial

The mean and standard deviation of the binomial are given by the following formulas:

$$M_b = np \qquad (7\text{–}8)$$

$$\text{S.D.}_b = \sqrt{npq} \qquad (7\text{–}9)$$

Where:

$M_b$ = mean of the binomial $x$ scale

$\text{S.D.}_b$ = standard deviation of the binomial $x$ scale

$n$ = number of trials

$p$ = probability of a specified outcome on any one trial, e.g., a head

$q = 1 - p$, the probability of the alternate outcome, e.g., a tail

Formulas 7–8 and 7–9 will be derived in the Mathematical Notes section below. Applying Formula 7–8 to the case where $n = 20$ gives $M_b = 20 \times \frac{1}{2} = 10$ and in Figure 7–1, it can be observed that $X = 10$ is the middle point in the distribution where the maximum expected frequency occurs. The standard deviation for $n = 20$ would be $\sqrt{20 \cdot \frac{1}{2} \cdot \frac{1}{2}} = \sqrt{5} = 2.24$.

## Using the Normal Curve Approximation to Determine the Probability of a Particular Binomial Outcome

With $n = 20$ and $p = \frac{1}{2}$, the normal curve approximates the binomial distribution with sufficient accuracy so that it can be used to obtain estimates of the probabilities of particular binomial outcomes. In Problem 19, the probability of obtaining a result of 10 rats turning left out of 15 under the assumption $(h_0)$ that the probability of a left turn is $\frac{1}{2}$ was computed by Formula 7–1 to be 0.09. Using the same formula, or the binomial coefficients given above for $n = 20$, the probability for exactly 15 rats out of 20 turning left would be $15504/1048576 = 0.0148$. 

The area under the normal curve will now be used to obtain an estimate of the probability just computed. The area under the normal curve involved (See Figure 7–1) will be that portion of the area under the curve between 14.5 and 15.5 since the score 15 covers an entire point, i.e., $15.0 \pm 0.5$. There is no area associated with $X = 15.0$ in the binomial, of course, since it is a discrete distribution. The scores all occur at integral values of $X$, with nothing in between. The sum of the probabilities for the binomial and the total area under the standard normal curve are both 1.0, however, so adding up the areas under the curve is equivalent to adding probabilities for discrete outcomes on the binomial.

To use the standard normal curve, however, everything must be converted to standard scores. Remembering that the standard score $Z$ is given by $(X - M)/$S.D., $X = 14.5$ and $15.5$ must be converted to standard scores by subtracting $M_b$ and then dividing by S.D.$_b$. For $n = 20$, it has already been determined above that $M_b = 10$ and S.D.$_b = 2.24$. The standard scores for the two end points of the desired interval, 14.5 to 15.5, are, therefore:

$$Z_{15.5} = \frac{15.5 - 10}{2.24} = \frac{5.5}{2.24} = 2.455$$

$$Z_{14.5} = \frac{14.5 - 10}{2.24} = \frac{4.5}{2.24} = 2.009$$

If these values are used to enter the table of normal curve values (See Appendix A) interpolation will give the following areas from $-\infty$ to $Z$:

for $Z = 2.455$    Area $= 0.9930$
for $Z = 2.009$    Area $= \underline{0.9777}$
Difference $= \underline{0.0153}$

This normal curve approximation value agrees closely with the 0.0148 value computed by the exact Formula 7–1. Both values round to 0.015. With $n = 20$, these approximations are sufficiently close for most practical purposes. As $n$ increases, the approximations improve. With $n$ less than 20, it is safer to use Formula 7–1.

### Using the Normal Curve Approximation to the Binomial to Make a Statistical Test of an Hypothesis

Since the normal curve approximation to the binomial may be used to give approximate probabilities for particular binomial outcomes, like Problem 19, it follows that normal curve approximation can also be used to test hypotheses, like Problem 20, provided $n$ is 20 or more.

Suppose the investigator performing the "crucial" experiment in Problem 20 (where 12 rats out of 14 jumped through the door with the circle) decided to repeat the experiment using 20 rats. The first experiment just failed to achieve significance at the 0.05 level. When the experiment was repeated with 20 rats, 15 out of 20 jumped through the door with the circle and 5 jumped through the door with the cross on it. Again, a two-tailed test should be run since $h_0$ would have been rejected with an extreme result in either direction.

The statistical test conducted with the binomial would require adding the probabilities for all terms representing 15 or more choices of either alternative. This would require adding the first six terms in the binomial expansion of $(\frac{1}{2} + \frac{1}{2})^{20}$ and then doubling this figure to take into account the last six terms as well. The first six binomial coefficients given above for $n = 20$ are: 1, 20, 190, 1140, 4845, and 15504. These sum to 21700. Dividing 21700 by 1,048,576, which is $2^{20}$, gives 0.0207. Doubling this figure gives 0.0414 as the probability of getting a result as extreme or more extreme than 15 out of 20 rats preferring one door or the other.

Solving this same problem using the normal curve approximation to the binomial distribution should give a probability reasonable close to the 0.0414 obtained using the binomial itself. Since 15 out of 20 rats jumped through the door with the circle on it, the area under the normal curve is needed for outcomes of 15 through 20, to take into account the probability of the particular outcome *plus* those outcomes that are more extreme. This means that the area under the curve to the right of a score of 14.5 must be obtained since the discrete score of 15 in the binomial is represented by the area from 14.5 to 15.5 under the continuous normal curve.

The standard score corresponding to a raw score of 14.5 must be determined before entering the normal curve table to find the area to the right of 14.5. These computations are shown as follows:

$$Z_{14.5} = \frac{14.5 - M_b}{\text{S.D.}_b}$$

$$= \frac{14.5 - np}{\sqrt{npq}}$$

$$= \frac{14.5 - 20 \cdot \frac{1}{2}}{\sqrt{20 \cdot \frac{1}{2} \cdot \frac{1}{2}}}$$

$$= \frac{14.5 - 10}{\sqrt{5}}$$

$$Z_{14.5} = \frac{4.5}{2.24} = 2.009$$

The standard score corresponding to 14.5 is 2.009. Entering the normal curve table with this value of $Z$ gives an area from $-\infty$ to $Z$ of 0.9777 or 0.0223 to the right of 14.5. Doubling this figure to 0.0446 gives the area to the right of 14.5 *plus* the corresponding area in the left tail of the distribution, i.e., below 5.5. The area below 5.5 approximates the probability for 15 or more rats preferring the door with the cross and the area above 14.5 approximates the probability for 15 or more rats preferring the door with the circle. Adding these together (by doubling one of them) gives an approximation to the probability of getting a result as extreme or more extreme than 15 rats preferring one or the other of the two possible choices.

The value of 0.0446 for the probability of 15 or more preferences in either one direction or the other computed from the normal curve approximation to the binomial compares favorably with the value of 0.0414 obtained by adding probabilities from the terms of the binomial itself. Since the obtained probability is less than 0.05, $h_0$ can be rejected at the five percent level of confidence.

Repetition of the experiment with a larger number of trials resulted in a statistically significant outcome even though the proportion of rats preferring the door with the circle on it actually dropped from $^{12}/_{14} = 0.86$ to $^{15}/_{20} = 0.75$. This comparison illustrates the fact that as the size of $n$ increases, rejection of $h_0$ can be achieved with a smaller and smaller departure from 50–50 in the choices between two alternatives.

*A Second Example.*  Suppose 100 randomly selected respondents are asked to state a preference for candidate A or candidate B in a forthcoming election. Of the 100 respondents, 57 favored candidate A and 43 favored candidate B. Can it be concluded that there is a preference for candidate A in the population from which this sample was selected?

The hypothesis to be tested, $h_0$, is that the probability of a choice of A is $\frac{1}{2}$ on any given trial. The normal curve approximation to the binomial will be applied because $n$ is well over 20. A $Z$ score is calculated for the point 56.5 since the area under the normal curve is to be obtained for all scores of 57 or more out of 100.

$$Z = \frac{56.5 - np}{\sqrt{npq}}$$

$$= \frac{56.5 - 50}{\sqrt{100 \cdot \frac{1}{2} \cdot \frac{1}{2}}}$$

$$Z = \frac{6.5}{5} = 1.30$$

From the normal curve table, the area from $-\infty$ to $Z$ for $Z = 1.30$ is 0.9032 which leaves 0.0968 as the area to the right of $Z = 1.30$. Taking into account the area in the other tail by doubling 0.0968 gives 0.1936 as the total area and the probability of obtaining a result as extreme as the one obtained or more extreme. This value is larger than all the possible significance level values, indicating that there is almost one chance in five of obtaining a result as extreme as this or more extreme when the true proportion equals $\frac{1}{2}$. Thus, even though 57 to 43 seems like a substantial difference in favor of candidate A, it does not prove to be even nearly statistically significant.

**Critical Values of** $Z$

In the previous examples using the normal curve approximation to the binomial, the probability associated with a particular outcome has been determined and this probability has then been compared with such criterion probability values as 0.05 and 0.01. An alternate way of determining whether the result is significant or not is to compare the obtained value of $Z$ wi'h the critical values of $Z$ associated with the criterion probabilities. For example, in the normal curve, the area to the right of $Z = 1.96$ is approximately 0.025. Doubling this gives 0.05. Hence, the critical value of $Z$ for the five percent level of significance with a two-tailed test is 1.96. If the obtained value of $Z$ is 1.96 or greater, the·result is "significant" and $h_0$ is rejected at the five percent level of confidence. If the obtained value of $Z$ is less than 1.96, the result is "not significant" and $h_0$ is not rejected. Table 7–2 gives the critical values

TABLE 7–2
Critical Values of $Z$

| Level of Confidence | Type of Test | Critical Value of Z |
|---|---|---|
| 0.10........ | Two-Tailed | 1.65 |
| 0.05........ | Two-Tailed | 1.96 |
| 0.05........ | One-Tailed | 1.65 |
| 0.02........ | Two-Tailed | 2.33 |
| 0.01........ | Two-Tailed | 2.58 |
| 0.01........ | One-Tailed | 2.33 |
| 0.002....... | Two-Tailed | 3.08 |
| 0.001....... | Two-Tailed | 3.30 |
| 0.001....... | One-Tailed | 3.08 |

of $Z$ for various levels and types of tests. The test of $h_0$ can be carried out either by computing the probability for the given outcome or by comparison of $Z$ with critical values of $Z$ when the normal curve approximation to the binomial is being used. When the binomial itself is being employed this alternative way of making the test is not available.

**Summary of Steps in Using the Normal Curve Approximation
to the Binomial to Make a Statistical Test of an Hypothesis**

1. Determine $h_0$, the hypothesis to be tested.
2. Decide whether a one-tailed test or a two-tailed test is to be used.
3. Determine $n$, the number of trials (usually 20 or more) and $x$, the number of trials in which a "success" occurred. The outcome labelled a "success" is chosen in such a way that $x$ is greater than $n/2$.

4.  Determine the mean of the distribution by Formula 7–8.
5.  Determine the standard deviation of the distribution by Formula 7–9.
6.  Subtract ½ point from $x$ and convert the resulting score to a standard score ($Z$ score). This is done by subtracting the mean to get a deviation score and then dividing the deviation score by the standard deviation.
7.  Enter the normal curve table with the $Z$ score obtained in step 6 and find the proportion of area under the normal curve beyond this value.
8.  If a one-tailed test is being conducted, let $p$ equal the probability obtained in step 7. If a two-tailed test is being conducted, let $p$ be twice the probability obtained in step 7.
9.  If $p$ is less than 0.05, reject $h_0$ at the five percent level of confidence. If $p$ is less than 0.01, reject $h_0$ at the one percent level of confidence. If $p$ is greater than 0.05, do not reject $h_0$.

## Mathematical Notes

In this section, Formulas 7–8 and 7–9 for the mean and standard deviation of the binomial will be developed from the definitions of these statistical concepts.

*Mean of the Binomial.* The mean may be defined as follows:

$$M = \frac{\sum_{i=1}^{k} f_i X_i}{N} = \sum_{i=1}^{k} X_i \left(\frac{f_i}{N}\right) \tag{7-10}$$

Where:

$k$ = the number of different scores
$X_i$ = a raw score
$f_i$ = the frequency for score $i$
$N$ = the number of cases

In Formula 7–10, the value $f_i/N$ is the relative frequency of score $X_i$, i.e., the ratio of the number of $X_i$ scores to the total number of scores. This can be taken as the probability of getting that particular score, $X_i$. Substituting the general term of the binomial, Formula 7–1, which gives the probability of a given score, $X$, for $(f_i/N)$ and modifying the notation to conform with the possible scores in the binomial, Formula 7–10 may be rewritten to give:

$$M_b = \sum_{x=0}^{n} x \cdot \left(\frac{n!}{x!(n-x)!} p^x q^{n-x}\right) \tag{7-11}$$

Since the values of $X$ in the binomial are 0, 1, 2, . . . , $n$ the notation in 7–11 is more convenient and conforms to the notation of Formula 7–1. Formula 7–11 involves $n + 1$ terms the first of which equals zero since when $x = 0$ is substituted in the expression, the result is zero. This permits elimination of the first term in the summation, making the limits $x = 1$ to $x = n$ instead of $x = 0$ to $x = n$. Also, the value of $x$ can be cancelled with the $x$ in $x!$ to give $(x - 1)!$ making 7–11 change to:

$$M_b = \sum_{x=1}^{n} \frac{n!}{(x-1)!(n-x)!} p^x q^{n-x}$$

Remove the constant $np$ outside the summation sign to give:

$$M_b = np \sum_{x=1}^{n} \frac{(n-1)!}{(x-1)!(n-x)!} p^{x-1} q^{n-x} \qquad (7\text{–}12)$$

Let $y = x - 1$; then $x = y + 1$ and substituting for $x$ in Formula 7–12 gives:

$$M_b = np \left( \sum_{y=0}^{n-1} \frac{(n-1)!}{y!(n-1-y)!} p^y q^{n-1-y} \right) \qquad (7\text{–}13)$$

The expression inside parentheses in Formula 7–13, however, is just the sum of the binomial terms for $(p + q)^{n-1}$. These terms must add up to 1, so this gives:

$$M_b = np \qquad (7\text{–}8)$$

*Standard Deviation of the Binomial.*    The raw score formula for the standard deviation is given by:

$$\text{S.D.} = \frac{1}{N} \sqrt{N\Sigma X^2 - (\Sigma X)^2} \qquad (4\text{–}2)$$

Taking into account that some scores are repeated and squaring both sides of Formula 4–2 leads to:

$$(\text{S.D.})^2 = \frac{\sum_{i=1}^{k} f_i X_i^2}{N} - \left( \frac{\sum_{i=1}^{k} f_i X_i}{N} \right)^2 \qquad (7\text{–}14)$$

As with the derivation for the mean of the binomial, the probability of a given score can be substituted for $f_i/N$, the relative frequency of score $X_i$. Also, the last term on the right of Formula 7–14 is the square of the mean, which for the binomial is $n^2 p^2$. Taking these facts into account, substituting $x$ as the binomial score, and remembering that the possible binomial scores are $x = 0, 1, 2, \ldots, n$, Formula 7–14 for the binomial may be rewritten as follows:

$$(\text{S.D.}_b)^2 = \left( \sum_{x=0}^{n} x^2 \cdot \frac{n!}{x!(n-x)!} p^x q^{n-x} \right) - n^2 p^2 \qquad (7\text{–}15)$$

The first term of the summation, for $x = 0$, will equal zero since zero times anything is still zero. Adjusting the summation limits for this fact, moving $np$ outside the summation sign and cancelling one $x$ from $x^2$ with the $x$ of $x!$ in the denominator, Formula 7–15 becomes:

$$(\text{S.D.}_b)^2 = \left( np \sum_{x=1}^{n} x \cdot \frac{(n-1)!}{(x-1)!(n-x)!} p^x q^{n-x} \right) - n^2 p^2 \qquad (7\text{–}16)$$

Let $y = x - 1$; then $x = y + 1$. Substituting for $x$ in 7–16 gives:

$$(\text{S.D.}_b)^2 = \left( np \sum_{y=0}^{n-1} (y+1) \cdot \frac{(n-1)!}{y!(n-1-y)!} p^y q^{n-1-y} \right) - n^2 p^2$$

$$(\text{S.D.}_b)^2 = \left( np \sum_{y=0}^{n-1} y \cdot \frac{(n-1)!}{y!(n-1-y)!} p^y q^{n-1-y} \right)$$

$$+ \left( np \sum_{y=0}^{n-1} \frac{(n-1)!}{y!(n-1-y)!} p^y q^{n-1-y} \right) - n^2 p^2 \qquad (7\text{–}17)$$

The first of the three terms to the right of the equal sign in Formula 7–17 above is like Formula 7–11 for the mean of the binomial except that $np$ appears in front of summation sign, $y$ is the variable symbol instead of $x$, and it is the sum of the binomial where there are $n - 1$ trials instead of $n$ trials. The mean of the binomial for $n - 1$ trials would be $(n - 1)p$ since with $n$ trials it is $np$. The first term on the right of Formula 7–17, therefore, becomes $np(n - 1)p$. The second term on the right is $np$ multiplied by the sum of the binomial terms for $n - 1$ trials, which equals 1, so this term is just $np$. Formula 7–17 reduces, therefore, to:

$$(\text{S.D.}_b)^2 = np(n-1)p + np \cdot 1 - n^2 p^2$$

$$(\text{S.D.}_b)^2 = n^2 p^2 - np^2 + np - n^2 p^2$$

$$(\text{S.D.}_b)^2 = -np^2 + np = np(1-p) = npq$$

$$\text{S.D.}_b = \sqrt{npq} \qquad (7\text{–}9)$$

## Problem 22
### USING THE BINOMIAL DISTRIBUTION, TEST THE SIGNIFICANCE OF THE DIFFERENCE BETWEEN CORRELATED FREQUENCIES

In one of the examples for Problem 21, a poll of 100 randomly selected respondents revealed that 57 favored candidate A and 43 favored candidate B. Although this did not prove to be a statistically significant difference in favor of candidate A, it would be a cause for concern among those supporting candidate B and a basis for guarded optimism on the part of supporters of candidate A.

On the basis of this set of results, supporters of candidate B might decide to repeat the poll with a larger number of cases and also to take a third poll two weeks later among the *same* individuals to see if there is any trend developing, either for or against their candidate. The results of this hypothetical study are shown in Table 7–3.

TABLE 7–3
Results of a Repeated Poll with the Same Subjects

|  |  | Third Poll | | |
| --- | --- | --- | --- | --- |
|  |  | A | B | Total |
| *Second Poll* | A......... | 103 | 12 | 115 |
|  | B......... | 2 | 83 | 85 |
|  | Total | 105 | 95 | 200 |

Two hundred subjects were randomly selected from the defined population and asked on two different occasions, two weeks apart, to express their preference for candidate A or candidate B. The first occasion of the second poll (the poll of 100 subjects was the first poll) gave 115 out of 200 in favor of candidate A, which is 57.5 percent, very close to the 57 percent favoring A in the first poll. When the poll was repeated two weeks later, however, only 105 out of 200 favored A, a drop to 52.5 percent. This result might lead B forces to take hope if the difference can be shown to represent a statistically significant change as opposed to a difference that could reasonably be attributed to mere random sampling variations.

This problem can be reduced to a problem of applying the binomial to a two-choice situation as follows. Consider the behavior of the respondents on both occasions. In Table 7–3 it can be seen that 103 cases chose A both times and 83 chose B both times. There is certainly no significant change in those cases. Two people, however, shifted from B on the second poll to A on the third and 12 shifted from A on the second poll to B on the third. This gives a 12–2 split in favor of shifting from A to B for the 14 people who changed from preferring one candidate to another.

If there is no trend in opinion, it could be assumed that a change from candidate A to B would be no more likely than a change from B to A. This would yield for $h_0$, the hypothesis to be tested, that the probability of a shift from A to B is ½. The binomial can be applied for $n$ trials, where $n$ is the number of changed votes, to see if the split is unbalanced enough to warrant rejection of $h_0$. Either the binomial itself can be used or the normal curve approximation to the binomial if $n$ is 20 or more.

In this case, $n$ is 14, far less than 20, so the normal curve approximation should not be used. The first three terms of the binomial for $( ½ + ½ )^{14}$ are given as follows:

$$\left(\frac{1}{2} + \frac{1}{2}\right)^{14} = \left(\frac{1}{2}\right)^{14} + \frac{14}{1}\left(\frac{1}{2}\right)^{13}\left(\frac{1}{2}\right)^{1} + \frac{14 \cdot 13}{2 \cdot 1}\left(\frac{1}{2}\right)^{12}\left(\frac{1}{2}\right)^{2} + \cdots$$

$$\left(\frac{1}{2} + \frac{1}{2}\right)^{14} = \frac{1}{4096} + \frac{14}{4096} + \frac{91}{4096} + \cdots$$

Summing the first three terms gives $^{106}\!/_{4096}$. This should be doubled to $^{212}\!/_{4096}$ to take into account the terms in the other tail, since a two-tailed test will be made here. This gives 0.051, which is not significant at the 0.05 level, being slightly larger than 0.05.

If a one-tailed test had been performed here, the result would have been significant at the 0.05 level, since only the terms on one side would have been considered, giving a probability of approximately 0.026 which is less than 0.05. A case could be made for using a one-tailed test here since the poll was being conducted by forces favorable to candidate B who are only interested in a trend favoring their candidate. If the question is being asked, "Is there a trend of any kind?", however, a two-tailed test would be called for. On the basis of the findings, supporters of candidate B might take heart that a trend may well be developing in favor of their candidate although it fails to reach significance at the five percent level for a two-tailed test.

This same type of analysis can be applied to a wide variety of situations where changes in opinion or response are being evaluated. The requirements are that the data be two-choice categorical, e.g., "yes"–"no", as opposed to continuous, and the same subjects must be evaluated twice. This method of analysis does not apply to the situation where two different samples of people are taken on the two occasions.

The approach described here in certain cases can lead to rather counterintuitive results. For example, if 2000 subjects had been tested instead of 200 and only 14 changed, giving the same split of 12–2 in favor of an A to B change, the level of significance would have been just the same as it was with the 200 cases. The subjects who do not change do not affect the outcome in any way. It is difficult to consider a change as significant or approaching significance when only 14 out of 2000 subjects register a change. Nevertheless, if 13 out of these 14 subjects were to change in one particular direction, the trend would be statistically significant at the one percent level of confidence with a two-tailed test.

## Summary of Steps in Using the Binomial Distribution to Test the Significance of the Difference between Correlated Frequencies

1. Prepare a four-fold table that contains the frequencies for the following: (a) those who gave response one on both occasions; (b) those who gave response two on both occasions; (c) those who gave response one on the first occasion and response two on the second occasion; (d) those who gave response two on the first occasion and response one on the second occasion.

2. Ignore frequencies $a$ and $b$ in Step 1 above, and consider only frequencies $c$ and $d$. Let $c + d = n$, and let the larger of the values $c$ and $d$ be $x$.

3. Using the values of $n$ and $x$ obtained in Step 2 above, treat the data as a test of an hypothesis, $h_0$, using the binomial distribution. Use the exact binomial test, as in Problem 20, or if $n$ is 20 or more, use the normal curve approximation (Problem 21).

4. If $h_0$ is rejected at a given level of significance by the test in Step 3, the difference between the frequencies is also significant at the same level. If $h_0$ is not rejected by the test in Step 3, the difference between frequencies is not significant.

## Exercises

7–1. According to a new visual theory developed by Professor X, placing a black dot at the center of a circle will make the circle appear larger. To test this hypothesis, an experiment was conducted in which 17 subjects were asked to state which of two circles appeared larger to them. The circles were actually equal in size but one had a black dot in it. Of the 17 subjects, 13 stated that the circle with the black dot appeared larger. Select an appropriate type of test, give $h_0$, carry out the test, state the result, and draw conclusions.

7–2. Eighteen randomly selected congressmen are polled concerning whether they intend to vote for a particular bill coming before the house. Of these 14 say they will vote no. State $h_0$ in terms of a vote outcome, select an appropriate test, make the test, and state your conclusions.

7-3. Although the normal curve approximation is not very good for $n = 17$, apply it to the specific outcome of a 13–4 split in Exercise 7–1 and compare with the probability obtained from the general term of the binomial.

7-4. Using the normal curve approximately, even though $n$ is only 18, find the probability for the 14–4 split in Exercise 7–2 and compare it with the corresponding value as computed by the general term of the binomial.

7-5. Determine the probability in Exercise 7–3 using the ordinate method and the normal curve. (See Mathematical Notes for Problem 15)

7-6. Determine the probability in Exercise 7–4 using the ordinate method and the normal curve. (See Mathematical notes for Problem 15)

7-7. A given manufacturing process is supposed to turn out plastic bags that can withstand a specified air pressure without breaking with only a 10 percent failure rate. Periodic tests are made to monitor the quality of production. The last sample of 10 bags yielded three bags that broke under pressure. Is the process out of control? Devise and carry out a test; draw conclusions.

7-8. A pair of dice is suspected of giving too many "snake eyes" (two ones). In a four-trial test, snake eyes turn up twice. Select a test, make it, state results and conclusions.

7-9. An audience fills out a questionnaire exploring their attitudes about India and then sees an information film on India followed by a repetition of the original questionnaire to check for changes in attitude. On one particular question, the breakdown of responses was as follows:

|           | Yes I | No I |
|-----------|-------|------|
| Yes II..... | 72    | 9    |
| No II..... | 16    | 48   |

Select a test, make it, state results, and give conclusions.

7-10. A physical educationist claims he can improve people's golf scores through hypnosis. To test this, 100 golfers play one game, receive hypnotic suggestion to do better and then play a second game. The results were as follows: 55 did better the second time, 8 had the same score, and 37 did worse. Evaluate the educationist's claim.

# chapter 8

# Testing Hypotheses with Continuously Measured Variables

When there are many possible outcomes from a particular experimental trial instead of just two, the binomial distribution is usually not an effective vehicle for making inferences concerning the accuracy of a statistical result. For example, if a group of students selected at random from a large population obtain a certain mean score on an achievement test, the binomial cannot be used to determine how well that sample mean approximates the mean of the population from which the sample was drawn. This chapter will consider other methods that are better adapted to the problems of statistical inference with such continuously measured variables. A full understanding of the theory of statistics upon which these methods are based would require considerably more mathematical background than has been assumed for this book. The formulas that are used, therefore, will in most cases have to be taken on faith by the student as far as this book is concerned. Interested students with sufficient mathematical background will be able to consult one or more of the many available books on mathematical statistics to satisfy their curiosity about the origins of these methods.

### The Sampling Distribution

One of the most important concepts in the theory of statistics is the notion of the sampling distribution. Each statistical constant, e.g., $M$, S.D., and others, has its own sampling distribution. A hypothetically correct, but impractical, method of constructing a sampling distribution for the mean would be the following. Consider an infinitely large population (or at least a very, very large one) with a true mean of $\mu$ (small Greek letter $mu$) and a true standard deviation of $\sigma$ (small Greek letter *sigma*). Draw a sample of size $N$ at random from this population and compute its mean to be $M_1$. Now draw a second sample of the same size, i.e., $N$, and find its mean, $M_2$. Repeat this process again and again to find $M_3, M_4, \ldots, M_q$ where $q$ is a very large number, always using the same sample size, $N$.

Suppose the population being treated here consisted of all males in the United States and the variable being measured were height in inches. For simplicity, assume that $\mu = 68$ inches and $\sigma = 3$ inches. If this were true, about 68 percent of the males in the U.S. would have heights between 65 inches and 71 inches, 95 percent would have heights between 62 inches and 74 inches, and 99 percent would have heights between 59 inches and 77 inches since height is approximately normally distributed in the population. Scores of individuals clearly scatter over a very wide range.

The mean scores of samples of size $N$, however, do not scatter as widely as the scores for individuals. This phenomenon is illustrated in Figure 8–1. Both distributions center about the same point, $\mu$, the popula-

FIGURE 8–1
Raw Score Distribution and Sampling Distribution

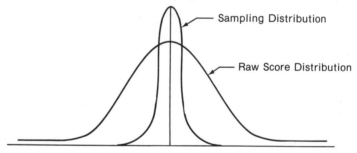

tion mean, but the variability of the sampling distribution is considerably smaller than the variability of the raw score distribution. In fact, it can be proved mathematically that the variability of the sampling distribution of the mean is given by:

$$\sigma_M = \frac{\sigma}{\sqrt{N}} \tag{8-1}$$

Where:

$\sigma_M$ = true standard deviation of the sampling distribution of means, the true standard error of the mean

$\sigma$ = true population standard deviation

$N$ = sample size for sample means making up the sampling distribution

Thus Formula 8–1 shows that the true standard deviation of the sampling distribution, called the "true standard error of the mean", is always smaller than $\sigma$ (except when $N=1$) and gets progressively smaller by comparison as $N$ increases in size. Since $\sigma$ is divided by the square root of $N$, however, it is clear that the law of diminishing returns operates. That is, as $N$ gets larger, it takes a bigger and bigger jump in $N$ to get the same amount of decrease in the size of $\sigma_M$.

The smaller $\sigma_M$ is, the more tightly sample means will be clustered about $\mu$, the population mean, which is at the center of both the raw score distribution and the sampling distribution. This means that as the size of the sample increases, the mean of the sample can be expected to be closer and closer to the true mean since *all* sample means tend to get closer and closer to the true mean as $N$ increases.

The actual size of $\sigma_M$ provides a means of determining how accurate the sample mean might be as an estimate of the population mean, $\mu$. With a normal population raw score distribution, the sampling distribution of means is normal in shape. Knowing that the sampling distribution is normal and knowing the value of $\sigma_M$ permits a rather precise specification of the limits within which sample values scatter about $\mu$, the true mean of the population. Thus 68 percent of sample means lie within the interval $\mu \pm 1.0\sigma_M$; 95 percent of the sample means lie within the interval $\mu \pm 1.96\sigma_M$; and 99 percent of the sample means lie within the interval $\mu \pm 2.58\sigma_M$. Of course these are theoretical values based on a theoretically derived sampling distribution. An empirically derived sampling distribution obtained in the manner described above might depart slightly from these precise figures. On the average, however, over many attempts the results should equal these theoretical values.

These facts permit some rather direct inferences concerning the accuracy of a sample mean as an estimate of the true mean. If the population is normal in form, then 95 percent of the sample means fall within $1.96\sigma_M$ of the true mean and 99 percent of the sample means fall within $2.58\sigma_M$ of the true mean. Suppose $\sigma_M$ were known to be 0.3 inches for the variable, height, in a certain population. Then, multiplying 0.3 by 1.96 and 2.58 gives 0.588 and 0.774, respectively, indicating that 95 percent of the sample means of size $N$ would be within 0.588 inches of the true mean and 99 percent of them would be within 0.774 inches of the true mean. Only five percent of the sample means would differ from the true mean by more than 0.588 inches and only one percent of the sample means would differ from the true mean by more than 0.774 inches.

The usual situation, of course, is that an investigator has only one sample and he does not know where the true mean is. He would like to use the mean of his sample to get some information about the location of the true mean. If he did have the kind of information given in the example above, however, he would be able to assert that his sample mean is not more than 0.588 inches or 0.744 inches away from the true mean, depending upon his willingness to take risks. If he uses 0.588, he has a five percent chance of being wrong; if he uses 0.774, he has a one percent chance of being wrong. In either case, however, he has placed a limit on the expected maximum discrepancy between the true mean and his sample mean and he also has an associated probability value to tell him how likely it is that he is correct. This type of statement certainly provides useful information that has a direct bearing on the accuracy with which the sample mean is estimating the population mean.

*Converting Sample Means to Z Scores.*   It has already been stated that the sampling distribution for the mean is normal in form provided that the samples are drawn randomly from a normally distributed population, that is, in such a way that each member of the population has an equally likely chance of being selected for the sample. In fact, the sampling distribution will approach normality as the size of the sample increases even if the population is not normally distributed. For normally distributed variables and all variables with large samples, therefore, the sampling distribution can be taken to be normal for all practical purposes. When sample means are converted to standard score $(Z)$ values with respect to the sampling distribution, then, it is possible to use the normal curve tables to determine the percentage of sample means having $Z$ scores in various parts of the sampling distribution. Converting a sample mean to a $Z$ score relative to the sampling distribution requires subtracting $\mu$, the population mean, from $M$, the sample mean, and a division of this difference by $\sigma_M$, the standard deviation of the sampling distribution. This is analogous to the conversion of a raw score to a standard score $(Z)$ by the formula $Z = (X - M)/\text{S.D.}$ In this case, the mean of a raw score distribution is subtracted from the raw score and the difference is divided by the standard deviation of the raw score distribution. If the raw score distribution is normal in form, the $Z$ scores derived from these raw scores are also normally distributed. Similarly, if the sample means are normally distributed, the $Z$ scores into which they are converted by the formula $Z = (M - \mu)/\sigma_M$ will also be normally distributed.

When $Z$ scores are normally distributed, the normal curve tables can be used to determine the percentage of $Z$ scores above or below a certain point in the distribution or between any two points. Taking a given sample mean, for example, and converting it to be a $Z$ score with respect to its own sampling distribution, by the formula $Z = (M - \mu)/\sigma_M$, it would be possible to determine from the normal curve tables for that value of $Z$ what proportion of sample means are larger than this one, or smaller than this one. The percentage of sample means falling between this one and the true mean, $\mu$, could also be determined using the normal curve tables.

Converting the sample means to normally distributed $Z$ scores with respect to the sampling distribution of means can be effected by means of the formula $Z = (M - \mu)/\sigma_M$. Unfortunately, this conversion cannot be carried out in most cases because neither $\mu$, the population mean, nor $\sigma_M$, the true standard error of the mean, is known. The value of $\sigma_M$ is not known because by Formula 8–1, $\sigma_M = \sigma/\sqrt{N}$, it is a function of $\sigma$, the population standard deviation which itself is unknown.

Not knowing $\mu$, the true population mean, proves to be less bothersome, however, than not knowing $\sigma$, the population standard deviation. When a particular statistical hypothesis about means is to be tested, it will ordinarily involve a specified hypothetical value for $\mu$, the population mean and central value in the normal sampling distribution. The variability of the sampling distribution, however, is not so specified by fixing upon a particular hypothesis to test. It is necessary, therefore, either to know what $\sigma$ is or to estimate it so that a true or estimated value of $\sigma_M$ can be obtained. Otherwise, it would not be possible to convert the sample mean to a standard score $(Z)$ relative to the sampling distribution.

### Estimating the Standard Error of the Mean

Formula 8–1 stated that the true standard error of the mean, $\sigma_M$, is given by $\sigma/\sqrt{N}$, that is, the true standard deviation of the population for the variable in question divided by the square root of the sample size. If $\sigma$ were known, then the value of $\sigma_M$ would become readily available since $N$ is always known. Unfortunately, however, $\sigma$ is not known since, in theory, at least, the population is infinite in size and it would be impossible to compute the standard deviation for an infinitely large population.

Although $\sigma$ is ordinarily unavailable, making it impossible to compute $\sigma_M$, it is possible to obtain an estimated value of $\sigma_M$, namely $s_M$, called the "standard error of the mean" by substituting an estimate of $\sigma$ in Formula 8–1. The estimated value of $\sigma$ is derived from the sample that has been drawn from the population, using the following formula:

$$s = \sqrt{\frac{\Sigma x^2}{N - 1}} \qquad (8\text{–}2)$$

Where:

$s$ = an estimate of the population standard deviation in variable $X$

$x$ = a deviation score in variable $X$

$N$ = the sample size

The only difference between Formula 8–2 and the formula for the standard deviation of a sample, S.D., is that the sum of the squared deviations is divided by $N - 1$ instead of by $N$. The effect of this is to make $s$ slightly larger than S.D., especially for small samples when $N - 1$ and $N$ are sufficiently different to affect the outcome. With large samples, S.D. and $s$ will be identical for all practical purposes.

The reason $s$ must be used as the estimate of $\sigma$ instead of the S.D. is that the S.D. is a "biased" estimate of $\sigma$. It can be shown mathematically that the average value of S.D. taken over a very large number of random samples will be slightly smaller than $\sigma$. It can also be shown that the slight correction involved in dividing $\Sigma x^2$ by $N - 1$ instead of $N$ is just sufficient to adjust for this bias so that the average value of $s$ over a very large number of samples will equal $\sigma$. Thus, $s$ is an "unbiased" estimate of $\sigma$. The sample mean, $M$, fortunately, is an "unbiased" estimate of $\mu$, the population mean, so no correction is necessary in estimating $\mu$.

Since the S.D. is sometimes known when $s$ is needed, and vice versa, it is convenient to change one to the other using one of the following two formulas:

$$s = \sqrt{\frac{\Sigma x^2}{N - 1}} = \sqrt{\frac{\Sigma x^2}{N}} \sqrt{\frac{N}{N - 1}} = (\text{S.D.}) \cdot \sqrt{\frac{N}{N - 1}} \qquad (8\text{-}3)$$

$$\text{S.D.} = \sqrt{\frac{\Sigma x^2}{N}} = \sqrt{\frac{\Sigma x^2}{N - 1}} \sqrt{\frac{N - 1}{N}} = s \cdot \sqrt{\frac{N - 1}{N}} \qquad (8\text{-}4)$$

Thus, to convert $s$ to S.D., multiply $s$ by the square root of $N - 1$ over $N$. To convert S.D. to $s$, multiply S.D. by the square root of $N$ over $N - 1$.

The standard error of the mean, an estimate of the true standard error of the mean, $\sigma_M$, then, can be obtained using the following formula:

$$s_M = \frac{s}{\sqrt{N}} \qquad (8\text{-}5)$$

Symbols are defined as before. An alternate mathematically equivalent formula for the standard error of the mean is given by:

$$s_M = \frac{\text{S.D.}}{\sqrt{N - 1}} \qquad (8\text{-}6)$$

That these two are equivalent can be shown as follows:

$$s_M = \frac{s}{\sqrt{N}} = \frac{\sqrt{\dfrac{\Sigma x^2}{N - 1}}}{\sqrt{N}} = \sqrt{\frac{\Sigma x^2}{N(N - 1)}} = \frac{\sqrt{\dfrac{\Sigma x^2}{N}}}{\sqrt{N - 1}} = \frac{\text{S.D.}}{\sqrt{N - 1}}$$

The choice of Formula 8–5 or 8–6 will depend on whether $s$ or S.D. is already known and sometimes on the convenience of taking a square root. If $N = 26$ obviously Formula 8–6 is easier to use, other things being equal.

*Converting Means to Z Scores Using* $s_M$. If $\mu_0$ is taken to be a hypothetical true mean of a population distribution and hence the center of the normal sampling distribution of means as well, it would be possible to convert mean scores to standard scores by the formula $t = (M - \mu_0)/s_M$. In this case, $\mu_0$, the hypothetical true mean, has been substituted for the unknown $\mu$, the true mean, and the estimated standard error of the mean, $s_M$, has been substituted for $\sigma_M$, the unknown true standard error of the mean. These standard scores are designated as $t$ *scores*, instead of $Z$ scores, to distinguish them from the standard scores that would be obtained by the formula $Z = (M - \mu)/\sigma_M$.

It has already been stated that converting sample means to $Z$ scores by the formula $(M - \mu)/\sigma_M$ gives scores that are normally distributed in most situations. What about $t$ scores? Are they normally distributed? Unfortunately, they are not normally distributed unless $\mu_0 = \mu$ and $s_M = \sigma_M$. Again, testing a certain hypothesis about the true mean, $\mu$, can be accomplished without knowing $\mu$, using $\mu_0$ for the purpose of calculating the standard score for a given sample mean. Distortions in the distribution of the $t$ scores brought about by the fact that $\mu_0$ does not equal $\mu$ do not interfere with the test of the hypothesis since if $\mu_0$ does equal $\mu$, the hypothesis being tested, there will be no distortion from this source. Distortions in the distribution of $t$ scores due to the fact that $s_M$ does not equal $\sigma_M$ are another matter, however. These distortions do affect the distribution of the $t$ scores in such a way that they no longer follow a normal distribution. When $s$, the sample estimate of $\sigma$, is too small, $s_M$ will be too small and $t$ will therefore be too large. When $s$, the sample estimate of $\sigma$, is too large, $s_M$ will be too large and $t$ will therefore be underestimated. Thus, when $s$ is used as an estimate of $\sigma$ in converting sample means to standard scores, i.e., $t$ scores, some of the $t$ values will be larger than they should be and some of them will be smaller than they should be as a consequence of errors in the values of $s$ as estimates of $\sigma$. The shape of the distribution of $t$ scores, therefore, is not normal in form. As a consequence, the normal curve tables cannot be used to determine the percentage of $t$ scores above or below a certain $t$ value.

*Student's* t *Distribution.* Means converted to standard scores using $t = (M - \mu_0)/s_M$ instead of $Z = (M - \mu)/\sigma_M$ yield scores that are not normal in form but instead follow the $t$ distribution. The $t$ distribution consists of a family of curves with a different curve for each value of $df$, the number of degrees of freedom. These curves approach more and more closely to the normal curve as the number of degrees of freedom $(df)$ increases. In computing a $t$ score for a sample mean relative to the sampling distribution of means, the number of degrees of freedom, $df$, equals $N - 1$, the number of cases in the sample minus one. Therefore, the particular $t$ distribution that is appropriate for describing the distribution of $t$ scores being considered comes closer and closer in form to the normal distribution as the size of the sample increases (since $N - 1$ is very close to $N$). The normal distribution, in fact, is a special case of the $t$ distribution when the $df$ is infinite, i.e., the sample is infinitely large. In practice, however, the particular $t$-distribution appropriate for a given size sample becomes very close to the normal distribution long before the sample size approaches infinity. Even with samples of size 50, or 49 degrees of freedom, the $t$ distribution is close to normal in shape. With a sample of size 100, or $df = 99$, there is very little difference between the $t$-distribution and the normal distribution.

Figure 8–2 shows a normal distribution and a $t$ distribution that illus-

FIGURE 8–2
Normal Curve and $t$-Distribution

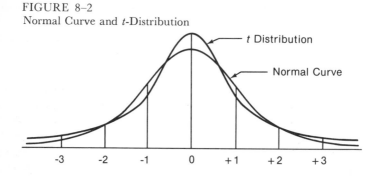

trate the difference between the two curves. The $t$ distribution is taller and narrower in the middle (more "leptokurtic") and has higher tails, due to the fact that some $t$ values are overestimated and others are underestimated as a consequence of using the approximate $s_M$ instead of the true $\sigma_M$ to compute the standard score. This means that the $t$ distribution has more area in the tails than the normal distribution. Both the standard $t$ distribution and the standard normal distribution have a total area under the curve equal to 1.00 but it is distributed differently in the two curves. In the two tails of the normal curve, above the point 1.96 and below $-1.96$, there is a total area of about 0.05, half in each tail. Beyond $\pm 2.58$, there is a total area of about 0.01, 0.005 in each tail. The corresponding figures for the $t$ distribution are larger, depending on the number of degrees of freedom. As the number of degrees of freedom increases, the areas in the tails of the $t$ distribution get smaller and smaller, approaching those for the normal curve as a limit.

The difference between the $t$ distribution and the normal curve is very apparent in the critical values associated with the commonly used levels of significance. The critical values of $t$ are shown in the Table of the $t$ Distribution (Appendix A, Table C). At the bottom of the table, with $df = \infty$, the normal curve critical values are given. These are approximately 1.65 for $p = 0.10$ (two-tailed test, or 0.05 for a one-tailed test), 1.96 for $p = 0.05$ (two-tailed test), 2.33 for $p = 0.02$ (two-tailed test, or 0.01 for a one-tailed test), and 2.58 for $p = 0.01$ (two tailed test). Notice that the critical values of $t$ required to achieve significance get larger and larger as the number of degrees of freedom (and consequently the sample size) gets smaller and smaller. With only 10 degrees of freedom, for example, the five and one percent level two-tailed critical values are large, $t = 2.23$ and 3.17, respectively, compared to the critical $Z$ values of 1.96 and 2.58 required for large samples.

To summarize, then, sample means from normally distributed populations themselves yield a normal distribution with a mean equal to the population mean, $\mu$, and a standard deviation equal to the population standard deviation, $\sigma$, divided by the square root of $N$, the sample size. This standard deviation of the sampling distribution is called the "true standard error of the mean," and is denoted by the symbol $\sigma_M$. Sample mean values may be converted to standard score values relative to this sampling distribution of means by the formula $Z = (M - \mu)/\sigma_M$. The normal curve tables may be used to find the percentage of cases above and below any such $Z$ values obtained from sample means. Since $\mu$ and $\sigma$ are not ordinarily known, however, it is usually impossible to use this formula to convert sample means to standard scores relative to the sampling distribution of means. Instead, the standard error of the mean is estimated by Formula 8–5 or 8–6, a hypothesized true mean, $\mu_0$, is selected, and sample means are converted to standard scores by the formula $t = (M - \mu_0)/s_M$. Even if $\mu = \mu_0$, $t$ scores still will not be normally distributed, however, due to distortions in the $t$ values resulting from the fact that $s_M$ is sometimes too large and sometimes too small. When $\mu = \mu_0$, however, the $t$ scores follow a known form, the $t$ distribution, which varies with the number of degrees of freedom. As the number of degrees of freedom increases, the $t$ distribution approaches the normal distribution more and more closely. Critical values of $t$ required for significance are larger than those required for $Z$ scores and the normal distribution. These values may be found by consulting the $t$ distribution table (Appendix A, Table C).

## Problem 23
### TEST THE HYPOTHESIS THAT THE MEAN OF THE POPULATION IS A CERTAIN VALUE

There are occasions in social science research where it may become important to test an hypothesis such as $\mu = C$, where $\mu$ is the population mean for a given variable and $C$ is some specified constant. Then, $\mu_0$, the hypothesized true mean, equals $C$. A particular theory might predict, for example, that the average number of trials to extinguish a particular response will be 50. To test the theory, the accuracy of this prediction can be tested. Thirty-seven subjects are trained until the specified response is established and then extinction trials are run to obliterate the response, counting the number of trials required for each subject.

With $N = 37$, the hypothesis will be tested using the $t$-distribution for $df = 36$, i.e., $N - 1$. The hypothesis to be tested is $h_0: \mu = 50$. Then if $\mu = 50$, the sampling distribution of means will be centered about 50 with a standard deviation that is estimated by $s_M = s/\sqrt{N}$ or $s_M = $ S.D.$/\sqrt{N - 1}$. Suppose in this particular experiment, the average number of extinction trials required to obliterate the response were 58 and the standard deviation of the sample of values were 15. The standard error of the mean then would be:

$$s_M = \frac{\text{S.D.}}{\sqrt{N - 1}} = \frac{15}{\sqrt{37 - 1}} = \frac{15}{6} = 2.5$$

The obtained sample mean value of 58 must be converted to a standard score for evaluation relative to the $t$ distribution with $df = 36$. This standard score is computed by the following formula:

$$t = \frac{M - \mu_0}{s_M} \tag{8-7}$$

The hypothesized true mean, 50 is substituted for $\mu_0$. The computed standard error of the mean, 2.5, is substituted for $s_M$ and the sample mean value, 58, is substituted for $M$ to give:

$$t = \frac{58 - 50}{2.5} = 3.2$$

This obtained standard score relative to the $t$ distribution must be compared with the critical values of $t$ obtainable from the Table of the $t$ Distribution (See Appendix A, Table C). If the one-percent level, two-tailed test, had been chosen in advance as the appropriate test, the critical value would be $t_{.01} = 2.72$.

The obtained value of $t$, 3.2, is larger than 2.72, the critical value of $t_{.01}$ for 36 degrees of freedom, so the obtained result represents a significant departure from expectation under $h_0$. It is therefore appropriate to reject $h_0$ at the one percent level of confidence. This would imply that $\mu \neq 50$ (is *not* equal to 50), although there is no statement as to what the correct value of $\mu$ actually is.

The reasoning is as follows. If $\mu = 50$, then 99 percent of all sample means from this distribution yield standard score values between $+2.72$ and $-2.72$. Only one time in a hundred does a randomly drawn sample have a computed standard $t$ score larger than 2.72 in absolute value. In this one experiment, however, a $t$ value of 3.2 was obtained. This is so unlikely (less than one chance in 100) under the assumption $\mu = 50$ that it must be concluded $\mu$ is *not* 50 but some other value, probably larger than 50, since the sample value was well above 50. Even though $h_0$ was rejected, however, there is always the possibility that it was a mistake to do so since there is some chance, although small, that a $t$ of 3.2 could occur even when $h_0$ is true.

In this particular experiment, rejection of $h_0$ represents a disconfirmation of the theory which predicted that the mean number of extinction trails would be 50. Since $h_0$: $\mu = 50$ was rejected there is less than one chance in 100 that the theory could be correct.

*Two Types of Errors.* In the example above, only one time in 100 would a value of $t$ larger in absolute value than 2.72 be obtained. It does occur, however, one time in 100. If this does happen by chance in a given experiment, the investigator will get an obtained value of $t$ greater than $+2.72$ or less than $-2.72$ and he will reject $h_0$ thinking it to be false on the basis of the fact that he had a large value of $t$. In rejecting $h_0$, however, he will be making an error, referred to as a "Type I error." He rejected $h_0$ when it was true. For the one percent level of confidence this happens on the average one time in every 100 experiments. For the five percent level of confidence, it happens five times in every 100 experiments.

A second type of error, called a "Type II error," is made when $h_0$ is false, i.e., $\mu$ is different from the hypothesized $\mu_0$, but the obtained $t$ does not equal or exceed the critical value of $t$ for that number of degrees of freedom. If $t$ does not exceed the critical value, $h_0$ cannot be rejected even if $h_0$ is false. If $h_0$ is not rejected when it is false, because $t$ is too small, a Type II error has been committed.

It is clear that there is a relationship between Type I and Type II errors. If the level of significance required is made stringent, e.g., 0.01 level, to reduce the Type I errors, Type II errors will increase. If the critical level is made easier to reduce Type II errors, Type I errors will increase. In selecting a significance level, the investigator must decide for himself which type of error he is more anxious to avoid. If he wishes to make sure he has identified a phenomenon of some importance before reporting it, he would do well to use a fairly stringent criterion of significance, e.g., the 0.01 level. If, on the other hand, he is more concerned that he does not miss something, he would choose a less stringent level, e.g., the 0.05 level.

## Summary of Steps in Testing the Hypothesis that the Mean of the Population is a Certain Value

1. Compute the mean of the sample, $M$.
2. Establish the value of $\mu_0$, the hypothesized true value of the mean to be tested.
3. Calculate the standard error of the mean, $s_M$, by Formula 8–5 or by Formula 8–6.
4. Convert the mean, $M$, to a standard $t$ score by subtracting $\mu_0$ from $M$, the sample mean, and dividing the resulting deviation score by $s_M$, the standard error of the mean.
5. With $N - 1$ degrees of freedom, where $N$ is the sample size, determine the critical values of $t$, $t_{.05}$ and $t_{.01}$, from the Table of the $t$ Distribution.
6. If the obtained $t$ score from Step 4 above is greater than $t_{.05}$ but less than $t_{.01}$, reject $\mu_0$ as the population mean at the five percent level of significance. If the obtained $t$ score from Step 4 above is greater than $t_{.01}$, reject $\mu_0$ as the population mean at the one percent level of significance. If the obtained $t$ score is less than $t_{.05}$, do not reject the hypothesis that $\mu_0$ is the population mean.

## Problem 24
### ESTABLISH A CONFIDENCE INTERVAL FOR THE TRUE MEAN

In Problem 23, a hypothesized value for $\mu$, the true mean, was available and the sample was used to investigate the tenability of this hypothesis. A more commonly encountered situation in the social sciences is that in which the mean of a sample is available but there is no hypothesis concerning where the true mean might be. In this situation, the sample mean, $M$, is taken as an estimate of the population mean, $\mu$, and the concern is with trying to establish how good the estimate is.

This problem is solved by setting up an interval about the sample mean that gives an idea of the limits within which the true mean could be expected to fall, together with an associated probability. The 95 percent confidence interval for the true mean, for example, is defined in such a way that if it is obtained for sample after sample, 95 percent of the time it will contain the true mean. The 99 percent confidence interval will contain the true mean 99 times out of 100 on the average if it is computed for a large number of samples. In practice, of course, the confidence interval, whether the 95 percent or 99 percent one, is computed only for a single sample and a statement is made that it contains the true mean. The statement may or may not be true in a given instance but if this is done over and over again most of the time the statement will be correct and the true mean *will* be contained within the confidence interval.

## Formulas for the 95 Percent and 99 Percent Confidence Intervals

The formulas for the 95 and 99 percent confidence intervals are as follows:

$$95 \text{ percent confidence interval: } M \pm t_{.05} \cdot s_M \qquad (8\text{–}8)$$

$$99 \text{ percent confidence interval: } M \pm t_{.01} \cdot s_M \qquad (8\text{–}9)$$

Thus, $s_M$ is computed by Formula 8–5 or 8–6, $s_M = \dfrac{s}{\sqrt{N}}$ or $\dfrac{\text{S.D.}}{\sqrt{N-1}}$

and multiplied by $t_{.05}$ or $t_{.01}$, for $N - 1$ degrees of freedom, taken from the Table of the $t$ Distribution (See Appendix A, Table C). These intervals, $t_{.05} \cdot s_M$ and $t_{.01} \cdot s_M$ are added to and subtracted from the sample mean, $M$, to provide intervals with $M$ at the middle, defined as the 95 and 99 percent confidence intervals, respectively.

Why do 95 and 99 percent of such intervals, respectively, contain the true mean on the average? It has already been established that sample means are distributed about the true mean as follows:

95 percent of sample means on the average are contained within the interval $\mu \pm t_{.05} \cdot s_M$

99 percent of the sample means on the average are contained within the interval $\mu \pm t_{.01} \cdot s_M$

Thus, the sample means are distributed about the true mean, $\mu$, while each confidence interval has the sample mean at its center. The interval within which 95 percent of the samples are contained is twice as wide as $t_{.05} \cdot s_M$, half above $\mu$, the true mean, and half below $\mu$. A particular sample drawn at random from the population could have a mean located anywhere in the sampling distribution, but 95 percent of them on the average are contained in the region $\mu \pm t_{.05} \cdot s_M$. Suppose that successive samples are drawn at random and a 95 percent confidence interval computed for each, the location of these confidence intervals relative to the true mean can be schematized as in Figure 8–3. Samples 1, 2, 3, 4, and $\infty$ have confidence intervals that contain the true mean $\mu$, although not at the center of the intervals. The sample means fall at the centers of the confidence intervals and of course the sample means do not all equal $\mu$, the true mean, which lies at the center of the sampling distribution.

The total width of the 95 percent confidence interval is the same as the total width of the interval containing 95 percent of the sample means in the sampling distribution, namely $t_{.05} \cdot s_M$. It is just that the confidence interval is centered about the sample mean whereas the sample means are distributed about the true mean. If a sample mean, however, is contained

FIGURE 8–3
Confidence Intervals and the True Mean

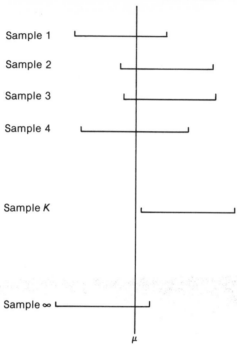

within the interval $\mu \pm t_{.05} \cdot s_M$ then the 95 percent confidence interval for the true mean obtained from that sample will actually contain $\mu$, the true mean, as with samples 1, 2, 3, 4, and $\infty$ in Figure 8–3. This is due to the fact that *no* sample within the interval $\mu \pm t_{.05} \cdot s_M$ is more than $t_{.05} \cdot s_M$ points away from $\mu$. Since the 95 percent confidence interval runs $t_{.05} \cdot s_M$ points on either side of the sample mean, the boundaries must extend beyond $\mu$. Only when the sample mean lies *outside* the interval $\mu \pm t_{.05} \cdot s_M$ will the 95 percent confidence interval fail to extend far enough to encompass $\mu$, as with sample $K$ in Figure 8–3.

Since 95 percent of the sample means lie in the interval $\mu \pm t_{.05} \cdot s_M$ on the average and all of the sample means in this region have confidence intervals large enough to encompass the true mean, it follows that only 5 percent of the time, on the average, will the true mean fail to be contained within the 95 percent confidence interval. The same line of reasoning dictates that only one percent of the time will the 99 percent confidence interval fail to encompass the true mean.

Computation of the confidence interval, then, provides a way of assessing the accuracy of the sample mean as an estimate of the true mean. It fixes an interval within which the true mean is likely to fall and provides a probability statement concerning the likelihood of being wrong in claiming that the true mean lies within that interval. In many cases, being able to tie down the location of $\mu$ to this degree is quite sufficient for the purposes at hand. If it is necessary to know only that $\mu$ is not less than 40, for example, and the 99 percent confidence interval runs from 43 to 47, there is little need to fix the location of $\mu$ any more precisely than it has been by means of this confidence interval.

It should be noted that the confidence interval can be used to solve Problem 23. After determining the 95 percent confidence interval, for example, $h_0$ will be rejected at the five percent level of significance provided that the hypothesized true mean lies outside the 95 percent confidence interval. If the 95 percent confidence interval contains the hypothesized true mean within its boundaries, $h_0$ is not rejected.

### An Example

To illustrate the application of these principles, a 99 percent confidence interval for the true mean will be computed given that $M = 20$, S.D. = 5, and $N = 16$. Then, using Formula 8–6:

$$s_M = \frac{\text{S.D.}}{\sqrt{N-1}} = \frac{5}{\sqrt{16-1}} = \frac{5}{3.87} = 1.29$$

From the Table of the $t$ Distribution (Appendix A, Table C) with $N - 1 = 15$ degrees of freedom, $t_{.01} = 2.95$. The 99 percent confidence interval, then, using Formula 8–9 is:

$$
\begin{aligned}
\text{99 percent confidence interval} &= M \pm t_{.01} \cdot s_M \\
&= 20 \pm (2.95)(1.29) \\
&= 20 \pm 3.80
\end{aligned}
$$

$$\text{99 percent confidence interval} = 16.8 \text{ to } 23.8$$

The true mean should fall within the interval 16.2 to 23.8. Ninety nine times out of 100 when this process is applied, on the average, the 99 percent confidence interval will contain the true mean if the variable in question is normally distributed or even approximately normal in form. To the extent that peculiarities in the parent population distribution result in distortions in the sampling distribution, some departures from this model can be anticipated.

### Summary of Steps in Establishing a Confidence Interval for the True Mean

1. Compute the mean of the sample, $M$.
2. Calculate the standard error of the mean, $s_M$, by Formula 8–5 or by Formula 8–6.

3. With $N - 1$ degrees of freedom, determine the critical values of $t$, $t_{.05}$ and $t_{.01}$, from the Table of the $t$ Distribution.
4. Multiply $s_M$ by $t_{.05}$ and by $t_{.01}$.
5. Add $t_{.05} \cdot s_M$ to $M$ and subtract it from $M$ to get the 95 percent confidence interval for the true mean, $\mu$.
6. Add $t_{.01} \cdot s_M$ to $M$ and subtract it from $M$ to get the 99 percent confidence interval for the true mean, $\mu$.

## Problem 25
## TEST THE SIGNIFICANCE OF A CORRELATION COEFFICIENT

When the correlation between two variables in the population is zero, the computed correlation coefficient between those two variables in a randomly drawn sample from that population will not necessarily equal zero. In fact, the correlation values in the samples yield a sampling distribution of correlation coefficients just as sample means yield a sampling distribution of means. A given sample correlation, $r$, can be converted to a $t$ score value relative to the sampling distribution of correlations where the true $r$ is zero by means of the following formula:

$$t = \frac{r\sqrt{N - 2}}{\sqrt{1 - r^2}} \tag{8-10}$$

Where:

$N$ = the number of cases in the sample, i.e., the number of pairs of scores

Formula 8–10 is appropriate for use in the situation where the investigator wishes to test whether his obtained sample correlation is significantly different from zero. In this case, $h_0$, the hypothesis to be tested, is that the true correlation is zero. The larger the sample correlation becomes, the bigger will be the $t$ value computed by Formula 8–10. If the computed $t$ value is larger than the critical value of $t$ taken from the Table of the t Distribution (Appendix A, Table C), $h_0$ is rejected and it is concluded that the true correlation is not equal to zero. Either a one-tailed or a two-tailed test can be used to test $h_0$ but a one-tailed test should be used only if it was perfectly clear in advance what sign the obtained correlation should have and only a correlation of that sign would be considered to be significant. If the intention is to reject $h_0$ with either a high positive or a high negative correlation, a two-tailed test must be made.

*Degrees of Freedom in Testing $r$.*   In the chapter on correlation, it was shown that the correlation coefficient is the slope of the regression line of best fit when standard scores are being plotted. Two constants must be specified to determine a straight line, the slope and the $Y$ intercept. This means that two degrees of freedom are lost in fitting a straight line to a sample of score pairs, one for each of these constants. The number of degrees of freedom that must be used in connection with Formula 8–10, therefore, is $N - 2$ instead of $N - 1$ which is used with means. The $t$ value at the 0.05 or 0.01 level for $N - 2$ df is taken from Table C in Appendix A to obtain the critical value of $t$ that must be equalled or exceeded to reject $h_0$. The $df$ for the mean is $N - 1$ because only one degree of freedom is lost in specifying the value of the mean. That is, any scores may be selected for $N - 1$ scores of the sample. The last score can always be chosen in such a way as to make the mean come out to be any specified value, no matter what the other scores are. In the case of the correlation coefficient, two such values must be retained to make the correlation coefficient come out to be a particular value.

### An Example

Given a correlation between two variables of 0.24 in a randomly selected sample of 66 subjects, determine if the correlation is significantly different from zero. Using Formula 8–10 gives:

$$t = \frac{0.24 \sqrt{66 - 2}}{\sqrt{1 - (0.24)^2}} = \frac{(0.24)(8)}{\sqrt{0.951}} = \frac{1.92}{0.975} = 1.97$$

With $N - 2 = 64$ $df$, $t_{.05}$ for a two tailed test is 2.00, so the $t$ value of 1.97 is not significant and $h_0$ cannot be rejected.

A somewhat larger correlation coefficient in this sample would have resulted in a $t$ value of 2.00 or larger which would have permitted rejecting $h_0$ if the five percent level had been selected in advance as the criterion. As it is, however, since $h_0$ could not be rejected, there is no statistical proof that the true correlation in the population is different from zero on the basis of the results from this random sample.

### Summary of Steps in Testing the Significance of a Correlation Coefficient

1. Compute the Pearson product-moment correlation coefficient, r.
2. Using $N - 2$ degrees of freedom, determine the critical values of $t$, $t_{.05}$ and $t_{.01}$, from the Table of the $t$-Distribution.
3. Substitute $r$ and $N$ in Formula 8–10 to obtain the $t$ score corresponding to this correlation coefficient.
4. If the $t$ score obtained in Step 3 above is greater than $t_{.05}$ but less than $t_{.01}$, reject at the five percent level of confidence the hypothesis that the true correlation is zero. If the $t$ score obtained in Step 3 above is greater than $t_{.01}$, reject at the one percent level of con-

fidence the hypothesis that the true correlation is zero. If the $t$ score is less than $t_{.05}$, do not reject the hypothesis that the true correlation is zero.

## Problem 26

### TEST THE SIGNIFICANCE OF THE DIFFERENCE BETWEEN TWO VARIANCE ESTIMATES WITH THE $F$ TEST

Given two populations with equal variances such that $\sigma_1^2 = \sigma_2^2$, random samples taken from these respective populations can be expected to have variance estimates that differ from one another, i.e., $s_1^2 \neq s_2^2$. A statistic, $F$, can be defined in this context as the ratio of two variance estimates as follows:

$$F = \frac{s_x^2}{s_y^2} \tag{8--11}$$

Where:

$F$ = the $F$ ratio

$s_x^2$ = the variance estimate for the sample that has the larger value of $s$

$s_y^2$ = the variance estimate for the sample that has the smaller value of $s$

Thus, for the two samples, $s^2$ is computed as $\Sigma x^2/(N-1)$ and the larger one is divided by the smaller one to get $F$ which in this situation is therefore always greater than or equal to 1.00.

If $\sigma_1^2 = \sigma_2^2$ and the estimates of these parameters, $s_1^2$ and $s_2^2$, are accurate, $F$ will equal one. In practice, however, random samples from the population produce variance estimates that differ from the true value in the population. The probability that a pair of samples will produce a particular $F$ value becomes smaller and smaller as the size of $F$ increases, other things being equal, when $\sigma_1^2 = \sigma_2^2$. If a large value of $F$ is obtained, for example, one must conclude either that a very rare thing has happened even though $\sigma_1^2 = \sigma_2^2$, or else it must be concluded that $\sigma_1^2 \neq \sigma_2^2$. A test of significance of the difference between two variance estimates with the $F$ test involves calculating $F$ by Formula 8--11 and then evaluating that $F$ to see if it is too large to continue entertaining the hypothesis, $h_0$, that $\sigma_1^2 = \sigma_2^2$.

How large $F$ can become with random sampling from populations where $\sigma_1 = \sigma_2$ depends on the sample size, just as the standard error of the mean depended on the sample size. More precisely, the distribution of $F$ can be shown mathematically to be a function of the number of degrees of freedom in the two samples, i.e., $(N_x - 1)$ and $(N_y - 1)$. Each combination of values $df_x$ (for the larger variance estimate) and $df_y$ (for the smaller variance estimate) results in its own distribution of $F$ such that there is a specified probability of getting that $F$ or a larger one when sampling at random from two populations with $\sigma_1^2 = \sigma_2^2$.

Two values from these distributions are given in Table E (See Appendix A), namely the five percent level value of $F$ (in light-face type) and the one percent level value of $F$ (in bold-face type). Table E is entered with the two degrees of freedom values, $df_x$ and $df_y$. The $df$ for the larger variance estimate is located across the top of Table E and the $df$ for the smaller variance estimate is located down the left hand column. At the intersection of this column and row are located $F_{.05}$ and $F_{.01}$ for that particular combination of degrees of freedom in the two samples. If $\sigma_1^2 = \sigma_2^2$, only five times in 100 on the average will the samples yield a value of $F$ as large or larger than $F_{.05}$. Only one time in 100 on the average will the samples yield a value of $F$ as large or larger than $F_{.01}$. If the obtained value of $F$ exceeds $F_{.05}$ for this particular combination of $df$ values for the two samples, the hypothesis, $h_0: \sigma_1^2 = \sigma_2^2$, can be rejected at the five percent level of significance. If the obtained $F$ exceeds $F_{.01}$, $h_0$ can be rejected at the one percent level of significance. If the obtained value of $F$ is less than $F_{.05}$, then $h_0$ is not rejected. As with the mean, $h_0$ is never "accepted", it is merely "not rejected".

*An Example.* Two methods of instruction with randomly selected groups have yielded mean achievement scores of $M_1 = 95$ and $M_2 = 99$, respectively. Other sample statistics were S.D.$_1 = 10$, S.D.$_2 = 7$, $N_1 = 50$, $N_2 = 65$. Before making a further statistical comparison that assumes $\sigma_1^2 = \sigma_2^2$, it is necessary to test the hypothesis, $h_0: \sigma_1^2 = \sigma_2^2$, to see if these other comparisons can be made legitimately in this situation. That is, it is necessary to test the validity of the assumption, $\sigma_1^2 = \sigma_2^2$, upon which the other statistical comparison is based.

Since in this case the S.D. is given for each sample, rather than $s$, the S.D. values will be converted to $s$ values by Formula 8–3:

$$s = \text{S.D.} \sqrt{\frac{N}{N - 1}} \qquad (8\text{–}3)$$

$$s^2 = (\text{S.D.})^2 \left(\frac{N}{N - 1}\right)$$

For the two samples:

$$s_1^2 = (10)^2 \cdot \left(\frac{50}{49}\right) = \frac{5000}{49} = 102.04$$

$$s_2^2 = (7)^2 \cdot \left(\frac{65}{64}\right) = \frac{49 \cdot 65}{64} = \frac{3185}{64} = 49.77$$

Since $s_1^2$ is larger than $s_2^2$, $s_1^2$ becomes $s_x^2$, $s_2^2$ becomes $s_y^2$ and Formula 8–11 gives:

$$F = \frac{s_x^2}{s_y^2} = \frac{102.04}{49.77} = 2.05$$

Entering Table E (Appendix A) with 49 degrees of freedom for the larger variance estimate (across top of table) and 64 degrees of freedom for the smaller variance estimate (down the left side), $F_{.05} = 1.54$ and $F_{.01} = 1.84$. Where necessary, interpolation can be used to obtain values of $F_{.05}$ and $F_{.01}$ for nontabled intermediate degrees of freedom. Since the obtained value of $F$, 2.05, is larger than 1.84, the $F_{.01}$ level value of $F$, if the one percent level of confidence had been chosen as the criterion, it would have been necessary to reject $h_0$ at that level and to conclude that $\sigma_1^2 \neq \sigma_2^2$. This would render technically inappropriate the application of further statistical comparisons with these data that rest upon the assumption that $\sigma_1^2 = \sigma_2^2$. In practice, however, such tests have been shown to be relatively insensitive to violations of this particular assumption.

### Summary of Steps in Testing the Significance of the Difference between Two Variance Estimates by Means of the F Test

1. Compute $s^2$, the estimated population variance, for each of two random samples, using the formula $s^2 = \Sigma x^2 / (N - 1)$.

2. Let the larger of these two variance estimates be designated as $s_x^2$ and the smaller variance estimate as $s_y^2$.

3. Substitute $s_x^2$ and $s_y^2$ into Formula 8–11 to obtain the value of $F$.

4. Using $N_x - 1$ degrees of freedom for the larger variance estimate and $N_y - 1$ degrees of freedom for the smaller variance estimate, enter the Table of the $F$ Distribution to find $F_{.05}$ and $F_{.01}$, the critical values of $F$.

5. If the value of $F$ obtained in Step 3 above is greater than $F_{0.05}$ but less than $F_{.01}$, reject at the five percent level of confidence the hypothesis that these two samples were drawn at random from populations with equal variances. If the value of $F$ obtained in Step 3 above is greater than $F_{.01}$, reject the same hypothesis at the one percent level of confidence. If the obtained value of $F$ is less than $F_{.05}$, do not reject the hypothesis that these two samples were drawn at random from populations with equal variances.

### Problem 27
### TEST THE SIGNIFICANCE OF THE DIFFERENCE BETWEEN TWO INDEPENDENT RANDOM SAMPLE MEANS USING THE $t$ TEST

One of the most common applications of statistics in the social sciences is to compare the means of two random samples, e.g., experimental and control groups, to determine if the difference is significant or not. In a typical experiment, individuals would be drawn from a common pool and would be randomly assigned to an experimental group (Group E) and a control group (Group C). Then, the experimental group (E) is given one type of treatment while the control group (C) is given another type of treatment. At the end of the experiment both groups, E and C, are tested on a common variable, presumably related to the treatments, and mean values computed. Then $M_E$, the mean of the experimental group, is compared with $M_C$, the mean of the control group, to see if the difference between the two groups is large enough to conclude that the two kinds of treatments had different effects on the experimental variable being studied.

## Sampling Distribution of Differences between Means

Given two populations that are normally distributed with means $\mu_1$ and $\mu_2$ and standard deviations of $\sigma_1$ and $\sigma_2$, draw two random samples of sizes $N_1$ and $N_2$ from the two populations and compute their means, $M_1$ and $M_2$. Let $D_1 = M_1 - M_2$, the difference between these two sample means. Draw two more samples of the same sizes, compute the sample means, and get $D_2$ as the difference between them. If this process is repeated over and over again to get values $D_1, D_2, D_3, \ldots, D_\infty$, the $D$ values can be used to form the sampling distribution of differences between means. This is a hypothetical distribution, of course, since it is not possible to draw an infinite number of sample pairs to get an infinite number of $D$ values. It can be proved mathematically, however, that this distribution is normal if the two populations from which the samples are drawn are normal. Even if the populations are not normally distributed, the sampling distribution of differences between means will come closer and closer to normality in shape as the sample sizes increase. As the mean of a single sample was converted to a standard score relative to the sampling distribution of means, so can the difference between two sample means be converted to a standard score relative to the sampling distribution of differences between means. Unless the true difference between means ($\mu_1 - \mu_2$) and the true standard deviations of the two distributions, $\sigma_1$ and $\sigma_2$, are known, the differences between means cannot be converted to normally distributed standard scores. Using an hypothesized true difference between means, however, and an estimated standard deviation of the sampling distribution of differences between means, $s_{D_M}$, it is possible to obtain standard scores, $t$ values, that follow the $t$ distribution.

The particular $t$ distribution that is appropriate for converting a given difference between means to a standard score depends on the number of degrees of freedom, as was the case with one sample and the sampling distribution of means. Whereas with one single sample and the sampling distribution of means, the number of degrees of freedom is $N - 1$, with two independent random samples, the number of degrees of freedom for the sampling distribution of differences between means is $N_1 + N_2 - 2$. This is just the sum of the degrees of freedom for the two separate samples, i.e., $(N_1 - 1) + (N_2 - 1) = N_1 + N_2 - 2$.

*The Null Hypothesis.* The center of the sampling distribution of differences between means is $\mu_1 - \mu_2$, the difference between population means. For most applications it is assumed that $\mu_1 = \mu_2$ or $\mu_1 - \mu_2 = 0$, i.e., there is *no* difference between the population means from which these samples have been drawn at random. This is the so-called null hypothesis, $h_0: \mu_1 = \mu_2$; or $h_0: \mu_1 - \mu_2 = 0$. When two sample means are being compared, e.g., the experimental group mean, $M_E$, and the control group mean, $M_C$, it is customary to test the null hypothesis, $h_0: \mu_1 - \mu_2 = 0$. This hypothesis assumes that there is no difference in the effect of the experimental and control treatments on the variable in question. The investigator may wish and expect to find $h_0$ to be false, but he assumes it as something to be tested rather than as something he believes to be true.

## The Standard Error of the Difference between Means

The standard deviation of the sampling distribution of the differences between means is estimated by the following formula:

$$s_{D_M} = \sqrt{\frac{\Sigma x_1^2 + \Sigma x_2^2}{N_1 + N_2 - 2} \left( \frac{N_1 + N_2}{N_1 \cdot N_2} \right)} \tag{8-12}$$

Where:

$s_{D_M} =$ the standard error of the difference between means

$\Sigma x_1^2 =$ the sum of the squared deviation scores for sample 1

$\Sigma x_2^2 =$ the sum of the squared deviation scores for sample 2

$N_1 =$ the number of cases in sample 1

$N_2 =$ the number of cases in sample 2

It should be remembered that $\Sigma x^2$ is difficult to compute directly because it involves fractional deviation scores. It may be expressed in terms of raw scores as:

$$\Sigma x^2 = \Sigma X^2 - \frac{(\Sigma X)^2}{N} \tag{8-13}$$

The sum of the squared deviations may be computed for each sample separately, using Formula 8–13, and the totals substituted into Formula 8–12 to find $s_{D_M}$.

In the typical situation, then, the sampling distribution of differences between means is presumed to center about zero, the hypothesized true difference between population means, and has a standard deviation (the standard error), $s_{D_M}$, estimated by Formula 8–12. The $t$ scores, computed by converting differences between sample means to standard scores, follows the $t$ distribution with $N_1 + N_2 - 2$ degrees of freedom.

*Testing the Null Hypothesis.* In order to test whether the null hypothesis, i.e., $h_0$: $\mu_1 - \mu_2 = 0$, is tenable or not, in comparing the means of two samples, the difference between two means is converted to a standard score as follows:

$$t = \frac{(M_1 - M_2) - 0}{s_{D_M}} \tag{8–14}$$

The difference $(M_1 - M_2)$ is a score in the sampling distribution of differences between means. The hypothesized mean of this distribution, zero, is subtracted from $(M_1 - M_2)$, to get a deviation score, and this deviation score is divided by $s_{D_M}$, the standard deviation of the sampling distribution of differences between means to convert the difference score to a standard score in the $t$ distribution. If the difference between means is small, other things being equal, $t$ in Formula 8–14 will be small, and if the difference between means is large, other things being equal, $t$ will be large.

If the value of $t$ computed using Formula 8–14 is larger than the critical values of $t$ given in the Table of the $t$ distribution (Appendix A, see Table C), the null hypothesis, $h_0$, will be rejected. If the value of $t$ computed by Formula 8–14 is smaller than the selected critical value of $t$, then the null hypothesis will not be rejected. The null hypothesis is *never* "accepted," only "not rejected." The critical value of $t$ will depend on the number of degrees of freedom, $N_1 + N_2 - 2$, and the type of test selected, e.g. two-tailed one percent level, one-tailed one percent level, and so on. Testing the null hypothesis in this way is referred to as "making a $t$ test of the difference between means."

In making a $t$ test, it will sometimes happen that a negative value of $t$ is obtained, e.g., when a larger mean is subtracted from a smaller mean. The $t$ distribution is symmetric, however, just as the normal curve is, so the area above a positive $t$ and the area below minus $t$ are the same. When a $t$ comes out negative, therefore, ignore the sign and treat it as though it were positive.

*An Example.* Subjects are drawn at random from a common pool to form an experimental group (E) and a control group (C). Group E receives visual imagery training for a pursuit rotor task while Group C receives a control type of training. In the pursuit rotor task, the subject tries to hold a stylus in electrical contact with a small metal dot that moves on a rotating turntable. The score is the number of seconds in contact. Both groups are then tested on a pursuit rotor task with the following results:

$$\text{Group C: } \Sigma X = 1433, \ \Sigma X^2 = 77435, \ N = 31$$

$$\text{Group E: } \Sigma X = 1033, \ \Sigma X^2 = 55525, \ N = 21$$

The means are $M_E = 1033/21 = 49.2$ and $M_C = 1433/31 = 46.2$. Thus, there is a difference, $M_E - M_C$, of 3 seconds on the average in favor of the experimental group, suggesting that the visual imagery training was effective. It remains to be tested, however, whether this is a statistically significant difference or whether it is one that could be attributed to random sampling fluctuations in the means.

Before the $t$ test can be applied to test the null hypothesis in this case, it must be shown that the variance estimates, $s_E^2$ and $s_C^2$, are not significantly different from each other since the $t$ test is based on the assumption that $\sigma_E^2 = \sigma_C^2$.

Formula 8–2 for $s$ requires a knowledge of $\Sigma x^2$, the sum of squares of the deviation scores. This value can be obtained for each sample using Formula 8–13:

For Group E:

$$\Sigma x^2 = \Sigma X^2 - \frac{(\Sigma X)^2}{N} = 55525 - \frac{(1033)^2}{21} = 4711$$

For Group C:

$$\Sigma x^2 = \Sigma X^2 - \frac{(\Sigma X)^2}{N} = 77435 - \frac{(1433)^2}{31} = 11{,}193$$

Using Formula 8–2, and squaring both sides,

$$s_E^2 = \frac{\Sigma x^2}{N-1} = \frac{4711}{20} = 235.5$$

$$s_C^2 = \frac{\Sigma x^2}{N-1} = \frac{11193}{30} = 373.1$$

Computing the $F$ ratio by means of Formula 8–11 gives:

$$F = \frac{s_x^2}{s_y^2} = \frac{373.1}{235.5} = 1.58$$

Entering Table E (See Appendix A) with 30 degrees of freedom for the larger variance estimate and 20 degrees of freedom for the smaller variance estimate yields $F_{.05} = 2.08$ as the value of $F$ required for significance at the five percent level. Since the obtained $F$ of 1.58 is well below the 2.08 required to reject $h_0$ at the five percent level of confidence, it is not unreasonable to entertain the idea that $\sigma_E^2 = \sigma_C^2$ so the $t$ test can proceed.

Having obtained a nonsignificant $F$ ratio for comparing variance estimates, the $t$ test will proceed by first calculating the standard error of the difference between means, $s_{D_M}$, the standard deviation of the sampling distribution of the differences between means, using Formula 8–12:

$$s_{D_M} = \sqrt{\frac{\Sigma x_1^2 + \Sigma x_2^2}{N_1 + N_2 - 2}\left(\frac{N_1 + N_2}{N_1 N_2}\right)} \qquad (8\text{–}12)$$

Since $\Sigma x^2$ has already been computed for both samples in the $F$-ratio test above, those values and the sample sizes can be substituted in Formula 8–12 to give:

$$s_{D_M} = \sqrt{\frac{4711 + 11193}{21 + 31 + 2}\left[\frac{21 + 31}{(21)(31)}\right]}$$

$$s_{D_M} = \sqrt{\frac{15904}{50} \cdot \frac{52}{651}} = \sqrt{(318.08)(.0799)}$$

$$s_{D_M} = \sqrt{25.41} = 5.04$$

Calculating a standard score, $t$; for the difference between means in the sampling distribution of differences between means, using Formula 8–14, gives:

$$t = \frac{(M_1 - M_2) - 0}{s_{D_M}} \qquad (8\text{–}14)$$

$$t = \frac{(49.2 - 46.2) - 0}{5.04} = \frac{3}{5.04} = 0.6$$

With $N_1 + N_2 - 2$, or 50, degrees of freedom, the Table of the $t$ distribution (Table C, Appendix A) shows $t_{.05} = 2.01$ for a two-tailed test. Since the experimental group (E) was expected to do better, a one-tailed test might have been justified. This would give $t_{.05} = 1.68$ for a one-tailed test. The obtained value of $t = 0.6$ is far below even the lower of these two critical values of $t$ so it is clearly not possible to reject the null hypothesis on the basis of this experiment. The visual imagery training, therefore, did not result in significantly superior performance for the experimental group.

*A Second Example.* Although the difference in the preceding example was not significant, it was in the predicted direction so the experimenter decides to repeat the experiment with larger samples to see if the trend holds and will reach significance with more cases. The data for the second study give the following results:

$$M_E = 50.1, \text{ S.D.}_E = 16, N_E = 50$$

$$M_C = 45.0, \text{ S.D.}_C = 20, N_C = 60$$

Applying Formula 8–3 to get variance estimates from the standard deviations:

$$s_E{}^2 = (\text{S.D.}_E)^2 \cdot \left(\frac{N}{N-1}\right) = (16)^2 \cdot \left(\frac{50}{49}\right) = (256)(1.020) = 261.1$$

$$s_C{}^2 = (20)^2 \cdot \left(\frac{60}{59}\right) = (400)(1.017) = 406.8$$

The $F$ ratio is $406.8/261.1 = 1.56$. Since $F_{.05}$ is 1.58, this result is still not significant although it is approaching the five percent level value rather closely. Many investigators would still consider the $t$ test applicable as long as the obtained $F$ fails to reach the one percent level of significance.

Since $(\text{S.D.})^2 = \Sigma x^2/N$, the value of $\Sigma x^2$ for each sample can be obtained by multiplying S.D.$^2$ by $N$ as follows:

$$(\Sigma x^2)_E = (\text{S.D.})_E{}^2 \cdot N_E = (16)^2(50) = (256)(50) = 12800$$

$$(\Sigma x^2)_C = (20)^2(60) = (400)(60) = 24000$$

Substituting these values in Formula 8–12 for $s_{DM}$ gives:

$$s_{DM} = \sqrt{\frac{\Sigma x_1{}^2 + \Sigma x_2{}^2}{N_1 + N_2 - 2}\left(\frac{N_1 + N_2}{N_1 N_2}\right)}$$

$$s_{DM} = \sqrt{\frac{12800 + 24000}{50 + 60 - 2}\left[\frac{50 + 60}{(50)(60)}\right]}$$

$$s_{DM} = \sqrt{\frac{36800}{108} \cdot \frac{110}{3000}} = \sqrt{(340.74)(0.0367)}$$

$$s_{DM} = \sqrt{12.505}$$

$$s_{DM} = 3.54$$

Computing the $t$ ratio by Formula 8–1 gives:

$$t = \frac{(M_1 - M_2) - 0}{s_{DM}} = \frac{50.1 - 45.0}{3.54} = \frac{5.1}{3.54} = 1.44$$

Since this value is still below even the large sample value of $t$ for a five percent level one-tailed test, i.e., $t_{.05} = 1.65$, the obtained value of $t$ is not significant and the null hypothesis cannot be rejected.

The obtained value of $t$ is pushing closer to significance, however, and if the trend holds, increasing the size of the samples would eventually result in a significant value of $t$. When samples as large as these fail to result in a significant difference between means, however, it must be concluded that the effect, if it exists at all, is certainly not very pronounced. Effects of experimental treatments that will make an important difference in practice would ordinarily be readily detected by samples as large as 50 in each group. On the other hand, even if there is only a tiny difference in true effect of diverse treatments, it can be detected as being statistically significant if the samples are made large enough. A statistically significant difference with large samples, therefore, may be of relatively little practical import.

## Summary of Steps in Testing the Significance of the Difference between Two Independent Random Samples Means Using the $t$ Test

1. Compute the means of the two independent random samples, $M_1$ and $M_2$.
2. Find the sum of the squared deviation scores, $\Sigma x^2$, for each of the two samples using Formula 8–13.
3. Find the standard error of the difference between means, $s_{DM}$, by substituting the sample sizes, $N_1$ and $N_2$, and the sums of squared deviation scores into Formula 8–12.
4. Obtain the $t$ score for this difference between means by dividing $(M_1 - M_2)$ by the standard error of the difference between means as called for in Formula 8–14.
5. Using $N_1 + N_2 - 2$ degrees of freedom, enter the Table of the $t$ distribution to obtain the critical values of $t$, $t_{.05}$ and $t_{.01}$.
6. If the $t$ score obtained in step 4 above is greater than $t_{.05}$ but less than $t_{.01}$, reject the null hypothesis at the five percent level of significance. If the $t$ score obtained in step 4 above is greater than $t_{.01}$, reject the null hypothesis at the one percent level of significance. If the obtained $t$ score is less than $t_{.05}$, do not reject the null hypothesis.

## Problem 28
### TEST THE DIFFERENCE BETWEEN TWO CORRELATED MEANS USING DIFFERENCE SCORES

In Problem 27, the significance of the difference between the means of two independent random samples was tested. In certain experimental situations it is better to use matched samples instead of independent samples. For example, if an experiment is to prove very costly per subject, it is important to obtain the necessary test of the null hypothesis with as few subjects as possible. One way to reduce the number of subjects necessary is to reduce the standard deviation of the sampling distribution. This can be done by using matched samples.

One type of matched sample design is to start with a pool of identical twin pairs. Members of the twin pairs are assigned at random to the experimental (E) and control (C) groups, respectively. It is intuitively evident that the difference between means of two such matched samples on the average would be less than the difference between the means of two independent random samples. When each person in one group has an identical twin in the other group, scores will be fairly comparable on any variable for matched individuals in the two groups, resulting in similar means. The *D* values making up the sampling distribution for matched samples would be less scattered about zero than for independent, random samples, giving a smaller standard error. With a smaller standard error, a given size difference between means becomes more significant. Hence matching experimental and control groups is a way of achieving the same level of significance with fewer cases, other things being equal. It increases the "precision" of the experiment. The actual difference due to the treatment stands out better because the background "noise" has been reduced by the matching process.

With independent, random sampling, the standard error of the sampling distribution must be reduced by increasing sample sizes. In many cases it is easier to do this, and less expensive, than it is to use matched groups. Getting a sample of twins is not easy and other types of matching may be less effective.

When groups of identical twins cannot be obtained but matched groups are desirable to reduce sampling error, it is often possible to obtain good results by matching subjects on one or more particularly relevant variables. If an experiment with two teaching methods is to be carried out that involves expensive equipment used on a concurrent basis, one way of proceeding to match groups would be to choose university student volunteers as subjects. Pairs of students of the same sex, approximate grade point average, major, and age could be chosen and assigned at random to the experimental (E) and control (C) groups, one member from each pair to group E and one to group C. This type of matching is usually not as good as having identical twins but it would probably function effectively to reduce the variability of the sampling distribution of differences between means in most cases.

### Distribution of Differences between Matched Pairs

When samples are matched, rather than independent, Formula 8–12 does not give a suitable estimate of the variability of the sampling distribution of differences between means because the matching process has reduced (hopefully) the average size of the differences between sample means. It is necessary, therefore, to approach the problem of testing the null hypothesis in a different manner when samples are matched.

An effective and simple way to test the difference between means of matched samples is to first compute the difference scores for each pair of matched subjects. That is, for each person in the experimental group, there is a person in the control group who is his twin or the person who is matched with him on one or more matching variables. The difference between the scores of these pairs of subjects on the experimental variable is obtained leading to a series of difference scores $d_1$, $d_2$, . . . $d_N$, where $N$ is the number of pairs of subjects, also the number of cases in each group. In a matched groups design, $N_E$ must, of course, equal $N_C$. These $d_i$ scores will sometimes be positive, sometimes negative. The signs are maintained and utilized in all calculations.

Assuming that the experimental variable being studied is normally distributed in the population from which the subjects are selected, the difference scores $d_1$, $d_2$, $d_3$, . . . $d_N$ taken from matched samples should also be normally distributed. Thus the distribution of difference scores can be treated as a normally distributed population of raw scores from which a sample, $d_1$, $d_2$, . . . , $d_N$, has been selected. This sample of difference scores, selected at random from the population, will have a mean value, $M_d$, the mean difference score. These difference scores will also have a standard deviation, S.D.$_d$, since some are larger than others.

It should be noted that the mean difference score, $M_d$, is also equal to the difference between the means of the two matched samples, this can be shown as follows:

$$M_E = \frac{1}{N} (X_{E_1} + X_{E_2} + \cdots + X_{E_N})$$

$$M_C = \frac{1}{N} (X_{C_1} + X_{C_2} + \cdots + X_{C_N})$$

$$M_E - M_C = \frac{1}{N} (X_{E_1} - X_{C_1}) + (X_{E_2} - X_{C_2}) + \cdots + (X_{E_N} - X_{C_N})$$

$$M_E - M_C = \frac{1}{N} (d_1 + d_2 + \cdots + d_N) = M_d$$

*Standard Error of the Mean Difference.* Since the difference scores $d_1$, $d_2$, . . . , $d_N$ function essentially as raw scores of a certain kind, they have a mean and standard deviation. It follows that they should also have a standard error of the mean. Formulas 8–5 and 8–6 gave two methods of estimating the standard error of the mean for ordinary raw scores:

$$s_M = \frac{s}{\sqrt{N}} = \frac{S.D.}{\sqrt{N-1}}$$

Comparable formulas for use with difference scores derived from matched samples are given by:

$$s_{d_M} = \frac{S.D._d}{\sqrt{N-1}} \tag{8–15}$$

$$s_{d_M} = \frac{s_d}{\sqrt{N}} \tag{8–16}$$

Where:

$s_{d_M}$ = the standard error of the mean for difference scores in matched samples

$s_d$ = the estimate of the population standard deviation of difference scores

S.D.$_d$ = the standard deviation of a sample of difference scores

Formulas 8–15 and 8–16 merely apply Formulas 8–5 and 8–6 to a sample of raw scores that are difference scores instead of the usual kind of raw scores.

*Testing the Null Hypothesis with Matched Samples.* If the means of the two populations from which the matched pairs are drawn are equal, i.e., $\mu_1 = \mu_2$, then the differences between pairs should average to zero if an infinitely large number of difference scores is used. The means of the samples of difference scores, therefore, are expected to distribute normally about zero as a mean when $\mu_1 = \mu_2$. Standard scores computed for these means of difference scores should be distributed according to the $t$ distribution when the standard deviation of the sampling distribution is estimated, as in the case of the means of samples of raw scores. Since one difference score becomes one raw score, the number of degrees of freedom will be $N - 1$, where $N$ is the number of differences. If the samples were not matched, i.e., independent of one another, the number of degrees of freedom would be twice as great.

To test the null hypothesis with matched samples, then, first take the difference scores between matched pairs of subjects, always subtracting in the same direction for all pairs, e.g., $E - C$ or $C - E$, and keeping algebraic signs on the $d$ values. Next, compute the average of the difference scores ($d$ values) as $M_d$. Obtain the standard error of the mean of the difference scores by Formula 8–15 or 8–16. This is just the same as applying Formula 8–5 or 8–6 on difference scores as raw scores. Then compute a standard score, $t$, for this mean difference score in the sampling distribution of means of difference scores as follows:

$$t = \frac{M_d - 0}{s_{d_M}} \tag{8-17}$$

Where:

$M_d$ = mean of the difference scores

$s_{d_M}$ = standard error of the mean for the difference scores

*Relationship of Problems 23 and 28.* In Problem 23, the mean of a sample of raw scores was evaluated to test whether it could have come from a population with a specified mean value, $\mu$. The present Problem

28 reduces to an example of Problem 23 with difference scores being used as raw scores and the population mean value, $\mu$, being zero. In Problem 28, then, the mean of the sample of difference scores is being evaluated to see if it could have come from a population with a mean of zero. The formulas for Problem 28 are identical to those for Problem 23 except that a subscript, $d$, has been added to make it clear that difference scores are being used as raw scores.

*An Example.*   Drug A is now used routinely in a certain large psychiatric in-patient facility to control symptoms in paranoid schizophrenic patients. A new drug, Drug B, has become available recently which the manufacturer claims to be superior to Drug A in controlling symptoms. Hospital officials decide to run a double blind study to compare the efficacy of the two drugs. A pool of matched pairs of patients is formed with individuals in the pairs matched for sex, body weight, severity of symptoms, and previous history of hospitalization. Members of the pairs are randomly assigned, by coin toss, to Groups E and C. Patients will be maintained on the drugs for a two-week period under intensive observation with extensive record keeping relative to their symptomatic behavior.

The drugs to be administered will be given by personnel who do not know who is being given which drug. The personnel evaluating the patients will not know which patients are receiving which drug. The observations over the entire two weeks are boiled down into overall ratings of severity of symptoms displayed. The ratings are converted to a scale from 1 to 9.

There were 10 pairs of patients. The overall symptom severity ratings for those receiving Drug A and Drug B are shown in the second two columns of Table 8–1. The first column gives the patient pair number. The difference scores, $d_i$, appear in the next to last column and the squares of these difference scores appear in the last column. The various column sums appear at the bottom of the table.

At the bottom of Table 8–1 the computations for testing the null hypothesis are shown. The mean difference score, $M_d$, is computed as the sum of the difference scores divided by $N$ and equals 1.1. The standard deviation of the difference scores, S.D.$_d$, is computed substituting $d$ for $X$ in the usual raw score formula for the standard deviation. It would also have been appropriate to calculate $s_d$ by the formula:

$$s_d = \sqrt{\frac{\Sigma(d - M_d)^2}{N - 1}} = \sqrt{\frac{\Sigma x_d^2}{N - 1}} \tag{8–18}$$

In formula 8–18, $\Sigma(d - M_d)^2$, or $\Sigma x_d^2$, is comparable to the $\Sigma x^2$ in Formula 8–13, i.e., a sum of deviation scores squared. In this situation, $d$ is a difference score, not a deviation score, hence $M_d$ must be

TABLE 8–1
Testing the Significance of the Difference between Means of Matched Samples

| Pair Number | Drug A | Drug B | d | $d^2$ |
|---|---|---|---|---|
| 1......... | 6 | 4 | 2 | 4 |
| 2......... | 7 | 6 | 1 | 1 |
| 3......... | 9 | 7 | 2 | 4 |
| 4......... | 6 | 7 | −1 | 1 |
| 5......... | 3 | 4 | −1 | 1 |
| 6......... | 4 | 2 | 2 | 4 |
| 7......... | 7 | 5 | 2 | 4 |
| 8......... | 2 | 1 | 1 | 1 |
| 9......... | 1 | 1 | 0 | 0 |
| 10........ | 8 | 5 | 3 | 9 |
| Σ......... | 53 | 42 | 11 | 29 |

$$M_d = \frac{\Sigma d}{N} = \frac{11}{10} = 1.1$$

$$\text{S.D.}_d = \frac{1}{N} \sqrt{N\Sigma X^2 - (\Sigma X)^2} = \frac{1}{N} \sqrt{N\Sigma d^2 - (\Sigma d)^2}$$

$$\text{S.D.}_d = \frac{1}{10} \sqrt{10(29) - (11)^2} = \frac{1}{10} \sqrt{290 - 121} = \frac{\sqrt{169}}{10} = \frac{13}{10}$$

$$\text{S.D.}_d = 1.3$$

$$s_{d_M} = \frac{\text{S.D.}_d}{\sqrt{N-1}} = \frac{1.3}{\sqrt{10-1}} = \frac{1.3}{\sqrt{9}} = \frac{1.3}{3} = .433$$

$$t = \frac{M_D - 0}{s_{d_M}} = \frac{1.1 - 0}{0.433} = 2.54*$$

$t_{.05}$ with 9 $df$ = 2.26

subtracted from $d$ to make it a deviation score. It is easier to compute the sum of deviation scores from raw scores as follows:

$$\Sigma x^2 = \Sigma X^2 - \frac{(\Sigma X)^2}{N} \qquad (8\text{–}13)$$

$$\Sigma(d - M_d)^2 = \Sigma x_d^2 = \Sigma d^2 - \frac{(\Sigma d)^2}{N} \qquad (8\text{–}19)$$

Formula 8–19 is obtained from Formula 8–13 merely by substituting $d$, the difference score, for $X$. In Table 8–1 the next step was to obtain $s_{d_M}$ using the value of S.D.$_d$ and Formula 8–15. Had $s_d$ been calculated instead of S.D.$_d$, Formula 8–16 would have been used to obtain $s_{d_M}$ instead of Formula 8–15. The only reason for using Formula 8–15 instead of 8–16 here was that $N$ is 10 and the square root of $(10 - 1)$ is easier to compute than the square root of 10.

Once the $M_d$ and $s_{d_M}$ values have been obtained, $t$ is computed as the ratio of these two, as shown in Table 8–1. The computed value of $t$ here was 2.54 which has an asterisk ($*$) placed beside it indicating a result significant at the five percent level. This is because $t_{.05}$ with 9 degrees of freedom is 2.26 for a two-tailed test and 2.54 is greater than 2.26. The $t_{.01}$ for a two-tailed test with 9 degrees of freedom is 3.25, greater than the obtained $t$ of 2.54, so this difference between means, or mean difference, was not large enough to reach significance at the one percent level. For a $t$ value greater than or equal to 3.25, two asterisks ($**$) would have been placed beside the $t$ value to indicate significance at the one percent level.

It should be noted that the number of degrees of freedom is cut in half with matched samples. This has the effect of requiring a larger value of $t$ to achieve significance. The effect of the matching must reduce the variability of the sampling distribution more than enough to compensate for the loss of degrees of freedom to be worthwhile. Matching groups on an irrelevant variable that has no effect on the main variable under consideration would be worse than no matching at all. In a test of teaching methods in an academic subject, for instance, it would probably be futile and perhaps even harmful to match groups on the basis of body weight.

### Summary of Steps in Testing the Difference between Two Correlated Means Using Difference Scores

1. Pair up the matched scores and take the difference between each pair of matched scores.
2. Compute the mean of these difference scores, $M_d$.
3. Compute the sum of the squared deviation scores, $\Sigma x_d^2$, for these difference scores using Formula 8–19.
4. Compute the estimated population standard deviation for these difference scores, $s_d$, using Formula 8–18.
5. Compute the standard error of the mean for these difference scores, $s_{d_M}$, using Formula 8–16.
6. Obtain the t-score by dividing the mean difference score, $M_d$, computed in Step 2 above, by the standard error of the mean for these difference scores, $s_{d_M}$, as called for in Formula 8–17.
7. Using $N - 1$ degrees of freedom, where N is the number of *pairs* of scores, enter the Table of the $t$ Distribution to find the critical values of $t$, $t_{.05}$ and $t_{.01}$.

8. If the *t* score obtained in Step 6 above is greater than $t_{.05}$ but less than $t_{.01}$, reject the null hypothesis at the five percent level of confidence. If the *t* score obtained in Step 6 above is greater than $t_{.01}$, reject the null hypothesis at the one percent level of confidence. If the obtained *t* score is less than $t_{.05}$, do not reject the null hypothesis.

## Problem 29
### TEST THE SIGNIFICANCE OF THE DIFFERENCE IN MEAN CHANGE FOR TWO INDEPENDENT RANDOM SAMPLES

In Problem 27, the difference between the means of two independent, random samples was evaluated to determine if it could reasonably be attributed to random sampling variations when $\mu_1 = \mu_2$. It is sometimes possible to improve the precision of an experiment of this type by taking before and after measures in both the experimental and control groups.

Consider an experiment in which two methods of teaching high jumping are being considered, the traditional straight-up scissors jump and the back flip method. With the design in Problem 27, men with no previous high jumping experience would be assigned at random to the two groups. Both groups would be trained but by different methods. At the end of the experiment a measure of each person's jumping skill would be obtained and the means of the two groups would be compared on this measure of jumping skill.

If the mean initial jumping ability happens by chance to be lower in one group than the other, as it will be in most cases, it will put the training method applied to that group at a disadvantage. The training will have to overcome the deficit of ability in the group in addition to achieving an increment in real performance greater than that in the other group.

One way of controlling to some extent for the difference in ability in the two randomly chosen groups is to compare how much one group *gains* in performance with how much the other group gains. In this design, each subject would be measured on his jumping performance *both before and after* training. The net improvement in performance would be the raw score for each person. The mean net change in performance would be computed for each group and this difference in means would be evaluated in the same way that was described in Problem 27. The raw score is, then, a change in score from pre- to post-experiment measurement instead of just a post-experiment measurement.

This approach does not eliminate all effects of random differences in ability in the two groups, but it does control such effects to some extent. The same increase in precision can be achieved by using larger samples. In any given situation, the experimenter has to evaluate the advantages and disadvantages of the various options open to him for increasing the probability of rejecting the null hypothesis if it is in fact false. Using matched samples, using before and after measurement, and increasing sample size are all ways of achieving this objective. *more accuracy*

*An Example.* Thirty-six male subjects were selected from a defined population and randomly assigned to an experimental group (E) and a control group (C). Two of the experimental group subjects failed to complete the experiment, leaving 16 subjects in group E and 18 in group C. Each subject was given three high jump trials at each level of the bar using any method he preferred and his highest successful jump was recorded as his initial score. These values are given in the $E_B$ column of Table 8–2 for the experimental group (*Before* scores) and in column $C_B$ of Table 8–2 for the control group. No subject had ever had training in high jumping. The experimental group was given a two weeks training course in high jumping with the "back flip" method and the control group was given a two weeks training course in high jumping by the "scissors" method. At the end of the training period, both groups of subjects were given three test jumps at each level and their highest successful jump was recorded. Those values are shown in columns $E_A$ and $C_A$, the *After* measures for the experimental and control groups, respectively. All measures are given in inches.

The differences between the *Before* and *After* measures are shown in columns $D_E$ and $D_C$ of Table 8–2 for the experimental and control groups, respectively. Once the D values are obtained, comparing their means becomes an example of Problem 27, testing the difference between independent sample means using the t test. The D score becomes a raw score for Problem 27. The calculations are shown at the bottom of Table 8–2.

The first step was to obtain the means of the change scores, $D_E$ and $D_C$, as $M_{D_E} = 6.31$ and $M_{D_C} = 4.56$. The next step was to compute the sum of the squared deviation scores (by Formula 8–13) for each group to be used in computing the standard error of the difference between means. These sums of squared deviations, 205.44 for Group E and 210.44 for Group C, were entered into Formula 8–12 to obtain the standard error of the difference between means, $s_{D_M} = 1.24$. Finally, t was computed by Formula 8–14, dividing the difference between means by the standard error, giving $t = 1.41$. With $N_1 + N_2 - 2$, or 32, degrees of freedom, $t_{.05}$ from the Table of the t Distribution (Appendix A, Table C) is 2.04 for a two-tailed test. Since the obtained value of t was less than 2.04, the null hypothesis cannot be rejected. This means that the experiment failed to establish a definite superiority of one training method over the other. The "back flip" method gave a larger average improvement but it was not statistically significant.

TABLE 8–2
Testing the Significance of the Difference between Independent, Random Sample
Mean Change Scores

| Subject | $E_B$ | $E_A$ | $D_E$ | $D_E^2$ | Subject | $C_B$ | $C_A$ | $D_C$ | $D_C^2$ |
|---|---|---|---|---|---|---|---|---|---|
| 1..... | 38 | 45 | 7 | 49 | 1..... | 48 | 44 | −4 | 16 |
| 2..... | 60 | 68 | 8 | 64 | 2..... | 50 | 55 | 5 | 25 |
| 3..... | 42 | 44 | 2 | 4 | 3..... | 46 | 53 | 7 | 49 |
| 4..... | 48 | 50 | 2 | 4 | 4..... | 45 | 55 | 10 | 100 |
| 5..... | 52 | 59 | 7 | 49 | 5..... | 57 | 62 | 5 | 25 |
| 6..... | 58 | 66 | 8 | 64 | 6..... | 52 | 58 | 6 | 36 |
| 7..... | 46 | 56 | 10 | 100 | 7..... | 38 | 45 | 7 | 49 |
| 8..... | 55 | 62 | 7 | 49 | 8..... | 42 | 40 | −2 | 4 |
| 9..... | 51 | 52 | 1 | 1 | 9..... | 47 | 50 | 3 | 9 |
| 10..... | 48 | 60 | 12 | 144 | 10..... | 36 | 40 | 4 | 16 |
| 11..... | 36 | 35 | −1 | 1 | 11..... | 39 | 43 | 4 | 16 |
| 12..... | 47 | 54 | 7 | 49 | 12..... | 54 | 60 | 6 | 36 |
| 13..... | 49 | 61 | 12 | 144 | 13..... | 50 | 60 | 10 | 100 |
| 14..... | 51 | 57 | 6 | 36 | 14..... | 61 | 66 | 5 | 25 |
| 15..... | 44 | 50 | 6 | 36 | 15..... | 43 | 50 | 7 | 49 |
| 16..... | 47 | 54 | 7 | 49 | 16..... | 50 | 52 | 2 | 4 |
|  |  |  |  |  | 17..... | 46 | 50 | 4 | 16 |
|  |  |  |  |  | 18..... | 48 | 51 | 3 | 9 |
| Σ................. |  |  | 101 | 843 | Σ................. |  |  | 82 | 584 |

$$M_{D_E} = \frac{\Sigma D_E}{N} = \frac{101}{16} = 6.31$$

$$M_{D_E} = \frac{\Sigma D_C}{N} = \frac{82}{18} = 4.56$$

$\Sigma x^2$ for Group E:

$$\Sigma x^2 = \Sigma X^2 - \frac{(\Sigma X)^2}{N} = \Sigma D_E^2 - \frac{(\Sigma D_E)^2}{N} = 843 - \frac{(101)^2}{16} = 843 - \frac{10201}{16}$$
$$= 843 - 637.56 = 205.44$$

$\Sigma x^2$ for Group C:

$$\Sigma x^2 = \Sigma X^2 - \frac{(\Sigma X)^2}{N} = \Sigma D_C^2 - \frac{(\Sigma D_C)^2}{N} = 584 - \frac{6724}{18} = 584 - 373.56 = 210.44$$

$$s_{D_M} = \sqrt{\frac{\Sigma x_1^2 + \Sigma x_2^2}{N_1 + N_2 - 2}\left(\frac{N_1 + N_2}{N_1 N_2}\right)}$$

$$s_{D_M} = \sqrt{\left(\frac{205.44 + 210.44}{16 + 18 - 2}\right)\left(\frac{16 + 18}{16 \cdot 18}\right)} = \sqrt{\left(\frac{415.88}{32}\right)\left(\frac{34}{288}\right)}$$

$$s_{D_M} = \sqrt{(12.996)(0.118)} = \sqrt{1.534}$$

$$s_{D_M} = 1.24$$

$$t = \frac{M_1 - M_2 - 0}{s_{D_M}} = \frac{M_{D_E} - M_{D_C} - 0}{s_{D_M}}$$

$$t = \frac{6.31 - 4.56 - 0}{1.24} = \frac{1.75}{1.24}$$

$$t = 1.41$$

With 32 *df*, $t_{.05} = 2.04$ (two-tailed test)

## Problem 30
## TEST THE SIGNIFICANCE OF THE DIFFERENCE IN
## MEAN CHANGE FOR TWO MATCHED SAMPLES

Just as it is possible to compare mean change scores with independent, random samples, so is it possible to compare mean change scores with matched groups. In either case, the net effect, as a rule, is to obtain a slight increase in the precision of the experiment where this technique is applicable. With independent, random samples, once the before and after measures were used to obtain change scores, means of the change scores were treated like means of raw scores and the significance of the difference between such means was evaluated by the methods of Problem 27.

When a before-and-after measures design is applied to matched samples, again the process calls for using the before and after measures to obtain differences or change scores. Once these change scores have been obtained, they are treated like raw scores and the difference between their means is evaluated by the methods of Problem 28. Problem 28 treated the significance of the difference between raw score means of matched samples using the *t* test. Problem 30, therefore, becomes an example of Problem 28 once before and after measures have been converted to change scores. It is customary to arrange the data so as to subtract the generally smaller scores from the generally larger scores so that the difference scores will be predominantly positive. The same direction of taking the difference, however, must be used over all subjects in both groups.

### An Example

Table 8–3 shows a completely worked out example of this type of problem. A new memory drug has been proposed as a method of accelerating the rate at which students can master college curricular material. It is suggested that under the influence of the drug, students can learn in three years what otherwise would take four years. If all students in state institutions finished in three years instead of four, this would save the taxpayers a good deal of money. Psychologist X is skeptical, however, and decides to run an experiment checking out these claims.

Thirteen pairs of identical twins are located for this experiment and a twin of each pair is assigned by coin toss to the experimental and control group. The experimental group will receive injections of the drug and the control group will receive placebo injections. A double-blind procedure will be employed to control for experimenter bias.

Subjects in both groups learn a list of nonsense syllables. The number of trials to obtain one perfect recitation of the list is recorded for each sub-

TABLE 8–3
Testing the Significance of the Difference between Means of Matched
Sample Change Scores

| Subject | $E_B$ | $E_A$ | $C_B$ | $C_A$ | $D_E$ | $D_C$ | $d$ | $d^2$ |
|---|---|---|---|---|---|---|---|---|
| 1.......... | 21 | 22 | 23 | 20 | −1 | 3 | 4 | 16 |
| 2.......... | 35 | 30 | 30 | 25 | 5 | 5 | 0 | 0 |
| 3.......... | 30 | 28 | 28 | 24 | 2 | 4 | 2 | 4 |
| 4.......... | 26 | 27 | 29 | 26 | −1 | 3 | 4 | 16 |
| 5.......... | 23 | 20 | 26 | 22 | 3 | 4 | 1 | 1 |
| 6.......... | 27 | 30 | 24 | 20 | −3 | 4 | 7 | 49 |
| 7.......... | 26 | 24 | 27 | 21 | 2 | 6 | 4 | 16 |
| 8.......... | 22 | 24 | 24 | 24 | −2 | 0 | 2 | 4 |
| 9.......... | 15 | 20 | 18 | 19 | −5 | −1 | 4 | 16 |
| 10.......... | 24 | 22 | 21 | 16 | 2 | 5 | 3 | 9 |
| 11.......... | 27 | 25 | 25 | 21 | 2 | 4 | 2 | 4 |
| 12.......... | 25 | 21 | 25 | 20 | 4 | 5 | 1 | 1 |
| 13.......... | 28 | 25 | 29 | 27 | 3 | 2 | −1 | 1 |
| Σ.......... | | | | | | | 33 | 137 |

$$M_d = \frac{\Sigma d}{N_2} = \frac{33}{13} = 2.54$$

$$\Sigma x^2 = \Sigma X^2 - \frac{(\Sigma X)^2}{N} = \Sigma d^2 - \frac{(\Sigma d)^2}{N} = 137 - \frac{(33)^2}{13}$$

$$\Sigma(d - M_d)^2 = \Sigma x_d^2 = 137 - 83.77 = 53.23$$

$$s_d = \sqrt{\frac{\Sigma x^2}{N-1}} = \sqrt{\frac{\Sigma(d-M_d)^2}{N-1}} = \sqrt{\frac{53.23}{12}} \doteq \sqrt{4.436} = 2.11$$

$$s_{d_M} = \frac{s_d}{\sqrt{N}} = \frac{2.11}{\sqrt{13}} = \frac{2.11}{3.61} = 0.58$$

$$t = \frac{M_d - 0}{s_{d_M}} = \frac{2.54}{0.58} = 4.38**$$

With 12 degrees of freedom, $t_{.01} = 3.06$, two-tailed test

ject. These values are given in Table 8–3 in column $E_B$ (*Before* measures)
for the experimental group and in column $C_B$ (*Before* measures) for the
control group. Following this, the injections are administered and after
a two hour interval both groups are tested on a new but comparable list
of nonsense syllables. Again the number of trials to reach one perfect
recitation of the entire list is recorded for each subject. These values are
given in column $E_A$ (*After* measures) for the experimental group and in
column $C_A$ (*After* measures) for the control group.

The change scores for the two groups will be more often positive than negative if the *after* scores are subtracted from the *before* scores. Column $D_E$ in Table 8–3 records these differences for the experimental group and column $D_C$ records these differences for the control group. In both groups, these difference scores are more often positive than negative, indicating that fewer trials were required to learn the list of nonsense syllables after the injections. The change scores are more on the positive side for the control group than they are for the experimental group, however, contrary to what would be expected on the basis of claims made for the drug.

From this point on, calculations proceed as with Problem 28. The $d$ scores in the $d$ column of Table 8–3 are obtained by subtracting the $D_E$ scores from the $D_C$ scores. If the $D_C$ scores had in general been larger, the subtraction would have been done the other way to make the $d$ scores predominately positive. The last column gives the squared $d$ values. The sums of the last two columns give $\Sigma d$ and $\Sigma d^2$, values needed for further calculations.

The mean of the difference scores, $M_d$, is the average of the difference scores, 2.54, as shown at the bottom of Table 8–3. The estimate of the population standard deviation, $s_d$, is calculated using Formula 8–2 but substituting the sum of squared deviation difference scores as $\Sigma x^2_d$. These calculations are shown at the bottom of Table 8–3 with 53.23 as $\Sigma x_d^2$ for difference scores and 2.11 as $s_d$, the estimated population standard deviation of the difference scores. The standard error of the mean for the difference scores is computed next, using Formula 8–16, yielding a value of 0.58. The $t$ ratio is then computed by dividing $M_d$, the mean difference score by $s_{d_M}$, the standard error of the mean of the difference scores, yielding a $t$ of 4.38. This value is marked by two asterisks (**) to indicate that it is significant at the one-percent level of confidence, two-tailed test, since $t_{.01}$ for a two-tailed test with 12 degrees of freedom (number of pairs, minus one) is only 3.06.

Had a one-tailed test been planned in advance here, under the assumption that the drug condition would at worst be equivalent to the no drug condition, it would not be theoretically correct to attribute statistical significance to these unexpected findings. If the investigator clearly intended in advance to make a two-tailed test, then it would be appropriate to claim significance for these findings. Taken at face value, the obtained outcome would suggest that the "memory" drug tended to depress performance although both groups did better on the second test, perhaps due to practice effect. In any event, the experiment would not lend scientific support for a social action program involving use of the drug to speed learning.

## Exercises

8-1.  An investor has been told by a friend, "Don't listen to stock brokers, they give bad advice." He decides to test this hypothesis. He picks 12 brokers at random and buys 100 shares of recommended stock through each broker. One year later he sells all the stocks. The number of points (rounded to nearest point) gained or lost on the different stocks are as follows: $-2$, $-10$, $+5$, $-7$, $-6$, $-8$, $-3$, $+1$, $+2$, $-12$, $-8$, $-4$. State $h_0$, carry out an appropriate test, and draw conclusions.

8-2.  A consulting psychologist represents to a firm that he can increase the industrial output per worker by more than 10 percent in one month using a special training program. The present output is 30 units per man a day. Fifteen workers were randomly selected to take the training program and were followed for one month. Their production rates at the end of one month were as follows: 25, 32, 35, 45, 40, 50, 47, 38, 36, 30, 37, 32, 35, 37, 35. State $h_0$, select an appropriate test, make the test, and state conclusions.

8-3.  Goaded by taunts of rival UCLA students, USC students decide to establish once and for all that they are not "dumb, rich kids." They draw a random sample of the USC student body and have the Wechsler Adult Intelligence Scale Administered to each student. They wish to establish a 95 percent confidence interval for the true mean. With a mean I.Q. score of 117 and a standard deviation of 8 what would the interval be with $N = 50$?

8-4.  Obtain the 99% confidence interval for $\mu$ when data from the sample are as follows: $\Sigma X = 512$; $\Sigma X^2 = 19{,}451$; $N = 17$.

8-5.  Given $N = 100$ and $r_{12} = 0.30$, test to see if $r$ is significant at the .01 level.

8-6.  With $N = 81$ and $r = 0.20$, test to see if the correlation coefficient is significant at the five percent level.

8-7.  Given the following data on two independent, random samples: $\Sigma X_1 = 226$, $\Sigma X_1^2 = 7128$, $N_1 = 20$, $\Sigma X_2 = 275$, $\Sigma X_2^2 = 8321$, $N_2 = 18$, test to see if these samples could have been drawn from populations with $\sigma_1{}^2 = \sigma_2{}^2$. Would it be appropriate to proceed with a $t$ test of the difference between means?

8-8.  Two independent, random samples, an experimental and control group, gave the following results: $M_E = 42$, S.D.$_E = 6.1$, $N_E = 26$; $M_C = 36$, S.D.$_C = 10.2$, $N_C = 17$. Make an $F$ test to see if these samples could have come from populations with equal variances. Is it appropriate to make a $t$ test of the difference between means?

8-9.  Test the difference between means for the data in Exercise 8–7. State $h_0$, select an appropriate test, make the test, and draw conclusions.

8-10.  Whether or not $F$ was significant in Exercise 8–8, test the difference between means. In the light of the $F$ test result in Exercise 8–8 and the $t$ test result, what conclusion would you draw?

8–11. Eight pairs of twins were recruited and assigned at random to an experimental (E) and a control group (C). Group E had a tape recorder that recited a poem over and over to them during times that their rapid eye movements indicated they were asleep. Group C had music of the same sound level instead of the poem. The next day, both groups learned the poem given on tape to the experimental group to one perfect recitation. The numbers of trials to reach one perfect recitation for the two groups are given below:

| Pair | Group E | Group C |
|---|---|---|
| 1......... | 25 | 28 |
| 2......... | 18 | 23 |
| 3......... | 23 | 23 |
| 4......... | 30 | 35 |
| 5......... | 28 | 30 |
| 6......... | 26 | 28 |
| 7......... | 25 | 29 |
| 8......... | 24 | 27 |

Make an appropriate statistical test of the difference between means and state conclusions.

8–12. Pairs of subjects are chosen from a given population that are matched on I.Q., age, sex, and education level. Group E, the experimental group, is subjected to a "memory course" while Group C, the control group is given intellectual tasks to perform equivalent in time and attention to the experimental tasks. Afterward, both groups are given a list of 20 names to remember on one presentation. The number of names retained were as follows:

| Pair | Group E | Group C |
|---|---|---|
| 1......... | 18 | 15 |
| 2......... | 16 | 17 |
| 3......... | 10 | 9 |
| 4......... | 10 | 9 |
| 5......... | 9 | 10 |
| 6......... | 13 | 8 |
| 7......... | 12 | 9 |
| 8......... | 13 | 10 |
| 9......... | 7 | 5 |
| 10......... | 11 | 8 |
| 11......... | 12 | 13 |
| 12......... | 13 | 13 |

State $h_0$, make an appropriate test, and draw conclusions.

8–13. The following data represent before and after measurements for experimental and control groups, using two independent, random samples. Make a *t* test and state the outcome.

| Pair | $E_B$ | $E_A$ | $C_B$ | $C_A$ |
|------|-------|-------|-------|-------|
| 1...... | 120 | 130 | 100 | 112 |
| 2...... | 115 | 125 | 97 | 95 |
| 3...... | 80 | 95 | 110 | 105 |
| 4...... | 95 | 95 | 118 | 116 |
| 5...... | 100 | 110 | 75 | 80 |
| 6...... | 110 | 105 | 83 | 86 |
| 7...... | 102 | 108 | 94 | 95 |
| 8...... | 98 | 104 | 96 | 97 |
| 9...... | 103 | 110 | 104 | 110 |
| 10...... | 85 | 94 | 106 | 103 |
| 11...... | 90 | 97 | 98 | 100 |
| 12...... | 99 | 106 | | |

8–14. Consider the first 10 subjects' (1–10) 0 and C scores in Exercise 6–13 as before and after scores for one independent, random sample. Make a *t* test of the difference in mean change scores and state the results.

8–15. Given the following data:

| Pair | $E_B$ | $E_A$ | $C_B$ | $C_A$ |
|------|-------|-------|-------|-------|
| 1....... | 22 | 32 | 24 | 23 |
| 2....... | 15 | 30 | 16 | 17 |
| 3....... | 35 | 40 | 32 | 31 |
| 4....... | 23 | 28 | 20 | 21 |
| 5....... | 25 | 24 | 28 | 30 |
| 6....... | 31 | 39 | 26 | 22 |
| 7....... | 26 | 26 | 30 | 34 |
| 8....... | 23 | 28 | 23 | 24 |
| 9....... | 16 | 30 | 17 | 19 |
| 10....... | 19 | 32 | 18 | 16 |
| 11....... | 20 | 28 | 22 | 20 |

These are before and after measures in an experiment with matched groups. Find $M_d$, $s_{d_M}$, $t$, and test the significance of the difference between means.

8–16. Identical twin calves are assigned at random to Group E and Group C, one twin pair member in each group. Calves in the experimental group are fed raw milk and twins in the control group are fed pasteurized milk. Weights are recorded before and after the experiment period as shown below. Test the significance of the mean change scores, state the results of the test, and draw conclusions.

| Pair | $E_B$ | $E_A$ | $C_B$ | $C_A$ |
|------|-------|-------|-------|-------|
| 1....... | 40 | 80 | 42 | 64 |
| 2....... | 35 | 71 | 38 | 70 |
| 3....... | 38 | 78 | 35 | 60 |
| 4....... | 45 | 92 | 43 | 72 |
| 5....... | 39 | 80 | 36 | 64 |
| 6....... | 42 | 78 | 40 | 59 |
| 7....... | 37 | 78 | 35 | 56 |
| 8....... | 39 | 82 | 39 | 70 |

chapter 9

# Analysis of Variance

In Chapter 8 several problems were presented that dealt with the statistical analysis of results from very simple experiments involving just two groups, an experimental group and a control group. Many experiments are conducted with much more complicated designs involving several groups of subjects. These experiments and the methods used to analyze their results can become very complicated, especially when the effects of more than one variable are being tested in the same experiment. A full discussion of these complicated analysis of variance (*anova*) designs is beyond the scope of this elementary treatment. Three of the simplest and most commonly used designs, however, will be presented. The object in this chapter, as elsewhere in the book, is to make available to the reader, in the most easily used and understood form, those methods of statistical analysis that he is most likely to encounter in reading the research literature or to find useful in designing a study of his own.

A characteristic feature of analysis of variance (*anova*) designs, especially the more complicated ones, is that they can seldom be applied to evaluate research study findings after the fact. That is, if the investigator does not plan the experiment ahead of time with a specific design of analysis in mind it is unlikely that his data will be suitable for use with any of the available methods. The moral to this story is that the investigator *must* decide *in advance* what method of analysis, i.e., what analysis of variance (*anova*) design he wishes to use and then to make sure that he plans to collect data that will be appropriate for analysis with that particular design. The three designs described here can be utilized to meet a substantial portion of the requirements of most social science experimental research situations. The three designs to be presented in this chapter are (1) one-way anova with equal size samples,

(2) one-way anova with different size samples, and (3) two-way anova. The first two designs represent a generalization of the *t* test to more than two groups and the third design permits the evaluation of the effects of two experimental factors simultaneously in contrast with the *t* test and the one-way anova designs which evaluate the effects of just one factor.

## Problem 31
## TEST THE DIFFERENCE BETWEEN MEANS OF SEVERAL SAMPLES OF EQUAL SIZE USING ONE-WAY ANOVA

With the *t* test it was possible to compare the means of an experimental and a control group to test the treatment effect of the experimental variable. It is often important, however, to have more than two groups. In testing the effect of a new drug, for example, it might be desirable to compare the effect of injecting the drug with the effect of injecting a placebo (e.g., distilled water). To control for the effects of suggestion, however, it would be important to have a third group which received no injection of any kind but otherwise was treated the same. It might turn out that getting an injection, regardless of whether it was the drug or distilled water, produced an effect. This would be shown by both the drug group and the distilled water group having higher means than the no-injection group but not higher than each other. This same type of design will also handle the experiment in which one or more additional groups are added to evaluate not just one drug against placebo and no injection conditions, but several. There can be any number of such groups.

This particular design requires first and foremost that subjects be drawn from a common pool and assigned at random to the different treatment subgroups, with an equal number of subjects, *n*, in each subgroup. The experimenter must plan, therefore, to ensure that these conditions are met. He cannot, for example, put out-patients in one group, in-patients in another, hospital attendants in another, and so on, in this type of experiment if he is to evaluate the effects of different kinds of treatment. If he did, the differences between means, if found to be significant, could be due to the different characteristics of the populations sampled for each subgroup rather than to the diverse effects of the different treatments. Failure to make random assignments of subjects to the different treatment groups from the same subject pool is a very common error in poorly designed experiments. In operational treatment facilities, for example, practitioners are loath to withhold favored types of treatment from patients who especially need it, in their view, and often insist on compromising the validity of the experiment by departing from the random assignment principle.

Just as the $t$ test assumed that the population variances of the two respective populations sampled are equal, so the one-way anova design assumes this for all the populations sampled. This may seem strange since the samples were supposed to be drawn from the same population. The treatment effects, however, could affect the variances of the groups on the experimental variable as well as the means. This type of design assumes that the variances of the groups are not significantly affected by the treatment effects while testing to see if the means are differently affected. Fortunately, however, this test is not very sensitive to differences in variances, particularly as the size of the samples increases. If the $F$ ratio for the two most discrepant population variance estimates from the different subgroups is significant at the 0.01 level, there could be a problem and the student should consult an advanced text on analysis of variance for ways for dealing with this situation. Otherwise, it would be reasonable to proceed with the application of this design for testing the differences between means.

In this design, it is also assumed that the population distribution of the experimental varible is normal in form. As with the assumption of equal variances, the one-way anova test is not particularly sensitive to departures from this assumption. As long as the distributions appear to be reasonable approximations to normal curves there is no problem. With U-shaped, truncated, flat, or otherwise seriously nonnormal distributions of the experimental variable, the student should plan to consult an advanced text for a more appropriate method of treating the data. This assumption implies, of course, that there must be enough variation in the scores to give a distribution. If there are only four or five possible scores on the experimental variable, the application of this design becomes rather questionable. It is desirable to have at least ten possible score values in the distribution of scores on the experimental variable to approximate adequately a continuous, normal distribution of scores.

## The One-Way Anova Test of Differences between Subgroup Means

The one-way anova test, then, is applied to the situation where several subsamples of equal size have been drawn at random from a common pool and subjected to different types of experimental treatment. The sum of scores and sum of squared scores for each subgroup are computed separately and then combined to give these totals for the entire group— (combining the subgroups). These data can be utilized to obtain two different estimates of the population variance, $\sigma^2$, for the experimental variable, a "within-groups" estimate and a "between-groups" estimate. These two variance estimates are compared by means of the $F$ ratio test. If $F$ is significant, the null hypothesis is rejected and the differences between means are presumed to exceed what would be expected if the samples were drawn from populations with equal means. If $F$ is not significant, $h_0$ is not rejected.

*Within-Groups Variance Estimate.* The variances of the population from which three samples are drawn at random is supposed to be the same for all three groups, $\sigma^2$. An estimate of $\sigma^2$ can be obtained from each group, i.e., $s_1^2$, $s_2^2$, $s_3^3$, by the formula $\Sigma x^2/(n-1)$ where $n$ is the number of cases in the subgroup. A better estimate of $\sigma^2$ can be obtained by combining the data from all three groups to get one overall estimate of $\sigma^2$. This is done by pooling the sums of squares of deviation scores and pooling the degrees of freedom, i.e.:

$$s^2 = \frac{\Sigma x_1^2 + \Sigma x_2^2 + \cdots + \Sigma x_k^2}{(n_1 - 1) + (n_2 - 1) + \cdots + (n_k - 1)} \tag{9-1}$$

An alternate way of pooling the sums of squares would be to just add all three groups together and take $\Sigma x^2$ for the total group and divide by $N-1$ where $N$ is the sum of $n_1 + n_2 + n_3$. This sum of squared deviation scores would be taken with respect to the overall mean, however, i.e., $x_i = X_i - M$, whereas in Formula 9-1, each deviation score is taken with respect to its own subgroup mean. Formula 9-1 gives a better estimate of $\sigma^2$ because the means of the subgroup populations are not presumed to be equal, as would be required to permit the use of deviation scores with respect to the overall mean (*grand mean*) of the three groups combined. By taking deviations with respect to subgroup means and pooling them across groups, the estimate of the population variance from the pooled data will not be inflated improperly by any differences that may exist between the means. Of course, if the differences between means are of negligible importance the two different ways of obtaining the pooled variance estimate would be roughly comparable. The method illustrated by Formula 9-1 is always used, however, to obtain the within-groups estimate of the population variance with respect to the experimental variable being measured. The one-way anova test of the difference between means will make use of this within-groups estimate of the population variance by comparing it to a second estimate of the population variance derived from information about the variability of the subgroup means.

*Between-Groups Variance Estimate.* Formula 8-5 in Chapter 8 provided an estimate of the variability of the sampling distribution of means, i.e., the standard error of the mean:

$$s_M = \frac{s}{\sqrt{n}} \tag{8-5}$$

Squaring both sides gives:

$$s_M^2 = \frac{s^2}{n} \tag{9-2}$$

Another estimate of $s_M^2$ can be obtained from the means of the subgroups in the experiment. This is done merely by calculating $s^2$, the variance estimate, based on the squared deviations of these means about the grand mean as follows:

$$s_M^2 = \frac{\sum_{i=1}^{k} (M_i - M)^2}{k - 1} \tag{9-3}$$

Where:

$M_i$ = mean of subgroup $i$
$M$ = grand mean
$k$ = number of subgroups

Substituting Formula 9–3 into Formula 9–2 gives:

$$\frac{s^2}{n} = \frac{\sum_{i=1}^{k} (M_i - M)^2}{k - 1}$$

Multiplying both sides by $n$ gives:

$$s^2 = \frac{n \sum_{i=1}^{k} (M_i - M)^2}{k - 1} \tag{9-4}$$

Formula 9–4 gives another estimate of the population variance derived from information about the variability of subgroup means. In Formula 9–3, $k - 1$ is the number of degrees of freedom in estimating the true variance of the distribution of means, $\sigma_M^2$, and $k - 1$ also represents the number of degrees of freedom in Formula 9–4 for the between-groups estimate of the population variance, $\sigma^2$. Since a variance estimate is a ratio of a sum of squares of deviation scores to a degrees of freedom value, it follows that the numerator of the expression on the right in Formula 9–4, $n \sum_{i=1}^{k} (M_i - M)^2$, represents the sum of squares for the between-groups estimate of $\sigma^2$, the true variance of the population with respect to the experimental variable.

*The F Ratio Test in the One-Way Anova.* Two ways of estimating $\sigma^2$, the true variance of the population with respect to the experimental variable, are available from the two methods just described, one based on variability of subjects about their own subgroup means (within-groups estimate of $\sigma^2$) and the other based on the variability of the subgroup means about the grand mean (between-groups estimate of $\sigma^2$). If there are significant treatment effects causing the means of the various experimental subgroups to vary more than randomly about the grand mean, then the between-groups estimate of $\sigma^2$ should be unduly inflated. The more pronounced the differences between sample means, the more inflated this variance estimate will become.

It follows, therefore, that if the between-groups estimate of $\sigma^2$ is significantly greater than the within-groups estimate of $\sigma^2$, the sample means must differ from each other more than would be expected of random samples from populations all having the same population mean, $\mu$.

To test whether the means of subgroups are significantly different from each other using the one-way anova, therefore, an $F$ ratio is obtained dividing the between-groups estimate of $\sigma^2$ by the within-groups estimate of $\sigma^2$. The significance of this $F$ is evaluated in the same manner as before (See Problem 26, Chapter 8).

The only difference between this application of $F$ and that in Problem 26 is that the between-groups estimate of $\sigma^2$ is *always* placed on top, whether it is larger or not. It almost always is larger. In case it is not, however, $F$ becomes less than 1.0 and is automatically not significant. This merely means that the variability of the group means is even less than what might be expected as a result of random sampling variations for samples drawn from populations with the same mean.

*An Example.* The application of the one-way anova with samples of equal size will be illustrated by means of a worked-out example. Table 9–1 shows the results of an hypothetical experiment in which an instructor tested the differential utility of three sets of selfinstruction programmed materials. Thirty students were selected from the instructor's classes to participate in the experiment. These subjects were assigned at random to groups 1, 2, and 3. Instructional materials sets were then numbered and assigned at random to groups 1, 2, and 3. Each group worked with its own set of instructional materials for one month. At the end of this period, all three groups took the same examination over subject matter all instructional materials sets were supposed to cover. Scores on this examination for the individuals in the three groups are given in Table 9–1 along with the squares of those scores. The sums of these columns are given just below the scores themselves in the row labeled $\Sigma$. The remainder of the calculations for making this test are divided into getting the within-groups variance estimate, the between-groups variance estimate, and computing the $F$ ratio.

To obtain the within-groups variance estimate, the first step is to compute the sum of squares of deviation scores for each separate subgroup with respect to its own subgroup mean, using the now familiar formula $\Sigma x^2 = \Sigma X^2 - (\Sigma X^2)/N$ where $N$ in this particular case means $n$, the size of the subgroup. These computed values $\Sigma x_1^2$, $\Sigma x_2^2$, $\Sigma x_3^2$, are given below Table 9–1 as 4010.0, 2562.5, and 4102.5. These are summed to obtain the within-groups sum of squares. The degrees of freedom are 9 in each group so the pooled $df$ for the within-groups sum of squares

TABLE 9–1
One-Way Anova with Samples of Equal Size

| Subject Number | Group 1 | | Group 2 | | Group 3 | |
|---|---|---|---|---|---|---|
| | $X$ | $X^2$ | $X$ | $X^2$ | $X$ | $X^2$ |
| 1......... | 45 | 2025 | 10 | 100 | 50 | 2500 |
| 2......... | 50 | 2500 | 55 | 3025 | 90 | 8100 |
| 3......... | 70 | 4900 | 35 | 1225 | 40 | 1600 |
| 4......... | 40 | 1600 | 50 | 2500 | 70 | 4900 |
| 5......... | 40 | 1600 | 65 | 4225 | 80 | 6400 |
| 6......... | 10 | 100 | 30 | 900 | 35 | 1225 |
| 7......... | 30 | 900 | 45 | 2025 | 25 | 625 |
| 8......... | 80 | 6400 | 40 | 1600 | 55 | 3025 |
| 9......... | 35 | 1225 | 20 | 400 | 75 | 5625 |
| 10........ | 20 | 400 | 25 | 625 | 45 | 2025 |
| $\Sigma$......... | 420 | 21650 | 375 | 16625 | 565 | 36025 |

$\Sigma\Sigma X = 1360$ $\qquad\qquad$ $\Sigma\Sigma X^2 = 74{,}300$

$N = n_1 + n_2 + n_3 = 10 + 10 + 10 = 30$

*Within-Groups Variance Estimate:*

$$\Sigma x^2 = \Sigma X^2 - \frac{(\Sigma X)^2}{n}$$

$$\Sigma x_1^2 = 21650 - \frac{(420)^2}{10} = 21650 - 17640 = 4010.0$$

$$\Sigma x_2^2 = 16625 - \frac{(375)^2}{10} = 16625 - 14062.5 = 2562.5$$

$$\Sigma x_3^2 = 36025 - \frac{(565)^2}{10} = 36025 - 31922.5 = 4102.5$$

Pooled Sum of Squares = $4010.0 + 2562.5 + 4102.5 = 10675.0$
Pooled Sum of Degrees of Freedom = $9 + 9 + 9 = 27$

Within Groups Variance Estimate = $s_w^2 = \dfrac{10675.0}{27} = 395.4$

*Between-Groups Variance Estimate:*

$M_1 = \dfrac{420}{10} = 42.0$ $\qquad\qquad$ $M_1^2 = 1764.00$

$M_2 = \dfrac{375}{10} = 37.5$ $\qquad\qquad$ $M_2^2 = 1406.25$

$M_3 = \dfrac{565}{10} = 56.5$ $\qquad\qquad$ $M_3^2 = 3192.25$

$\Sigma$........ $136.0$ $\qquad\qquad\qquad$ $\Sigma = 6362.50$

$$\sum_{i=1}^{3} (M_i - M)^2 = 6362.5 - \frac{(136)^2}{3} = 6362.5 - 6165.3 = 197.2$$

TABLE 9–1—(*continued*)

$$n \sum_{i=1}^{3} (M_i - M)^2 = 10(197.2) = 1972.0$$

Degrees of Freedom $= k - 1 = 2$

Between-Groups Variance Estimate $= s_B^2 = \dfrac{1972.0}{2} = 986.0$

*F Ratio:*

$$F = \frac{s_B^2}{s_w^2} = \frac{986.0}{395.4} = 2.49 \text{ N.S.}$$

$F_{.05} = 3.35 \qquad\qquad F_{.01} = 5.49$

is 27. Dividing the sum of squares by the *df* gives $s_w^2 = 10675.0/27 = 395.4$ as the within groups estimate of $\sigma^2$.

To obtain the between-groups variance estimate, the subsample mean values and their squares are computed. The sums of these values are used to obtain $\sum_{i=1}^{3} (M_i - M_t)^2$, the sum of the squared deviations of the subsample means, $M_i$, from the grand mean, $M$. This value, 197.2, must be multiplied by 10, the subsample size $(n)$, to get the sum of squares for the between-groups estimate of $\sigma^2$, as shown in Formula 9–4. The sum of squares between groups, 1972.0, is divided by $(k - 1) = 2$, the number of *df*, to obtain the between-groups variance estimate, $s_B^2 = 986.0$.

Finally, to obtain $F$, the between-groups variance estimate is divided by the within-groups variance estimate to give $F = s_B^2/s_w^2 = 2.49$. Entering Table E (Appendix A) with 2 *df* for the larger variance estimate and 27 *df* for the smaller variance estimate, $F_{.05} = 3.35$ and $F_{.01} = 5.49$ are obtained as the values of $F$ required for significance at the five and one percent levels, respectively. Since the obtained value of $F$ was only 2.49, it is labelled N.S. for "not significant". As a consequence of the insignificant $F$ ratio, $h_0$, the null hypothesis, cannot be rejected on the basis of these results. If there are in fact differences between means induced as a consequence of the treatment effects, this test does not show them to be statistically significant. The fairly large numerical difference between the means and the size of $F$ obtained in relation to $F_{.05}$ might lead the investigator to hope that a repetition of the experiment with larger samples could yield significant results. With sample sizes of only 10, however, the differences between the teaching materials are not large enough to yield a statistically significant effect.

The results of a one-way anova are typically summarized in a table similar to that shown in Table 9–2. The total sum of squares and degrees

TABLE 9–2
One-Way Anova Summary for Data in Table 9–1

| Source of Variation | Sum of Squares | Degrees of Freedom | Mean Square (Variance Estimate) | F |
|---|---|---|---|---|
| Between-Groups......... | 1972.0 | 2 | 986.0 | 2.49 |
| Within-Groups......... | 10675.0 | 27 | 395.4 | |
| Total............... | 12647.0 | 29 | | |

$F_{.05} = 3.35; F_{.01} = 5.49$

of freedom, obtained in Table 9–2 by adding the values for between-groups and within-groups, should check with the values obtained independently for all three groups combined. The *df* of 29 checks because the *df* for the total group would be $N - 1$ or 29 since the three subgroups with $n = 10$ add up to $N = 30$. The sum of squares for the total group may be obtained using sums of the columns in Table 9–1 as follows:

$$\Sigma X = 420 + 375 + 565 = 1360$$

$$\Sigma X^2 = 21650 + 16625 + 36025 = 74300$$

$$\Sigma x^2 = \Sigma X^2 - \frac{(\Sigma X)^2}{N} = 74300 - \frac{(1360)^2}{30}$$

$$\Sigma x^2 = 74300 - 61653.3 = 12646.7$$

This value differs from that in Table 9–2 by only 0.3, a discrepancy attributable to rounding errors. In doing a one-way anova this check should always be made to guard aganst errors.

### *t* Tests Following the One-Way Anova *F* Test

It would have been possible to compare the mean of each subsample with the mean of each other subsample in the example given above using the *t* test (Problem 27). This practice is discouraged in favor of the one-way anova approach because in making many *t* tests a certain percentage of them can be expected to achieve significance even sampling at random from populations with equal means and variances. In fact, out of 100 *t* tests, roughly five should achieve significance at the five percent level by chance alone. With seven subgroups, for example, one significant *t* test at the 0.05 level could be expected by chance comparing each subgroup mean with each other subgroup mean since there are $(7 \times 6)/2 = 21$ such comparisons.

It is most appropriate to use the one-way anova to determine if there is an overall significant *F* ratio before examining differences between individual pairs of subgroup means for significance. If *F* is not significant, it is not statistically acceptable to claim significance for individual pair subgroup mean comparisons that exceed $t_{.05}$ or some other chosen level of confidence. If $F$ is significant, then it would be appropriate to make individual $t$ tests to determine the significance of differences between pairs of subgroup means. It is always important to remember, however, that the more $t$ tests are made, the greater is the possibility that a spuriously significant $t$ will be obtained. The investigator must exercise suitable caution in interpreting his results. In fact, where many such $t$ tests are being made, higher values than the usual $t_{.05}$ and $t_{.01}$ critical values are required for significance. Procedures for determining these higher critical values will be described in connection with Problem 32 later in this chapter.

Where $F$ is not significant, some investigators will proceed to make individual $t$ tests for the large subgroup mean pair differences even though it is not strictly correct statistically to do so. They can sometimes obtain some hypotheses for further investigation in this manner. For example, if one pair of means out of a large number of mean pairs gives a significant $t$ while the other pairs do not, the investigator may be encouraged to set up another experiment using just those two treatments to see if the phenomenon is replicable, i.e., whether or not he will obtain a statistically significant $t$ test again with new samples.

*Other Applications of the One-Way Anova.* The one-way anova can also be applied to certain nonexperimental studies in which subjects are not assigned at random to the different subgroups from the same pool. For example, it can be used to compare different populations to see if their means vary with respect to some variable of interest. Random samples of males could be drawn from the general population within several widely separated states of the United States. Measurements of height could be taken and the one-way anova used to investigate the variability of means in height of the subgroups. If $F$ proves to be significant, an investigation could be launched into the reasons for mean differences in height of the male populations of the different states. In this application of the one-way anova, however, there is no experimental intervention, giving different treatment to the various subgroups followed by measurement of the posttreatment state. If any such variation in treatment takes place, with the expectation of evaluating the effect of the different treatment by the one-way anova, random assignment of the subjects from a common pool to the different subgroups is an absolute necessity.

## Summary of Steps in Testing the Difference between Means of Several Samples of Equal Size Using One-Way Anova

1. Compute the sum of raw scores, $\Sigma X$, and the sum of raw scores squared, $\Sigma X^2$, for each sample separately.

2.  Obtain the overall sum of raw scores, $\Sigma\Sigma X$, and the overall sum of raw scores squared, $\Sigma\Sigma X^2$.
3.  Compute the mean of each sample, $M_i$, and the overall mean, $M$.
4.  Find the sum of squares of deviation scores within each sample separately using the totals obtained in Step 1 above and the formula $\Sigma x^2 = \Sigma X^2 - (\Sigma X)^2/n$ where $n$ is the number of cases in each sample.
5.  Obtain the between-groups sum of squares by summing the squared differences between sample means and the overall mean, $\Sigma(M_i - M)^2$, and multiplying by $n$, the common sample size.
6.  Obtain the between-groups variance estimate, $s_B^2$, by dividing the between-groups sum of squares from Step 5 by $k - 1$ where $k$ is the number of samples.
7.  Obtain the within-groups sum of squares by adding up the sum of squared deviation scores for the separate samples, computed in Step 4 above.
8.  Obtain the within-groups variance estimate, $s_W^2$, by dividing the within-groups sum of squares from Step 7 above by $k(n - 1)$, the sum of the degrees of freedom for the separate samples.
9.  Verify that the between-groups sum of squares from Step 5 plus the within-groups sum of squares from Step 7 equals the total group sum of squares within the limits of rounding error. The total group sum of squares is obtained from the totals computed in Step 2 above by the formula $\Sigma\Sigma X^2 - (\Sigma\Sigma X)^2/N$ where $N$ equals $k$ times $n$.
10. Divide the between-groups variance estimate, $s_B^2$, by the within-groups variance estimate, $s_W^2$, to obtain the $F$ ratio or $F$ value.
11. Enter the Table of the $F$ Distribution with $k - 1$ and $k(n - 1)$ degrees of freedom for the between-groups and within-groups variance estimates, respectively, to obtain the critical values of $F$, namely, $F_{.05}$ and $F_{.01}$.
12. If the value of $F$ computed in Step 10 above is greater than $F_{.05}$ but less than $F_{.01}$, reject the null hypothesis at the five percent level of confidence. If the value of $F$ computed in Step 10 above is greater than $F_{.01}$, reject the null hypothesis at the one percent level of confidence. If the obtained value of $F$ is less than $F_{.05}$, do not reject the null hypothesis.
13. If the null hypothesis is rejected in Step 10 above, $t$ tests may be conducted to determine which pairs of sample means are significantly different from each other.

### Mathematical Notes

The fact that the total sum of squares must be equal to the within-groups plus the between-groups sums of squares will be proved in this section. Begin with the formula for a deviation score about the grand mean, $M$:

$$x = X_{ij} - M$$

$$x = (X_{ij} - M_i) + (M_i - M)$$

$$x = x_{ij} + d_i$$

$$x^2 = x_{ij}^2 + d_i^2 + 2x_i d_i$$

Summing both sides over the entire total group, i.e., from 1 to $N$ while breaking up the sums on the right side in the subgroups separately gives:

$$\sum_1^N x^2 = \sum_{j=1}^n x_{1j}^2 + \sum_{j=1}^n x_{2j}^2 + \sum_{j=1}^n x_{3j}^2 + \cdots \sum_{j=1}^n x_{kj}^2 +$$

$$nd_1^2 + nd_2^2 + nd_3^2 +$$

$$2\sum_{j=1}^n x_{1j}d_1 + 2\sum_{j=1}^n x_{2j}d_2 + \cdots + 2\sum_{j=1}^n x_{kj}d_k \quad (9\text{--}5)$$

$$\sum_1^N x^2 = \text{Within Sum of Squares} + \text{Between Sum of Squares} +$$

$$2d_1 \sum_{j=1}^n x_{1j} + 2d_2 \sum_{j=1}^n x_{2j} + \cdots + 2d_k \sum_{j=1}^n x_{kj} \quad (9\text{--}6)$$

The first row of terms on the right of Formula 9–5 is just the summation of the sums of squares of the deviation scores within groups, i.e., the within-groups sum of squares. The second row of terms in Formula 9–5 is $n\Sigma d_i^2$ or $n\Sigma(M_i - M)^2$ which by Formula 9–4 is the between-groups sum of squares. The third row of terms in Formula 9–5 reduces to the second row of terms in Formula 9–6 because the $d$ values are constants with respect to the sums over subgroups and hence can be taken outside the summation signs. What is left is the sum of deviation scores which is zero in each case so the second row of terms in Formula 9–6 vanishes, leaving:

$$\sum_1^N x^2 = \text{Within } SSq + \text{Between } SSq$$

## Problem 32
### TEST THE DIFFERENCE BETWEEN MEANS OF SEVERAL SAMPLES OF UNEQUAL SIZE USING ONE-WAY ANOVA

Many research situations arise in which it becomes desirable to test the significance of differences among means of several samples of unequal size. This could occur in a planned experiment where the subgroup sizes were originally equal but became unequal due to dropouts. Such dropouts should be replaced, but this is not always possible. In such cases, the method of Problem 32 would be used to test the mean differences instead of that illustrated in Problem 31.

An even more common application of Problem 32 is that to data that are naturally occurring rather than experimentally obtained. The different subsamples may not consist of subjects randomly assigned from a common pool who are then to be subjected to different experimental treatments. Rather, they may represent samples from basically different populations that are being compared on some variable of interest. For example, an investigator might be interested in the relationship of blood levels of potassium and area of the country where the individual resides. He might be able to obtain results of blood tests made on draftees inducted at different centers throughout the country. By sifting these records he could come up with several samples of individuals from particular localities. The available data on potassium blood level could then be used as the "experimental" variable in a one-way anova. These samples would very likely be of different size, however. He could throw out data at random to make all samples of equal size in order to use the method of Problem 31, but it would be preferable to use the method of Problem 32 which permits the subsamples to vary in size.

### Formulas for Calculating Variance Estimates for One-Way Anova with Subgroups of Unequal Size

Formula 9–1 gave the method of determining the within-groups estimate of the population variance, $\sigma^2$, for the one-way anova with samples of equal size. This same formula also applies when the samples are of unequal size:

$$s^2 = \frac{\Sigma x_1^2 + \Sigma x_2^2 + \cdots + \Sigma x_k^2}{(n_1 - 1) + (n_2 - 1) + \cdots + (n_k - 1)} \qquad (9\text{–}7)$$

The only difference is that the values of the samples sizes, $n_1$, $n_2$, . . . , $n_k$, however, are different when Problem 32 is being worked out whereas they are all equal for Formula 9–1 and Problem 31. In Formula 9–7 for Problem 32, the implied limits on the summations are from 1 to $n_i$, so these change with the sample size. In Formula 9–1 for Problem 31, the limits on the summations are all the same because all the sample sizes are equal.

In both Problems 31 and 32, then, the sums of squares for the within-groups variance estimate are obtained by pooling the sums of squares of deviations from the various subgroups. The degrees of freedom are also pooled. The population variance estimate in both problems is the ratio of the pooled sum of squares to the pooled degrees of freedom.

*Between Groups Variance Estimate.* The formula for the between-groups variance estimate in the one-way anova with subgroups of unequal size varies slightly from Formula 9–4 which applies to subgroups of equal size. In Formula 9–4, each squared discrepancy between a subgroup mean and the grand mean is effectively weighted by its sample size, *n*. Since *n* is the same for all groups, it becomes a constant and moves outside the summation sign. When the samples are of unequal size, however, each squared discrepancy is weighted by its own subsample size, $n_i$, which puts the *n* back inside the summation sign since it varies with the sample as shown in Formula 9–8:

$$s^2 = \frac{\sum_{i=1}^{k} n_i(M_i - M)^2}{k - 1} \tag{9-8}$$

The degrees of freedom are the same for the between-groups variance estimate in the one-way anova both for samples of equal size and samples of unequal size, namely, $k - 1$, where *k* is the number of groups.

*An Example.* The procedures for doing a one-way anova with samples of unequal size will be illustrated by means of a worked-out example. Table 9–3 gives the raw scores and squared raw scores for four samples of male inductees from four widely scattered regions of the country. The raw scores represent scaled potassium blood levels for these recruits. The question is whether these subgroups could be considered to be random samples from populations that have equal population mean potassium blood levels. This question will be answered by doing a one-way anova for samples of unequal size. As with the one-way anova for groups of equal size, this test assumes essentially normally distributed populations with equal variances for the different groups with respect to the experimental variable while entertaining the possibility that the population means will be different.

The sums of the columns in Table 9–3 give $\Sigma X$ and $\Sigma X^2$ for the subgroups. The overall sums for the groups combined are given as $\Sigma\Sigma X$ and $\Sigma\Sigma X^2$. The computations for the variance estimates are given at the bottom of Table 9–3. The within groups variance estimate is obtained in the same way that it was obtained in Table 9–1 for the one-way anova with groups of equal size (Problem 31).

For the between-groups variance estimate each deviation score of a subgroup mean from the grand mean is calculated and multiplied by the subgroup sample size, $n_i$. These figures are summed to obtain the between groups sum of squares according to Formula 9–8. The between groups sum of squares is then divided by the between-groups degrees of

TABLE 9-3
One-Way Anova with Samples of Unequal Size

| Subject Number | Group 1 | | Group 2 | | Group 3 | | Group 4 | |
|---|---|---|---|---|---|---|---|---|
| | $X$ | $X^2$ | $X$ | $X^2$ | $X$ | $X^2$ | $X$ | $X^2$ |
| 1.......... | 8 | 64 | 16 | 256 | 14 | 196 | 27 | 729 |
| 2.......... | 13 | 169 | 19 | 361 | 10 | 100 | 15 | 225 |
| 3.......... | 6 | 36 | 8 | 64 | 10 | 100 | 20 | 400 |
| 4.......... | 16 | 256 | 16 | 256 | 2 | 4 | 19 | 361 |
| 5.......... | 1 | 1 | 18 | 324 | 9 | 81 | 10 | 100 |
| 6.......... | 14 | 196 | 20 | 400 | 3 | 9 | 17 | 289 |
| 7.......... | 12 | 144 | 17 | 289 | 16 | 256 | 21 | 441 |
| 8.......... | 14 | 196 | 30 | 900 | 4 | 16 | 20 | 400 |
| 9.......... | 21 | 441 | 18 | 324 | 17 | 289 | 32 | 1024 |
| 10.......... | 15 | 225 | 24 | 576 | 9 | 81 | 24 | 576 |
| 11.......... | 13 | 169 | 11 | 121 | 20 | 400 | | |
| 12.......... | 25 | 625 | 20 | 400 | 8 | 64 | | |
| 13.......... | 10 | 100 | 26 | 676 | 12 | 144 | | |
| 14.......... | 19 | 361 | 19 | 361 | 11 | 121 | | |
| 15.......... | 14 | 196 | 15 | 225 | | | | |
| $\Sigma$.......... | 201 | 3179 | 277 | 5533 | 145 | 1861 | 205 | 4545 |

$\Sigma\Sigma X = 828$    $\Sigma\Sigma X^2 = 15118$

$N = n_1 + n_2 + n_3 + n_4 = 15 + 15 + 14 + 10 = 54$

*Within-Groups Variance Estimate:*

$$\Sigma x^2 = \Sigma X^2 - \frac{(\Sigma X)^2}{n}$$

$$\Sigma x_1^2 = 3179 - \frac{(201)^2}{15} = 3179 - 2693.4 = 485.6$$

$$\Sigma x_2^2 = 5533 - \frac{(277)^2}{15} = 5533 - 5115.3 = 417.7$$

$$\Sigma x_3^2 = 1861 - \frac{(145)^2}{14} = 1861 - 1501.8 = 359.2$$

$$\Sigma x_4^1 = 4545 - \frac{(205)^2}{10} = 4545 - 4205.5 = 342.5$$

Pooled Sum of Squares = $485.6 + 417.7 + 359.2 + 342.5 = 1605.0$
Pooled Sum of Degrees of Freedom = $14 + 14 + 13 + 9 = 50$

Within-Groups Variance Estimate = $s_w^2 = \dfrac{1605.0}{50} = 32.1$

*Between-Groups Variance Estimate:*

$$M_1 = \frac{201}{15} = 13.40; \ M_2 = \frac{277}{15} = 18.47; \ M_3 = \frac{145}{14} = 10.36; \ M_4 = \frac{205}{10} = 20.50;$$

TABLE 9–3—*(continued)*

$$M = \frac{828}{54} = 15.3$$

$n_1(M_1 - M)^2 = 15(13.40 - 15.3)^2 = 15(-1.93)^2 = 15(3.725) = 55.87$
$n_2(M_2 - M)^2 = 15(18.47 - 15.3)^2 = 15(3.14)^2 = 15(9.860) = 147.89$
$n_3(M_3 - M)^2 = 14(10.36 - 15.3)^2 = 14(-4.97)^2 = 14(24.711) = 345.81$
$n_4(M_4 - M)^2 = 10(20.50 - 15.3)^2 = 10(5.17)^2 = 10(26.729) = 267.29$

$$\sum_{i=1}^{4} n_i(M_i - M)^2 = 55.87 + 147.89 + 345.81 + 267.29 = 816.86$$

$$\text{Degrees of Freedom} = (k - 1) = (4 - 1) = 3$$

$$\text{Between-Groups Variance Estimate} = s_B^2 = \frac{816.9}{3} = 272.3$$

*F Ratio:*

$$F = \frac{s_B^2}{s_w^2} = \frac{272.3}{32.1} = 8.48**$$

$$F_{.05} = 2.79 \qquad F_{.01} = 4.20$$

freedom, one less than the number of groups. The results of this one-way anova are summarized in Table 9–4.

TABLE 9–4
One-Way Anova Summary for Data of Table 9–3

| Source of Variation | Sum of Squares | Degrees of Freedom | Mean Square (variance estimate) | F |
|---|---|---|---|---|
| Between Groups....... | 816.9 | 3 | 272.3 | 8.45** |
| Within Groups........ | 1605.0 | 50 | 32.1 | |
| Total................ | 2421.9 | 53 | | |

The total sum of squares must be checked against the value obtained using $\Sigma X$ and $\Sigma X^2$ for the entire group as follows:

$$\Sigma x^2 = \Sigma X^2 - \frac{(\Sigma X)^2}{N} = 15118 - \frac{(828)^2}{54} = 15118 - 12696 = 2422.0$$

The difference between 2422 and the value of 2421.9 shown in Table 9–4 is due to rounding error.

The total degrees of freedom shown in Table 9–4 checks since 53 equals $N - 1$. The F ratio has two asterisks (**) beside it since it exceeds the $F_{.01}$ value with 3 and 50 *df*, respectively, for the larger and smaller variance estimates.

*Checking Subgroup Variance Estimates.* Since $F$ was significant, it will be necessary to examine the $s^2$ values for the various subgroups to see if they are close enough to each other in size to make the $F$ test appropriate. Using the $\Sigma x_i^2$ values from Table 9–3, these variance estimates may be obtained as follows:

$$s_1^2 = \frac{\Sigma x_1^2}{n_1 - 1} = \frac{485.6}{14} = 32.4$$

$$s_2^2 = \frac{\Sigma x_2^2}{n_2 - 1} = \frac{417.7}{14} = 29.8$$

$$s_3^2 = \frac{\Sigma x_3^2}{n_3 - 1} = \frac{359.2}{13} = 27.6$$

$$s_4^2 = \frac{\Sigma x_4^2}{n_4 - 1} = \frac{342.5}{9} = 38.1$$

The largest $F$ would be $s_4^2/s_3^2 = 1.4$. For 9 and 13 $df$, respectively, for the larger and smaller variance estimate, $F_{.05}$ is 2.72 so the null hypothesis, $h_0$, cannot be rejected as far as variances are concerned. Since there is no proof that the population variances are unequal, the one-way anova to test differences between means is appropriate as far as the assumption of equal variances is concerned. In this case with an $F$ twice as large as that required for significance at the 0.01 level, differences in $s_i^2$ values would have to be very substantial to cast doubt on the conclusion that the means differ significantly from each other.

*Comparing Means by the* t *Test.* Since the $F$ ratio test for the one-way anova turned out to be statistically significant for the data in Table 9–3, it is appropriate to compare individual pairs of means by the $t$ test using the method of Problem 27. The computations for these $t$ tests are shown in Table 9–5. The formula for $s_{D_M}$ in Table 9–5 differs from the usual one, Formula 8–12 for Problem 27. In the denominator in Table 9–5 $s_w^2$ has been substituted for $(\Sigma x_1^2 + \Sigma x_1^2)/(n_1 + n_2 - 2)$ in Formula 8–12 for $s_{D_M}$. Both these expressions are within-groups variance estimates. In the one-way anova, if more than two groups are available, a better estimate of $\sigma^2$ can be obtained by pooling all the sums of squares and dividing by the pooled degrees of freedom. This better estimate, $s_w^2$, is then used in the formula for $s_{D_M}$ shown in Table 9–5 as the denominator of the $t$ ratio.

Three of the $t$ test comparisons in Table 9–5, 1–4, 2–3, and 3–4, yielded differences between means significant at the 0.01 level, indicated by two asterisks after the $t$ value. The sign on the $t$ value is ignored since the $t$ distribution is symmetrical. One additional comparison, 1–2, gave a statistically significant difference between means, but at the 0.05 level. This is indicated in Table 9–5 by one asterisk after the $t$ value.

TABLE 9–5

*t*-Tests of Differences between Subgroup
Means for the Data in Table 9–3

$$t = \frac{M_1 - M_2 - 0}{\sqrt{s_w{}^2 \left( \dfrac{n_1 + n_2}{n_1 n_2} \right)}}$$

$$t_{12} = \frac{13.4 - 18.5 - 0}{\sqrt{32.1 \left( \dfrac{15 + 15}{15 \cdot 15} \right)}} = \frac{-5.1}{\sqrt{4.27}} = -2.46*$$

$$t_{13} = \frac{13.4 - 10.4 - 0}{\sqrt{32.1 \left( \dfrac{15 + 14}{15 \cdot 14} \right)}} = \frac{3.0}{\sqrt{4.43}} = 1.43$$

$$t_{14} = \frac{13.4 - 20.5 - 0}{\sqrt{32.1 \left( \dfrac{15 + 10}{15 \cdot 10} \right)}} = \frac{-7.1}{\sqrt{5.35}} = -3.07**$$

$$t_{23} = \frac{18.5 - 10.4 - 0}{\sqrt{32.1 \left( \dfrac{15 + 14}{15 \cdot 14} \right)}} = \frac{8.1}{\sqrt{4.43}} = 3.86**$$

$$t_{24} = \frac{18.5 - 20.5 - 0}{\sqrt{32.1 \left( \dfrac{15 + 10}{15 \cdot 10} \right)}} = \frac{-2.0}{\sqrt{5.35}} = -0.87$$

$$t_{34} = \frac{10.4 - 20.5 - 0}{\sqrt{32.1 \left( \dfrac{14 + 10}{14 \cdot 10} \right)}} = \frac{-10.1}{\sqrt{5.50}} = -4.30**$$

$df = 22, t_{.05} = 2.07, t_{.01} = 2.82$

$df = 23, t_{.05} = 2.07, t_{.01} = 2.81$

$df = 27, t_{.05} = 2.05, t_{.01} = 2.77$

$df = 28, t_{.05} = 2.05, t_{.01} = 2.76$

The two remaining comparisons failed to yield statistically significant differences between means.

Unfortunately, these *t* tests and their apparent levels of significance cannot be taken at face value. The *t* test is basically designed for the situation where only one comparison between two available means is to be made. In this particular case, with four treatment groups, a total of six *t* tests were made. The more *t* tests that are made, the greater the likelihood of getting one that exceeds the usual significance level by chance. With several groups, then, just making all the possible $t$-tests and reporting those that achieved statistical significance at the 0.05 and/or the 0.01 level is not acceptable.

If an investigator decides in advance that he is going to test only one particular pair of means out of several possible pairs, if $F$ is significant he can proceed to use the $t$ test in the usual way. This would not involve multiplying the chances of getting a significant $t$ when the null hypothesis is true. If he is planning to test all the possible mean differences, however, the ordinary significance levels for $t$ are too low. This underestimation is a function of the number of $t$ tests to be made. As the number of groups increases, the possible comparisons multiply rapidly and as a consequence so do the opportunities to obtain significant results by chance.

A method for dealing with this problem has been worked out by Scheffé[1] which determines how big $t$ must be to achieve significance where all possible comparisons are being made. This test also makes allowances for comparisons involving averages of subgroups. This multiplies the number of comparisons requiring a value of $t$ for significance that may be higher than it need be if the investigator had no idea of investigating means of combined groups. If an obtained value exceeds the critical Scheffé $t'$ value, however, it is certainly safe to treat the difference as statistically significant.

*The Scheffé Test.*    To determine a value $t'_{.05}$ which must be exceeded by an obtained $t$ for it to be considered significant at the 0.05 level when all possible comparisons are being made in the one-way anova, define $t'$ as follows:

$$t'_{.05} = \sqrt{(k - 1)F_{.05}} \qquad (9\text{--}9)$$

Where:

$t'_{.05}$ = the critical value of $t$

$k$ = the number of groups in the one-way anova

$F_{.05}$ = the $F$ ratio required for significance with $k - 1$ as the $df$ for the larger variance estimate and $N - k$ as the $df$ for the smaller variance estimate

Any and all obtained $t$ ratios that exceed $t'_{.05}$ are regarded as significant at the 0.05 level at least. This test applies to the one-way anova with equal size subgroups as well. If a one percent level value of $t$ is desired, use the following formula:

$$t'_{.01} = \sqrt{(k - 1)F_{.01}} \qquad (9\text{--}10)$$

The definitions of terms in Formula 9–10 parallel those for Formula 9–9.

---

[1] Scheffé, H., "A method for judging all contrasts in the analysis of variance." *Biometrika* **40** (1953), 87–104.

For the data in Tables 9–3, 9–4, and 9–5, the $F_{.05}$ and $F_{.01}$ values for Formulas 9–9 and 9–10 would be 2.79 and 4.20 using 3 and 50 degrees of freedom for the larger and smaller variance estimates, respectively. Substituting these values of $F$ in Formulas 9–9 and 9–10 gives:

$$t'_{.05} = \sqrt{3(2.79)} = \sqrt{8.37} = 2.89$$

$$t'_{.01} = \sqrt{3(4.20)} = \sqrt{12.60} = 3.55$$

Comparing these values of $t'_{.05}$ and $t'_{.01}$ with the $t_{.05}$ and $t_{.01}$ values required for significance at the 0.05 and 0.01 level for the $t$ test, as shown in Table 9–5, it can be seen that with $k = 4$ the Scheffé $t'_{.05}$ value is not dissimilar to the $t$ test $t_{.01}$ value. The Scheffé $t'_{.01}$ value, of course, is considerably larger than the $t$ test $t_{.01}$ value.

In Table 9–5, the comparisons 2–3 and 3–4 still remain significant at the 0.01 level when compared against $t'_{.01} = 3.55$, but the 1–4 $t$ value of 3.07 drops down to being significant at only the 0.05 level. The 1–2 value of $t = -2.46$ drops from being significant at the 0.05 level to not being significant at all.

To summarize, if only one $t$ test is to be made, the significance levels shown in Table 9–5 represent appropriate standards for making that evaluation. If all possible $t$ tests are to be made, including those for averages of all possible combined subgroups, the Scheffé $t'_{.05}$ and $t'_{.01}$ values give fair criteria of significance for the obtained $t$ values. If only some of the comparisons are actually to be considered, the appropriate significance levels for $t$ would lie somewhere in between. Using the Scheffé values, however, is a conservative approach that will guard against being too aggressive in rejecting the null hypothesis at a specified level of confidence.

## Summary of Steps in Testing the Difference between Means of Several Samples of Unequal Size Using One-Way Anova

1. Compute the sum of raw scores, $\Sigma X$, and the sum of raw scores squared, $\Sigma X^2$, for each sample separately.
2. Obtain the overall sum of raw scores, $\Sigma\Sigma X$, and the overall sum of raw scores squared, $\Sigma\Sigma X^2$.
3. Compute the mean of each sample, $M_i$, and the overall mean, $M$.
4. Find the sum of squares of deviation scores within each sample separately using the totals obtained in Step 1 above and the formula $\Sigma x^2 = \Sigma X^2 - (\Sigma X)^2/n$ where $n$ is the sample size.
5. Obtain the between-groups sum of squares as follows. First, for each sample, compute $n_i(M_i - M)^2$, where $n_i$ is the sample size. Add these values over the $k$ samples.

6. Obtain the between-groups variance estimate, $s_B^2$, by dividing the between-groups sum of squares from Step 5 by $k - 1$ where $k$ is the number of samples.

7. Obtain the within-groups sum of squares by adding up the sum of squared deviation scores for the separate samples, computed in Step 4 above.

8. Obtain the within-groups variance estimate, $s_W^2$, by dividing the within-groups sum of squares from Step 7 above by $\sum_{i=1}^{k} (n_i - 1)$, the sum of the degrees of freedom for the $k$ separate samples.

9. Verify that the between-groups sum of squares from Step 5 plus the within-groups sum of squares from Step 7 equals the total group sum of squares within the limits of rounding error. The total group sum of squares is obtained from the totals computed in Step 2 above by the formula $\Sigma\Sigma X^2 - (\Sigma\Sigma X)^2 / N$ where $N$ is the sum of the separate sample sizes, or the overall sample size.

10. Divide the between-groups variance estimate, $s_B^2$, by the within-groups variance estimate, $s_W^2$, to obtain the $F$ value.

11. Enter the Table of the $F$ Distribution with $k - 1$ and $\sum_{i=1}^{k} (n_i - 1)$ degrees of freedom for the between-groups and within groups variance estimates, respectively, to obtain the critical values of $F$, namely, $F_{.05}$ and $F_{.01}$.

12. If the value of $F$ computed in Step 10 above is greater than $F_{.05}$ but less than $F_{.01}$, reject the null hypothesis at the five percent level of confidence. If the value of $F$ computed in Step 10 above is greater than $F_{.01}$, reject the null hypothesis at the one percent level of confidence. If the obtained value of $F$ is less than $F_{.05}$, do not reject the null hypothesis.

13. If the null hypothesis is rejected in Step 10 above, $t$ tests may be conducted to determine which pairs of sample means are significantly different from each other.

## Mathematical Notes

The representation of the total sum of squares as the sum of the within-groups sum of squares plus the between-groups sum of squares was shown in the Mathematical Notes section of Problem 31 for the case with equal size subsamples. Inspection of Formula 9–5 in that section will reveal readily that a similar mathematical development can be used to show that these sums of squares also add up for the case of subgroups of unequal size. The between-groups sum of squares is calculated as the numerator of Formula 9–8, however, instead of Formula 9–4 which requires equal subgroup sample sizes. In Formula 9–5 each $n$ will be subscripted to represent the $n$ of the given subgroup, and these values may differ in size. The within-groups sum of squares is obtained in the same way as with subsamples of equal size although the upper limit on the summation will differ from subgroup to subgroup, (i.e., $n_1$, $n_2$, . . . , $n_k$) instead of being just $n$ for every group.

## Problem 33
### TEST THE EFFECTS OF TWO FACTORS SIMULTANEOUSLY USING THE TWO-WAY ANOVA

The head of a college department has three available instructors from whom to choose the individual to be assigned regularly to teach an important undergraduate course. He is also concerned with finding out whether the lecture or discussion method produces greater student achievement on a standardized examination to be taken by students at the end of the course to rate his department nationally. The department head decides to conduct an experiment to answer both questions at the same time.

Sixty students are selected to participate in the experiment. These students' names are written on slips of paper that are mixed thoroughly and placed in a hat. Names are drawn blindly from the hat one at a time for six separate experimental groups until all subjects have been assigned to one of the groups, giving 10 subjects per group. Each of the three instructors will teach two groups, one by the lecture method and the other by the discussion method, giving a total of six groups. The postcourse national examination scores for the students in these six groups are shown in Table 9–6. These scores are scaled scores with a mean of 10 and a standard deviation of 3 based on national norms. No score higher than 19 or lower than 1 is given. The scaled scores themselves are the $X$ scores and their squares are given under the heading $X^2$.

### Computing Procedures for the Two-Way Anova

The sums of the columns in Table 9–6 are given in the rows designated by the summation symbol $\Sigma$. Thus the sum of the $X$ scores for the first group is 93 and the sum of $X^2$ is 953.

In addition to showing $\Sigma X$ and $\Sigma X^2$ for each of the six subgroups, Table 9–6 also shows the double sums ($\Sigma\Sigma$) by rows and columns as well as the overall sum of $X$ and $X^2$. For example, 365 and 4831 give $\Sigma X$ and $\Sigma X^2$ for row 1, i.e., the sums of those values for the three groups

TABLE 9–6
Two-Way Analysis of Variance

| | Subject | Instructor A | | Instructor B | | Instructor C | | $\Sigma\Sigma X$ | $\Sigma\Sigma X^2$ |
|---|---|---|---|---|---|---|---|---|---|
| | | $X$ | $X^2$ | $X$ | $X^2$ | $X$ | $X^2$ | | |
| *Discussion* | 1...... | 10 | 100 | 7 | 49 | 15 | 225 | | |
| | 2...... | 13 | 169 | 10 | 100 | 10 | 100 | | |
| | 3...... | 12 | 144 | 12 | 144 | 18 | 324 | | |
| | 4...... | 2 | 4 | 19 | 361 | 14 | 196 | | |
| | 5...... | 7 | 49 | 15 | 225 | 13 | 169 | | |
| | 6...... | 11 | 121 | 14 | 196 | 14 | 196 | | |
| | 7...... | 9 | 81 | 16 | 256 | 9 | 81 | | |
| | 8...... | 8 | 64 | 13 | 169 | 17 | 289 | | |
| | 9...... | 10 | 100 | 14 | 196 | 12 | 144 | | |
| | 10...... | 11 | 121 | 17 | 289 | 13 | 169 | | |
| | $\Sigma$....... | 93 | 953 | 137 | 1985 | 135 | 1893 | 365 | 4831 |
| *Lecture* | 1...... | 18 | 324 | 10 | 100 | 6 | 36 | | |
| | 2...... | 15 | 225 | 13 | 169 | 14 | 196 | | |
| | 3...... | 13 | 169 | 4 | 16 | 7 | 49 | | |
| | 4...... | 14 | 196 | 7 | 49 | 5 | 25 | | |
| | 5...... | 11 | 121 | 1 | 1 | 2 | 4 | | |
| | 6...... | 6 | 36 | 14 | 196 | 13 | 169 | | |
| | 7...... | 13 | 169 | 9 | 81 | 10 | 100 | | |
| | 8...... | 10 | 100 | 10 | 100 | 9 | 81 | | |
| | 9...... | 16 | 256 | 8 | 64 | 8 | 64 | | |
| | 10...... | 14 | 196 | 12 | 144 | 12 | 144 | | |
| | $\Sigma$....... | 130 | 1792 | 88 | 920 | 86 | 868 | 304 | 3580 |
| | $\Sigma\Sigma$...... | 223 | 2745 | 225 | 2905 | 221 | 2761 | 669 | 8411 |

$$N = kn = 6 \cdot 10 = 60$$

*Within-Groups Variance Estimate:*

$$\Sigma x^2 = \Sigma X^2 - \frac{(\Sigma X)^2}{n}$$

$$\Sigma x_1^2 = \quad 953 - \frac{(93)^2}{10} = \quad 953 - \quad 864.9 = \qquad 88.1$$

$$\Sigma x_2^2 = 1985 - \frac{(137)^2}{10} = 1985 - 1876.9 = \qquad 108.1$$

$$\Sigma x_3^2 = 1893 - \frac{(135)^2}{10} = 1893 - 1822.5 = \qquad 70.5$$

$$\Sigma x_4^2 = 1792 - \frac{(130)^2}{10} = 1792 - 1690.0 = \qquad 102.0$$

$$\Sigma x_5^2 = \quad 920 - \frac{(88)^2}{10} = \quad 920 - \quad 774.4 = \qquad 145.6$$

$$\Sigma x_6^2 = \quad 868 - \frac{(86)^2}{10} = \quad 868 - \quad 739.6 = \qquad 128.4$$

TABLE 9–6 (*continued*)

Pooled Sum of Squares $= \Sigma\Sigma x_i^2 =$    642.7

Pooled Sum of Degrees of Freedom $= k(n - 1) =$    54

Within-Groups Variance Estimate $= s_w^2 = \dfrac{642.7}{54} = 11.90$

*Between-Rows Variance Estimate:*

$$M_{R_1} = \frac{365}{30} = 12.17 \qquad M_{R_1}^2 = 148.11$$

$$M_{R_2} = \frac{304}{30} = 10.13 \qquad M_{R_2}^2 = 102.62$$

$$\Sigma = \ldots\ldots 22.30 \qquad \Sigma = 250.73$$

$$\Sigma(M_{R_i} - M)^2 = 250.73 - \frac{(22.30)^2}{2} = 250.73 - 248.64 = 2.09$$

$$nc\Sigma(M_{R_i} - M)^2 = 10 \cdot 3 \cdot (2.09) = 62.70$$
$$df = r - 1 = 2 - 1 = 1$$

Between-Rows Variance Estimate $= s_R^2 = \dfrac{62.70}{1} = 62.70$

*Between-Columns Variance Estimate:*

$$M_{C_1} = \frac{223}{20} = 11.15 \qquad M_{C_1}^2 = 124.32$$

$$M_{C_2} = \frac{225}{20} = 11.25 \qquad M_{C_2}^2 = 126.56$$

$$M_{C_3} = \frac{221}{20} = 11.05 \qquad M_{C_3}^2 = 122.10$$

$$\Sigma = \ldots\ldots 33.45 \qquad \Sigma = 372.98$$

$$\Sigma(M_{C_i} - M)^2 = 372.98 - \frac{(33.45)^2}{3} = 372.98 - 372.97 = 0.01$$

$$nr\Sigma(M_{C_i} - M)^2 = 10 \cdot 2(0.01) = 0.20$$
$$df = c - 1 = 3 - 1 = 2$$

Between-Columns Variance Estimate $= s_C^2 = \dfrac{0.20}{2} = 0.10$

*Total Sum of Squares:*

$$\sum_1^N x^2 = \Sigma X^2 - \frac{(\Sigma X)^2}{N} = 8411 - \frac{(669)^2}{60} = 951.65$$

TABLE 9–6 (*continued*)

*Between-Groups Sum of Squares:*

$$M_1 = \frac{93}{10} = 9.3 \qquad M_1{}^2 = 86.49$$

$$M_2 = \frac{137}{10} = 13.7 \qquad M_2{}^2 = 187.69$$

$$M_3 = \frac{135}{10} = 13.5 \qquad M_3{}^2 = 182.25$$

$$M_4 = \frac{130}{10} = 13.0 \qquad M_4{}^2 = 169.00$$

$$M_5 = \frac{88}{10} = 8.8 \qquad M_5{}^2 = 77.44$$

$$M_6 = \frac{86}{10} = 8.6 \qquad M_6{}^2 = 73.96$$

$$\Sigma = \ldots \ldots 66.9 \qquad \Sigma = 776.83$$

$$\Sigma(M_i - M)^2 = 776.83 - \frac{(66.9)^2}{6} = 776.83 - 745.935 = 30.895$$

$$n\Sigma(M_i - M)^2 = 10(30.895) = 308.95$$

Within-$SSq$ + Between-$SSq$ = 642.7 + 308.95 = 951.65

Between-$SSq$ − Row-$SSq$ − Column-$SSq$ = 308.95 − 62.7 − 0.2 = 246.05

Interaction-$SSq$ = 246.0

*Interaction Variance Estimate:*

Interaction $SSq$ = 246.0

Interaction $df = (r - 1)(c - 1) = (2 - 1)(3 - 1) = 2$

Interaction Variance Estimate $= s_{R \times C}{}^2 = \dfrac{246.0}{2} = 123.0$

*Independent Check on the Interaction Sum of Squares:*

$(M_i - M_{R_i} - M_{C_i} + M)^2$
$( 9.30 - 12.17 - 11.15 + 11.15)^2 = (-2.87)^2 = 8.24$
$(13.70 - 12.17 - 11.25 + 11.15)^2 = (1.43)^2 = 2.04$
$(13.50 - 12.17 - 11.05 + 11.15)^2 = (1.43)^2 = 2.04$
$(13.00 - 10.13 - 11.15 + 11.15)^2 = (2.87)^2 = 8.24$
$( 8.80 - 10.13 - 11.25 + 11.15)^2 = (-1.43)^2 = 2.04$
$( 8.60 - 10.13 - 11.05 + 11.15)^2 = (-1.43)^2 = 2.04$
$\Sigma \ldots \ldots \ldots \ldots \ldots \ldots \ldots \ldots \ldots \ldots \ldots \ldots \ldots 24.64$

$n\Sigma(M_i - M_{R_i} - M_{C_i} + M)^2 = 10(24.64) = 246.4$

that are getting the discussion method. The sums for row 2 are $\Sigma X = 304$ and $\Sigma X^2 = 3580$, the result of adding the individual sums for the three groups getting the lecture method. The column sums for the first column, i.e., for Instructor A, in Table 9–6, are $\Sigma X = 223$ and $\Sigma X^2 = 2745$. For Instructor B, column two, the sums are $\Sigma X = 225$ and $\Sigma X^2 = 2905$; for Instructor C, column three, $\Sigma X = 221$ and $\Sigma X^2 = 2761$. Adding up the $\Sigma X$ and $\Sigma X^2$ values by rows or by columns gives the total $\Sigma\Sigma X = 669$ and $\Sigma\Sigma X^2 = 8411$ for the entire sample comprising all six subgroups with $N = 60$.

*Within-Groups Variance Estimate.* Just as with the one-way anova in Problems 31 and 32, a within-groups estimate of the population variance is computed with the two-way anova. As with the one-way anova, the sum of squared deviations relative to their own subgroup means is computed for each group separately. These sums of squares are pooled, i.e., $\Sigma\Sigma x_i^2$, to get the sum of squares within groups. The $k - 1$ degrees of freedom from each group are also pooled to get $k(n - 1)$ $df$. The pooled sum of squares is divided by the pooled $df$ to get $s_w^2$, the within groups estimate of the population variance, $\sigma^2$. As with the one-way anova, $\sigma^2$ is assumed to be equal for each subgroup. The computations leading to the determination that $s_w^2 = 11.90$ for the data in Table 9–6 are shown below the table.

*Between-Rows Variance Estimate.* In the one-way anova, variations of subgroup means from the grand mean provided the basis for another variance estimate, the between-groups variance estimate. In the two-way anova, the groups in each row can be pooled to provide two groups that will give an estimate of the population variance by the same method used to get the between-groups variance estimate in the one-way anova. When the three subgroups having the discussion method (row 1) are combined, $\Sigma X = 365$; this gives $M_{R_1} = {}^{365}\!/_{30} = 12.17$ for the mean of the first row. Since three groups were combined, the number of cases in the combined group is 30. The corresponding mean for the second row is $M_{R_2} = {}^{304}\!/_{30} = 10.13$. The sum of squares of deviations of these two means about their average, $M$, the grand mean, is found in the usual way, leading to a value of 2.09. This sum of squares of deviation scores must be multiplied by the number of cases in the combined group, i.e., $(n \times c) = 10 \times 3 = 30$, or the number of subjects per subgroup times the number of columns, to get the sum of squares for the between-rows variance estimate. This is analogous to the numerator for Formula 9–4 which gives the between-groups sum of squares for the one-way anova. There, however, each unit group has $n$ cases, so $\Sigma(M_i - M)^2$ is multiplied by $n$. Here, each unit group has $nc$ cases, so $\Sigma(M_i - M)^2$ is multiplied by $nc$ to get the between-rows sum of squares. The calculations to obtain this sum of squares are shown below Table 9–6, giving a value of 62.70. The degrees of freedom for this variance estimate are $(r - 1)$

where $r$ is the number of rows, i.e., $(2 - 1)$ or 1. The between-rows variance estimate, therefore, obtained by dividing the sum of squares by the $df$, is $62.70/1 = 62.70$.

*Between-Columns Variance Estimate.*    Just as the subgroups in each row were combined with the data in Table 9–6 to form two larger groups to obtain an estimate of the population variance from the variation of row means about the grand mean, so can the subgroups in each column be combined to form three larger groups. The variation in the means of these columns about the grand mean can be used to provide yet another estimate of the population variance. As with the between-rows variance estimate, the basic method of obtaining this between-columns variance estimate is very similar to that used to obtain the between-groups variance estimate in the one-way anova. The computations are shown below the data in Table 9–6. The sum of squared deviations of the column means about the grand mean are obtained by computing column means and then using the usual formula $\Sigma x^2 = \Sigma X^2 - (\Sigma X)^2/N$. In this case, as shown in Table 9–6, $\Sigma x^2$ for columns = 0.01. To obtain the between-columns sum of squares, the sum of squared deviations is then multiplied by the number of cases in each pooled group, i.e., $n \cdot r$, the number of cases per group by the number of rows, or 20. In Table 9–6 this value is 0.20. The number of degrees of freedom is the number of columns minus 1, or 2. The between-columns variance estimate, $s_c^2$, is obtained by dividing the between-columns sum of squares by the between columns $df$, giving in Table 9–6 $s_c^2 = 0.20/2 = 0.10$.

*Interaction Variance Estimate.*    In the one way anova, the total sum of squares equalled the sum of the within-groups. sum of squares plus the between-groups sum of squares, except for rounding errors. This same principle holds for the two-way anova. The total sum of squares and within-groups sum of squares are calculated in the same way for both the two-way and the one-way anova. The between-groups sum of squares for the two-way anova is calculated the same as for the one-way anova if all subgroups are treated as they are in the one-way anova, without regard to row and column position. For each subgroup, $M_i$ and $M_i^2$ are obtained. The sums of these values are inserted in the usual formula for the sum of squared deviations and this total is multiplied by $n$, the subgroup size. In Table 9–6, the computations for these sums of squares yield the following:

$$\begin{array}{r} \text{Within-Groups Sum of Squares} = 642.70 \\ \text{Between-Groups Sum of Squares} = \underline{308.95} \\ 951.65 \end{array}$$

Addition of these two sums of squares gives a figure that equals the independently calculated total sum of squares, 951.65.

The between-groups sum of squares in the two-way anova is not just due to one experimental factor, however. Differences between column-means as well as differences between row-means contribute to the between-groups sum of squares. Adding the between-rows sum of squares, 62.70, and the between-columns sum of squares, 0.20, gives a total, however, of only 62.90 when the between-groups sum of squares is 308.95. Subtracting the between-rows and between-columns sums of squares from the between-groups sum of squares, $308.95 - 62.7 - 0.2 = 246.05$, gives 246.05 as a "left over" sum of squares not attributable to row or column effects. In this particular experiment this means that there are variations among the subgroup means that cannot be attributed to differences between teachers, per se, or to differences between teaching methods, per se. There must be something else operating.

This "something else" is the *interaction* effect and this left over sum of squares is called the "interaction sum of squares." In this experiment, if there is a significant interaction effect, what it means is that some teachers do better with one method than they do with the other, or vice versa. If there were no interaction effect, the interaction sum of squares would be zero and it would not matter which teacher is paired with which method. The interaction sum of squares will seldom be zero, of course, but if it does not lead to a statistically significant result, no interaction effect can be claimed.

The interaction sum of squares also can lead to yet another estimate of the population variance, $\sigma^2$, when divided by the number of degrees of freedom. The *df* for the interaction sum of squares is $(r - 1)(c - 1)$, the product of the number of rows minus one times the number of columns minus one. Calculations for the interaction variance estimate, $s^2_{R \times C}$, are shown for the data in Table 9–6 below the calculations for the between-groups sum of squares as follows: $s^2_{R \times C} = {}^{246}\!/_2 = 123.0$.

*Check on the Interaction Sum of Squares.* The interaction sum of squares should also be calculated by another method to insure that the method shown in Table 9–6 has given correct results. The deviation of a subgroup mean from the grand mean, $(M_i - M)$, can be viewed as consisting of a row effect, $M_{R_i} - M$, a column effect, $M_{C_i} - M$, and an interaction effect. To get the part of the deviation that is due to interaction, subtract out the other two effects as follows:

$$(M_i - M) - (M_{R_i} - M) - (M_{C_i} - M)$$

Removing parentheses and cancelling $M$s gives:

$$(M_i - M_{R_i} - M_{C_i} + M) \tag{9–11}$$

Formula 9–11 gives the deviation due to interaction. Squaring and summing over all such values for all subgroups gives:

$$\sum_{i=1}^{k} (M_i - M_{R_i} - M_{C_i} + M)^2 \tag{9–12}$$

As with other sums of squared deviation scores involving variations of means of groups, the sum of squared deviations must be multiplied by the number of cases in the group for which the mean is computed to obtain the sum of squares. This gives for the interaction sum of squares:

$$n \cdot \sum_{i=1}^{k} (M_i - M_{R_i} - M_{C_i} + M)^2 \qquad (9\text{-}13)$$

Using Formula 9–12 on the six subgroups for the data in Table 9–6 gives the results shown at the bottom of Table 9–6 under the title "Independent Check on the Interaction Sum of Squares." Multiplying 24.64 by $n = 10$, as called for by Formula 9–13, gives 246.4 as the interaction sum of squares. This figure agrees within rounding error with the interaction sum of squares, 246.0, obtained by subtracting the between-columns and between-rows sum of squares from the between-groups sum of squares (see Table 9–6).

*Summary of the Two-Way Anova Results.*   As with the one-way anova, it is customary to summarize the results of the two-way anova in a table that shows the various sums of squares, degrees of freedom, variance estimates, and $F$ ratio values. This summary for the data in Table 9–6 is shown in Table 9–7.

TABLE 9–7
Summary of Two-Way Anova for Data in Table 9–6

| Source of Variation | Sum of Squares | Degrees of Freedom | Mean Square (variance estimate) | F |
|---|---|---|---|---|
| Between Rows............. | 62.7 | 1 | 62.7 | 5.3* |
| Between Columns.......... | 0.2 | 2 | 0.1 | 0.0 |
| Interaction............... | 246.0 | 2 | 123.0 | 10.3** |
| Within Groups............ | 642.7 | 54 | 11.9 | |
| Total.................... | 951.6 | 59 | | |

Rows: $F_{.05} = 4.02$; $F_{.01} = 7.12$
Columns: $F_{.05} = 3.17$; $F_{.01} = 5.01$
Interaction: $F_{.05} = 3.17$; $F_{.01} = 5.01$

As with the one-way anova, the within-groups variance estimate is used as the denominator for the $F$ ratio tests. The between-rows, between-columns, and interaction variance estimates are divided, respectively, by the within-groups variance estimate to obtain the $F$ ratios for testing these various effects. As Table 9–7 shows, the between-rows $F$ ratio was significant at the 0.05 level, the interaction $F$ ratio was significant at the 0.01 level, and the between-columns $F$ ratio was not significant. This would suggest that: (1) the discussion method was, in general, more effective than the lecture method; (2) there were no overall differences among teachers; and (3) some teachers do better with one teaching method than with the other.

If the overall $F$ ratio is significant, means of rows or columns may be tested, pair by pair, to locate significant differences. As with the one-way anova, multiple comparisons increase the probability of obtaining a result significant by chance if the ordinary $t$ test is applied. For this reason, it is recommended that some procedure, such as the Scheffé method, be applied to decide how large the $t$ ratio must be to be regarded as significant when several pairs of mean differences are being tested. The method described in connection with Problem 32 will serve for this purpose.

## Applications of the Two-Way Anova

The method just described is applicable to a wide variety of experiments many of which do not involve random selection of cases for each subgroup from the same common pool. One of the factors, rows or columns, might involve distinct categories of individuals. For example, random samples of males and females might be given several different treatments, or people of high, medium, and low intelligence might each be given four different study methods. In this case, four random samples of size $n$ would be drawn from people of high intelligence, four samples of size $n$ from people of medium intelligence, and four samples of size $n$ from people of low intelligence.

It is also possible to have quantitative levels both ways or categories both ways. People of high, medium, and low intelligence who are either high or low in weight could be tested on pain thresholds and a two-way anova run on the scores. Or, brown-eyed and blue-eyed people who are males or females could be tested on a visual acuity task to provide data for a two-way anova.

*Special Cases of the Two-Way Anova.* There are certain classes of experiments occasionally encountered that might appear on the surface to fit the model described here but which actually require a more complicated statistical treatment. If the conditions selected for the treatment in either rows or columns, or both, in the two-way anova represent a random selection from a large number of possibilities, the data must be treated in a different manner. Suppose a random selection of six major university varsity broad jumpers each made ten jumps with each of two types of track shoes under properly randomized and spaced conditions. The jump lengths would be the $X$ values. This would give a six row, two column two-way anova but the data should not be analyzed by the methods described in connection with Problem 33 since one of the two factors is based on a random selection of conditions (people, in this case). For experiments of this kind, the $F$ ratios do not necessarily use the within-groups variance estimate in the denominator. The reader will find appropriate methods for dealing with this situation in more advanced texts.

## Summary of Steps in Testing the Effects of Two Factors Simultaneously Using the Two-Way Anova

1.  Compute the sum of the raw scores, $\Sigma X$, and the sum of the squared raw scores, $\Sigma X^2$, for each sample separately. There are $k = (r \times c)$ samples, where $r$ is the number of rows and $c$ is the number of columns.
2.  Using the totals from Step 1 above, find the sum of the squared deviation scores, $\Sigma x^2$, for each sample separately by means of the formula $\Sigma x^2 = \Sigma X^2 - (\Sigma X)^2/n$, where $n$ is the number of cases in each sample.
3.  Add up the sums of squared deviation scores obtained in Step 2 for all samples to get the within groups sum of squares.
4.  Divide the within-groups sum of squares from Step 3 by the sum of the degrees of freedom for the $k$ separate samples, i.e., $k(n-1)$, to get the within-groups variance estimate, $s_W^2$.
5.  Using the totals from Step 1 above, combine the $\Sigma X$ values for each row and compute a row-mean for each row, $M_{r_i}$.
6.  Using the totals from Step 1 above, combine the $\Sigma X$ values for each column and compute a column-mean for each column, $M_{c_i}$.
7.  Using the totals from Step 1 above, combine the $\Sigma X$ values for all samples and compute a grand overall mean, $M$.
8.  Compute the row sum of squares by summing the values $(M_{r_i} - M)^2$ for each row and multiplying the total by $nc$, the sample size times the number of columns.
9.  Divide the row sum of squares obtained in Step 8 by $r-1$, the degrees of freedom for rows, to obtain the between-rows variance estimate, $s_R^2$.
10. Compute the column sum of squares by summing the values $(M_{c_i} - M)^2$ for each column and multiplying the total by $nr$, the sample size times the number of rows.
11. Divide the column sum of squares obtained in Step 10 by $c-1$, the degrees of freedom for columns, to obtain the between-columns variance estimate, $s_C^2$.
12. Using the totals obtained in Step 1 above, calculate the mean of each sample, $M_i$.

13. Obtain the between-groups sum of squares by adding up the $(M_i - M)^2$ values for all $k$ samples and multiplying the total by $n$, the sample size.

14. Using the totals from Step 1 above, calculate the total sum of squares by the formula $\Sigma x^2 = \Sigma\Sigma X^2 - (\Sigma\Sigma X)^2/N$, where $N$ is the number of cases in all samples combined and the double summations are over all $k$ samples.

15. Add the within-groups sum of squares from Step 3 to the between-groups sum of squares obtained in Step 13 and check to see if this sum equals the total sum of squares obtained in Step 14.

16. Calculate the interaction sum of squares by subtracting the between-rows sum of squares (Step 8) and the between-columns sum of squares (Step 10) from the between-groups sum of squares obtained in Step 13.

17. Check the interaction sum of squares independently using Formula 9–13.

18. Calculate the interaction variance estimate, $s^2_{R \times C}$, by dividing the interaction sum of squares by $(r-1)(c-1)$, the number of degrees of freedom for interaction.

19. Calculate the $F$ ratios by dividing the between-rows, between-columns, and interaction variance estimates by the within-groups variance estimate.

20. Enter the Table of the $F$ Distribution with the appropriate numbers of degrees of freedom to find the critical values of $F$, $F_{.05}$ and $F_{.01}$, for each of these $F$ ratio tests.

21. Determine the significance of each effect by comparing the obtained values of $F$ with the critical values of $F$.

### Mathematical Notes

It will be shown in this section that the total sum of squares for the two-way anova can be obtained by adding up the within-groups, between-rows, between-columns, and interaction sums of squares. As with the one-way anova, begin with a deviation score about the grand mean:

$$x = X_{ij} - M$$

$$x = X_{ij} + (M_i - M_i) + (M_{R_i} - M_{R_i}) + (M_{C_i} - M_{C_i}) - M$$

$$x = (X_{ij} - M_i) + (M_{R_i} - M) + (M_{C_i} - M) + (M_i - M_{R_i} - M_{C_i} + M)$$

Squaring both sides and summing from 1 to $N$ gives:

$$\sum_1^N x^2 = \sum_1^N (X_{ij} - M_i)^2 + \sum_1^N (M_{R_i} - M)^2 + \sum_1^N (M_{C_i} - M)^2$$
$$+ \sum_1^N (M_i - M_{R_i} - M_{C_i} + M)^2 + \text{cross product terms} \quad (9\text{--}14)$$

The first term on the right of Formula 9–14 is the within-groups sum of squares, since it is just $\sum_1^n x_1{}^2 + \sum_1^n x_2{}^2 + \cdots + \sum_1^n x_k{}^2$ (See Table 9–6). The second term on the right equals $nc \cdot \sum_1^r (M_{R_i} - M)^2$ since there are $nc$ cases that have the same $(M_{R_i} - M)$ value and there are only $r$ such values. This expression is the same as that for the between-rows sum of squares shown in Table 9–6. The third term on the right can be rewritten as $nr \cdot \sum_1^c (M_{C_i} - M)^2$ since there are $nr$ cases that have the same $(M_{C_i} - M)$ value and there are only $c$ such values. This expression is the same as that for the between-columns sum of squares shown in Table 9–6. The fourth term on the right of Formula 9–14 can be rewritten as $n \cdot \sum_1^k (M_i - M_R - M_C + M)^2$ since all the cases in the same subgroup will have the same $(M_i - M_R - M_C - M)^2$ value. This expression is the same as Formula 9–13, the sum of squares for the interaction variance estimate.

The total sum of squares on the left of Formula 9–14 equals the sum of the sums of squares within-groups, between-rows, between-columns, and for interaction plus the cross product terms. For the equation to hold as expected, these cross product terms must be equal to zero. To see that these cross product terms are zero, it is necessary to consider them within each subgroup. The first term with the others yields cross products that involve a constant times a deviation score, as with the one-way anova. When the constant is removed outside the summation sign, a sum of deviation scores is left inside and this equals zero. The other cross products involve cross product terms that are identical for every case within a subgroup, giving $n$ times that cross product for the group. For the $2 \cdot \sum_1^N (M_{R_i} - M)(M_{C_i} - M)$ cross product term this will give:

$$2[n(M_{R_1} - M)(M_{C_1} - M) + n(M_{R_1} - M)(M_{C_2} - M)$$
$$+ n(M_{R_1} - M)(M_{C_3} - M) + n(M_{R_2} - M)(M_{C_1} - M)$$
$$+ n(M_{R_2} - M)(M_{C_2} - M) + (M_{R_2} - M)(M_{C_3} - M)]$$

Combining terms:

$$2[n(M_{R_1} - M)(M_{C_1} - M + M_{C_2} - M + M_{C_3} - M)$$
$$n(M_{R_2} - M)(M_{C_1} - M + M_{C_2} - M + M_{C_3} - M)]$$

Since $M_{C_1} + M_{C_2} + M_{C_3} = 3M$, this term equals zero. The remaining cross product terms can be shown to equal zero in a similar fashion.

## Exercises

9-1. The data given below repreent ad libitum consumption totals in cc of 10 percent alcohol solution, 20 percent alcohol solution, and distilled water, respectively, in a specified time period by three groups of rats maintained on the same deficiency diet. Test to see if there are differences among these means at the 0.05 level using the one-way anova.

| Individual | 10% Alcohol | 20% Alcohol | Water |
|---|---|---|---|
| 1............ | 40 | 26 | 13 |
| 2............ | 45 | 20 | 16 |
| 3............ | 38 | 30 | 14 |
| 4............ | 37 | 24 | 20 |
| 5............ | 36 | 25 | 15 |
| 6............ | 30 | 19 | 17 |
| 7............ | 33 | 27 | 10 |
| 8............ | 38 | 25 | 14 |
| 9............ | 39 | 26 | 16 |
| 10............ | 35 | 26 | 15 |

9-2. A personality test was administered to samples of male (Group AM) and female (Group AF) applicants for employment in a given firm as well as to volunteer male (Group EM) and female (Group EF) employees, who were promised the test results would not affect their present or future opportunities with the company. Emotional stability scores on the test were converted to Stanine Scores. These scores for the four groups are given below. Perform a one-way anova to test for differences between means.

| Individual | AM | AF | EM | EF |
|---|---|---|---|---|
| 1........ | 8 | 6 | 7 | 5 |
| 2........ | 4 | 8 | 5 | 2 |
| 3........ | 9 | 6 | 4 | 6 |
| 4........ | 7 | 7 | 3 | 4 |
| 5........ | 9 | 4 | 4 | 9 |
| 6........ | 2 | 6 | 6 | 1 |
| 7........ | 5 | 3 | 2 | 4 |
| 8........ | 8 | 5 | 1 | 3 |
| 9........ | 6 | 8 | 8 | 3 |
| 10........ | 7 | 7 | 5 | 7 |
| 11........ | 4 | 2 | 6 | 4 |
| 12........ | 7 | 7 | 5 | 4 |

9-3. Check to see if the assumption of equal variances for the $F$ test in Exercise 9-1 is tenable.

9-4. Check to see if the assumption of equal variances for the $F$ test in Exercise 9-2 is tenable.

9-5. If it is appropriate, make individual $t$ tests for pairs of means in Exercise 9-1. What are the Scheffé $t'_{.05}$ and $t'_{.01}$ values?

9-6.   If it is appropriate, make individual $t$ tests for pairs of means in Exercise 9-2. What are the Scheffé $t'_{.05}$ and $t'_{.01}$ values?

9-7.   The data below represent scaled and rounded off blood levels of a certain hormone for normals, (A); nonpsychiatric patients, (B); and hospitalized schizophrenics, (C). Using the one-way anova test for differences among the means.

| Individual | A | B | C |
|---|---|---|---|
| 1......... | 13 | 12 | 9 |
| 2......... | 12 | 12 | 3 |
| 3......... | 12 | 10 | 8 |
| 4......... | 14 | 11 | 6 |
| 5......... | 16 | 14 | 11 |
| 6......... | 13 | 13 | 7 |
| 7......... | 9 | 10 | 5 |
| 8......... | 11 | 8 | 9 |
| 9......... | 12 | 13 | 8 |
| 10........ | 13 | 11 | 9 |
| 11........ | 14 | 12 | |
| 12........ | 13 | | |

9-8.   Rats are randomly assigned to four conditions, A, B, C, D. The rats are placed in Skinner Boxes. In group A, the rat receives a food pellet every time it presses the bar. In Group B, the rat receives a food pellet every other time it presses the bar; in groups C and D, the reinforcement rate drops to one in three and one in four, respectively. The data below give the number of bar presses in a given time period after the start of the experiment. Two rats in group B and one rat in group D died before completion of the experiment, leading to unequal numbers of subjects in each group. Test for differences among means using the one-way anova.

| Individual | A | B | C | D |
|---|---|---|---|---|
| 1......... | 8 | 13 | 14 | 18 |
| 2......... | 10 | 13 | 16 | 16 |
| 3......... | 14 | 16 | 20 | 20 |
| 4......... | 7 | 14 | 17 | 20 |
| 5......... | 3 | 12 | 10 | 19 |
| 6......... | 9 | 14 | 13 | 21 |
| 7......... | 10 | 6 | 16 | 19 |
| 8......... | 10 | 10 | 15 | 25 |
| 9......... | 11 | | 15 | 13 |
| 10........ | 8 | | 16 | |

9-9.   If it is appropriate, carry out individual $t$ tests for pairs of means in Exercise 9-7. What are the Scheffé $t'_{.05}$ and $t'_{.01}$ values?

9-10.  If it is appropriate, carry out individual $t$ tests for pairs of means in Exercise 9-8. What are the Scheffé $t'_{.05}$ and $t'_{.01}$ values?

9-11.  An investigator wishes to test the effect of three weight reducing programs and also to determine if sex is a factor to be considered.

He draws at random three samples of women and three random samples of men from the population being studied. The three weight reducing programs are Diet (D), Exercise (E), and Diet plus Exercise (DE). Each sample is also either male (M) or female (F). The number of pounds lost per subject in the trial period is shown below for each of the six samples. Using the two-way anova, test the main effects (row and column differences) and the interaction effect.

| Individual | MD | ME | MDE | FD | FE | FDE |
|---|---|---|---|---|---|---|
| 1 | 13 | 11 | 20 | 8 | 6 | 6 |
| 2 | 8 | 14 | 25 | 18 | 7 | 10 |
| 3 | 10 | 11 | 16 | 17 | 12 | 10 |
| 4 | 13 | 12 | 15 | 22 | 9 | 13 |
| 5 | 13 | 10 | 12 | 16 | 6 | 18 |
| 6 | 14 | 10 | 9 | 14 | 7 | 11 |
| 7 | 12 | 5 | 17 | 12 | 8 | 12 |
| 8 | 13 | 8 | 14 | 16 | 4 | 14 |
| 9 | 18 | 13 | 15 | 17 | 6 | 13 |
| 10 | 12 | 11 | 16 | 16 | 2 | 11 |

9–12.  In a certain coed liberal arts college, a controversy has arisen about the relative scholarly capacities of males and females and sorority and fraternity members as opposed to nonorganization students. To test these hypotheses, four random samples of 12 cases each are taken from senior fraternity men (MO), senior nonfraternity men (MN), senior sorority women (FS), and senior nonsorority women (FN). Grade point averages for these students are computed and rounded off to one decimal place and multiplied by 10 to give the $X$ scores below. Using the two-way anova, test the main effects for rows (male vs female), columns (org. vs nonorg.), and interaction.

| Individual | MO | MN | FO | FN |
|---|---|---|---|---|
| 1 | 24 | 29 | 31 | 29 |
| 2 | 25 | 28 | 27 | 35 |
| 3 | 20 | 32 | 30 | 33 |
| 4 | 27 | 23 | 26 | 31 |
| 5 | 37 | 29 | 29 | 39 |
| 6 | 30 | 27 | 27 | 27 |
| 7 | 23 | 30 | 33 | 32 |
| 8 | 27 | 26 | 25 | 30 |
| 9 | 26 | 36 | 28 | 21 |
| 10 | 31 | 30 | 30 | 32 |
| 11 | 28 | 29 | 35 | 31 |
| 12 | 25 | 33 | 28 | 30 |

9–13.  If justified, make $t$ tests for the data in Exercise 9–11, computing the Scheffé critical values $t'_{.05}$ and $t'_{.01}$. Interpret the results for the entire experiment.

# chapter 10

# Comparing Frequencies by Means of Chi Square

Many situations arise in social science research where it becomes important to compare two sets of frequencies for the purpose of ascertaining if they are significantly different from each other. One of these situations involves comparing obtained frequencies in a frequency distribution with theoretically determined frequency values to see if the discrepancies between the two are within tolerable limits. In another type of application, the distribution of native-born American caucasians into various blood types, for example, might be compared with the numbers of native-born American blacks in these same categories to test some hypothesis about inheritance. Wherever it is necessary to compare two sets of frequencies there is a good chance that chi square may be an appropriate statistic.

## Formula for Chi Square

The formula for chi square is the following:

$$\chi^2 = \sum \frac{(f_0 - f_t)^2}{f_t} \tag{10-1}$$

Where:

$$\chi^2 = \text{chi square}$$
$$f_0 = \text{an obtained frequency}$$
$$f_t = \text{a theoretical frequency}$$

In the application of chi square where obtained frequencies are compared with theoretically determined frequencies, the $f_0$ values are the actual obtained frequencies in the distribution and the $f_t$ values are the corresponding theoretical frequencies. In applying Formula 10-1 to a problem of this kind, the difference between each obtained

236

and theoretical frequency is squared and divided by the theoretical frequency $f_t$. These values are determined for each category and summed to give the overall chi square value for the entire frequency distribution.

## The Distribution of Chi Square

Examination of Formula 10–1 readily reveals that $\chi^2$ would be zero if the obtained frequencies for a given distribution exactly equalled the theoretical frequencies, since all the $(f_0 - f_t)$ values would be zero. If the obtained frequencies depart to some extent from the theoretical frequencies, as they would in almost every case even if the sample is drawn at random from a normally distributed population, then chi square would be some positive number greater than zero. The squaring of $(f_0 - f_t)$ removes the negative sign from any negative $(f_0 - f_t)$ value. It can be seen that chi square will increase in size as the departure of the $f_0$ and the $f_t$ values become more pronounced.

A distribution of chi square values for a particular number of class intervals could be developed empirically as follows. Draw a sample of 200 cases from a very large, normally distributed population and determine the number of cases falling into each of a specified set of class intervals. All cases falling above or below the end intervals will be placed in the end intervals. Then, determine the theoretical normal curve frequencies by the methods outlined in Chapter 5 and compute the chi square value by Formula 10–1. Repeat this experiment, taking another 200 cases, and compute another chi square. Do this over and over again a very large number of times. Set up a frequency distribution that shows how many times each chi square value was obtained. Convert this to a smoothed curve which can be used to determine the proportion of chi square values that fall above, below, or between any two specified chi square values.

This process would be analogous to an empirical determination of the sampling distribution of means, as described in Chapter 8. In practice, of course, the sampling distribution of means is not determined empirically in such a manner. In the case of chi square, it is also unnecessary to determine the distribution of chi square empirically in the manner specified. It is possible to prove mathematically that under proper conditions the distribution takes a particular mathematical form. This distribution is called the "Chi Square Distribution."

The normal distribution was shown to be a family of curves, requiring specification of $M$ and S.D. to designate a particular member of the family. The $t$ distribution was also described in Chapter 8 as a family of curves requiring specification of $M$, S.D., and $df$, the degrees of freedom, to designate a particular $t$ distribution. Of course it is the standard normal curve, the one with $M=0$ and S.D.$=1$, that is tabled in the back of the book. In the Table of the $t$ Distribution, selected values of $t$ are given from the standard $t$ distributions for different values of $df$, the degrees of freedom.

The chi square distribution also is a family of curves, not just one curve. There is a different curve for each value of $df$, the number of degrees of freedom. In the experiment described above as a method for empirically determining the chi square distribution, $df$, the number of degrees of freedom, was held constant for each determination of $\chi^2$ and hence the total obtained distribution would be appropriate for that particular value of $df$ associated with that situation. The method of determining the number of degrees of freedom for an application of chi square varies from situation to situation. In comparing obtained and normal curve frequencies, the number of degrees of freedom, $df$, depends on the number of class intervals in the distribution (see Problem F).

A different theoretical chi square distribution can be determined mathematically for each value of $df$. Table D (Appendix A) gives the critical values of chi square for the various numbers of degrees of freedom. A specified percentage of obtained chi square values will be larger than a critical value by chance. For example, with 10 degrees of freedom, the value of chi square in Table D for $P = 0.05$ is 18.307. In testing normal curve frequencies against obtained sample frequencies where the $df$ value is 10, only five percent of the time will the obtained chi square value be larger than 18.307 if the samples are drawn at random from a normally distributed population.

*Testing the Null Hypothesis.* In making an application of chi square, the typical procedure is to determine the number of degrees of freedom in the situation at hand, look up in Table D the value of chi square for the 0.05 or 0.01 level, or whatever standard is being used, and then determine if the computed chi square is bigger than the critical value, e.g., $\chi^2_{.05}$ or $\chi^2_{.01}$. If the obtained $\chi^2$ is greater than or equal to $\chi^2_{.05}$, for example, for that number of degrees of freedom, the null hypothesis is rejected at the 0.05 level. If the obtained chi square is less than $\chi^2_{.05}$, the null hypothesis is not rejected. Of course, the one percent level value of $\chi^2$, $\chi^2_{.01}$, could be used as the critical value, or any other tabled value of $\chi^2$. The most commonly used critical values are, of course, $\chi^2_{.05}$ and $\chi^2_{.01}$. Since the differences $(f_0 - f_t)$ are squared, obscuring whether $f_0$ was greater than $f_t$, or vice versa, $\chi^2$ is always positive and the test is always a two-tailed test.

Calculating $\chi^2$ in the proper fashion and determining the correct number of degrees of freedom involve certain complications depending on the particular application of chi square. Some of the most commonly encountered contingencies will be described in connection with the individual problems to be presented below.

## Problem 34
## TEST THE DIFFERENCE BETWEEN OBTAINED AND
## THEORETICAL FREQUENCY DISTRIBUTION VALUES

When observations are being taken at random from a specified population, they can be expected to fall into certain categories with frequencies based on those in the population. Or, if observations are being determined according to a certain model, the distribution of those observations into categories should follow a pattern implied by the model. The chi square test can be used to determine whether or not the obtained distribution of observations into the available categories conforms to expectations, based on the theoretical distribution or model. As the discrepancies between obtained and expected frequencies increases, the computed chi square will increase in size. When the computed chi square exceeds the critical chi square value for the particular situation being investigated, then it must be concluded that the obtained observations are not distributing themselves in accordance with the theoretical distribution or model being tested.

*An Example.* A college instructor has been using multiple-choice questions with five possible answers for the examinations in his large introductory course. His intention in writing the questions is to avoid any systematic preference for *a*, *b*, *c*, *d*, or *e* as the correct answer position. He decides to conduct an investigation to see if in fact he has successfully avoided any systematic bias in his choice of correct and incorrect answer positions. He counts 435 items in his question file with the following distribution of correct answers: $a = 89$, $b = 72$, $c = 95$, $d = 105$, $e = 74$. If the response alternative positions were equally preferred, there would have been 435/5, or 87, questions with correct answers in each of the five response positions. Clearly, the numbers of questions in the various categories depart from the theoretical expected value of 87 in each category. The question to be answered by the chi square test is whether these departures are large enough to represent a statistically significant bias.

Applying Formula 10–1 to this problem gives:

$$\chi^2 = \frac{(89-87)^2}{87} + \frac{(72-87)^2}{87} + \frac{(95-87)^2}{87} + \frac{(105-87)^2}{87} + \frac{(74-87)^2}{87}$$

$$\chi^2 = \frac{4}{87} + \frac{225}{87} + \frac{64}{87} + \frac{324}{87} + \frac{169}{87}$$

$$\chi^2 = \frac{786}{87} = 9.03 = 9.0$$

*Degrees of Freedom.* In this situation there are five categories of response into which the observations are distributed. The only constraint on these frequencies is that they must add up to 435, the total number of items. No matter what frequencies occur in four of the cells, the total may be kept equal to 435 by making the last cell frequency to be 435 minus the sum of the other cell frequencies. Thus, once four cell frequencies are determined, the last cell frequency is fixed by the requirement that the cell frequencies must add up to 435. Since four cell frequencies can vary freely, there are four degrees of freedom for this chi square test. Entering Table D (Appendix A), the critical values of $\chi^2$ with four *df* are found to be:

$$\chi^2_{.05} = 9.488 \quad \text{and} \quad \chi^2_{.01} = 13.277.$$

*Interpretation of Results.*  Since the obtained $\chi^2$ value of 9.0 was less than $\chi^2_{.05} = 9.488$, the five percent level critical value of chi square for four degrees of freedom, it is not possible to reject the null hypothesis in this situation. It cannot be concluded, therefore, that the instructor is departing from random choice of correct answer position in writing his test items. The closeness of the obtained chi square value to the five percent level critical value of chi square would tend to suggest, however, that there is a strong possibility he may be biased in his choices. In particular, the frequency for the *d* category is elevated so he should probably make a conscious effort to use it somewhat less frequently than he has in the past while using *b* and *e* somewhat more often.

## Summary of Steps in Testing the Difference between Obtained and Theoretical Frequency Distribution Values

1.  Determine the theoretical frequency for each category in the frequency distribution, making certain that the sum of these values equals the sum of the actual obtained frequencies.
2.  If any of the theoretical frequencies is less than 10, if possible combine adjacent categories in both theoretical and obtained frequency distributions so as to eliminate categories with small theoretical frequencies. Do not compute chi square by this method if categories cannot be combined to eliminate all theoretical frequencies less than five.
3.  Take the difference between the theoretical and the obtained frequency for each category in the frequency distribution.
4.  For each category, square the difference obtained in Step 3 above and divide it by the theoretical frequency value.
5.  Add up the values computed in Step 4 to get chi square, $\chi^2$.
6.  Using the appropriate number of degrees of freedom, usually one less than the number of categories in the frequency distribution, find the critical values of chi square, $\chi^2_{.05}$ and $\chi^2_{.01}$.
7.  If the value of $\chi^2$ obtained in Step 5 above is greater than $\chi^2_{.05}$ but less than $\chi^2_{.01}$, reject the null hypothesis at the five percent

level of confidence. If the obtained value of $\chi^2$ is greater than $\chi^2_{0.01}$, reject the null hypothesis at the one percent level of confidence. If the obtained value of $x^2$ is less than $\chi^2_{.05}$, do not reject the null hypothesis.

## Problem 35
## TEST THE ASSOCIATION BETWEEN TWO VARIABLES USING FREQUENCIES IN A FOUR-FOLD TABLE

When variables are measured on a continuous, normally distributed scale, the best way to determine whether or not they are related to each other is to compute the product-moment correlation coefficient between them, assuming that the regression line of best fit is a straight line. Many variables of interest, however, are not continuously measured on a normally distributed scale. In fact many such variables are dichotomous, i.e., in only two categories. Such two-category variables include sex membership, high school graduate vs not, Republican vs Democrat, serviceman vs civilian, employed vs not, and so on. Many social science research projects involve the study of such dichotomous variables in relation to each other and to questionnaire data, themselves often in dichotomous form. Chi square is an appropriate statistic for determining whether or not there is a significant association between such variables.

*An Example.* Table 10–1 shows a worked-out example of the application of $\chi^2$ to a fourfold table. The data represent frequencies of *yes* and *no* replies to a questionnaire item by male and female respondents in a randomly selected sample of voters. It is clear that men answered this question very differently from women, but the question remains whether the difference is significant. The null hypothesis, $h_0$, in this situation would be that there is no association between sex of respondent and opinion on this issue as manifested by replies to a questionnaire item. Chi square can be used to determine if the frequencies of reply in the *yes* and *no* categories vary significantly or not between men and women.

*Determining the Theoretical Frequencies.* If there is no difference between men and women in their replies, the split between *yes* and *no* replies should be the same in both groups. The $f_t$ values are determined in such a way as to make this occur. This can be done in various ways. The simplest procedure is to multiply the column total by the row total for the cell in question and then divide by the overall $N$. In Table 10–1, therefore, the $f_t$ for the upper left-hand cell is found by $(91 \times 94)/178 = 48.1$; the upper right-hand cell $f_t$ is given by $(87 \times 94)/178 = 45.9$; the lower left cell $f_t$ is $(91 \times 84)/178 = 42.9$; and the lower right cell $f_t$ is $(87 \times 84)/178 = 41.1$.

The method of obtaining the $f_t$ values described above is just a method of making the *yes–no* splits equal proportionally in the males and females

TABLE 10–1
Computing Chi Square in a Four-Fold Table

| | Obtained Frequencies $(f_0)$ | | |
| | Males | Females | Total |
|---|---|---|---|
| Yes.............. | 32 | 62 | 94 |
| No.............. | 59 | 25 | 84 |
| Total............ | 91 | 87 | 178 |

| | Expected Frequencies $(f_t)$ | | |
| | Males | Females | Total |
|---|---|---|---|
| Yes.............. | 48.1 | 45.9 | 94.0 |
| No.............. | 42.9 | 41.1 | 84.0 |
| Total............ | 91.0 | 87.0 | 178.0 |

$$\chi^2 = \sum \frac{(f_0 - f_t)^2}{f_t}$$

$$\chi^2 = \frac{(32.0 - 48.1)^2}{48.1} + \frac{(59.0 - 42.9)^2}{42.9} + \frac{(62.0 - 45.9)^2}{45.9} + \frac{(25.0 - 41.1)^2}{41.1}$$

$$\chi^2 = \frac{(-16.1)^2}{48.1} + \frac{(16.1)^2}{42.9} + \frac{(16.1)^2}{45.9} + \frac{(-16.1)^2}{41.1}$$

$$\chi^2 = 5.39 + 6.04 + 5.65 + 6.31$$

$$\chi^2 = 23.39$$

$$\chi^2 = 23.4**$$

With 1 $df$, $\chi^2_{.05} = 3.84$, $\chi^2_{.01} = 6.64$

columns, although the frequencies themselves are not equated since the number of males overall is different from the number of females. The ratios 48.1/42.9, 45.9/41.1, and 94/84 all equal the same value, 1.12. Also, 48.1/91.0 = 45.9/87.0 = 94.0/178.0 and 42.9/91 = 41.1/87.0 = 84/178.0. Corresponding ratio relationships among the $f_t$ values in Table 10–1 hold for the rows. That is, 48.1/45.9 = 42.9/41.1, 48.1/94.0 = 42.9/84.0 = 91.0/178.0, and 45.9/94.0 = 41.1/84.0 = 87.0/178.0. These proportionality relationships provide the basis for obtaining the correct $f_t$ for any cell by multiplying the row and column totals for a cell and dividing the product by the overall $N$.

Once the table of $f_t$ values is determined, as shown in Table 10–1, the computation of chi square proceeds by taking the difference between the $f_0$ and $f_t$ values for each cell. These differences are squared, divided by the $f_t$ values and summed to obtain $\chi^2$, as shown at the bottom of

Table 10–1. The value of $\chi^2$ is ordinarily rounded to one decimal place unless it is very close to a critical value. In this example, $\chi^2 = 23.4$, a substantial value for a four-fold table.

*Determining the Number of Degrees of Freedom.* Before $\chi^2_{.05}$ and $\chi^2_{.01}$ can be determined from the table of chi square (Table D, Appendix A), it is necessary to determine the number of degrees of freedom for the particular situation being studied. In the case of a four-fold table, it is always true that $df = 1$. It can be seen intuitively that this is true by considering the table of frequencies in Table 10–1. With the row and column totals fixed, the specification of any one of the four cell frequencies automatically determines what the others must be since the other cell frequencies are then determinable as the difference between the row or column total and the already fixed cell frequency. Since only one cell frequency can be specified without fixing the others, there is only one degree of freedom in the four-fold table situation. In the example shown in Table 10–1, $\chi^2$ is significant beyond the one percent level of confidence since with one degree of freedom, $\chi^2_{.01}$ is 6.64 and the obtained $\chi^2$ is 23.4. It is marked with two asterisks (**) in Table 10–1 to indicate significance at the 0.01 level.

*When Theoretical Frequencies Are Small.* As with other comparisons of obtained and theoretical frequencies, the use of chi square in the four-fold table situation requires theoretical frequencies of adequate size. Ten is a good minimum to use as a guide. In planning a research investigation, every effort should be made to insure that no $f_t$ will be less than 10 if chi square is to be used to evaluate the results.

There is a gradual deterioration in the fit of the chi square distribution to the data as the size of the smallest $f_t$ value decreases. This deterioration does not take place suddenly at 10 as a sharp dividing point. It begins slowly long before 10 and continues to increase in severity as the expected frequency for a cell drops below 10. Some investigators feel that it is still appropriate to apply chi square as long as no theoretical frequency is less than five. In the case where one or more $f_t$ values are between 5.0 and 10.0, it is common to apply Yates' correction[1] which involves subtracting 0.5 from the absolute value (ignoring the sign) of each $(f_o - f_t)$ value. This has the effect of reducing the overall size of the obtained chi square, making it more difficult to reject $h_o$. This correction is applied only in $2 \times 2$ tables. In Table 10–1, if it were necessary to apply Yates' correction, all the $(f_o - f_t)$ values would be 15.6 (after removing the sign of the difference) instead of 16.1. Of course, no $f_t$ value in Table 10–1 is less than 10 so Yates' correction is unnecessary.

When the $f_t$ value for any cell drops below 5.0, chi square definitely

---

[1] F. Yates, "Contingency tables involving small numbers and the $\chi^2$ test." *Suppl. J. Royal Stat. Soc.,* 1 1934, 217–235.

should *not* be computed by the methods described here. It is possible, however, to carry out an "exact" chi square test when one (or more) of the $f_t$ values in a four-fold table is less than 5.0. The reader should consult a more advanced text if such a test is needed. It should be noted that these restrictions on the use of the chi square test apply to the theoretical frequencies, not to the obtained frequencies. Some $f_o$ values can be small without invalidating the chi square test as long as the $f_t$ values are large enough.

## Summary of Steps in Testing the Association between Two Variables Using Frequencies in a Four-Fold Table

1.  Arrange the obtained frequencies in a four-fold table for the four possible outcomes, e.g., high on both variables, high on one but low on two, low on one but high on two, and low on both variables.
2.  Determine the theoretical frequency for each cell multiplying the column sum by the row sum in the four-fold table and dividing this product by the sum of the frequencies in all four cells, $N$.
3.  For each cell, take the difference between the obtained and theoretical frequencies, square the difference, and divide by the theoretical frequency.
4.  Sum the four values obtained in Step 3 above to obtain the value of $\chi^2$.
5.  If any theoretical frequency obtained in Step 2 above is between 5.0 and 10.0, reduce the absolute size of the difference between all four obtained and theoretical frequencies in Step 3 above by 0.5 before squaring (Yates' correction). If any theoretical frequency is less than 5.0, do not compute $\chi^2$ by this method.
6.  Using one degree of freedom, determine the critical values of chi square, $\chi^2_{.05}$ and $\chi^2_{.01}$.
7.  Test the value of $\chi^2$ obtained in Step 4 above against the critical values of $\chi^2$ to determine whether the null hypothesis is to be rejected or not.

## Problem 36
## TEST THE ASSOCIATION BETWEEN TWO VARIABLES USING FREQUENCIES IN LARGER CONTINGENCY TABLES

The four-fold table in Problem 35 had only two categories for each variable, e.g., *yes–no* response for *males* and *females*. Many tables occur in which one or both variables have more than two categories. Responses to questions, for example, could fall into *yes, no,* or *?,* instead of just *yes* or *no*. Political affiliation could be classified as *Democrat, Republican,* or *Independent*. For these multiple category investigations of association between noncontinuous variables, chi square is often the only practical statistic to use.

*An Example.* In a large organization with members representing a wide range of educational backgrounds, an anonymous questionnaire is circulated asking for political affiliation and amount of educated completed, among other things. One hundred and ninety completed questionnaires are returned out of 250 mailed out. The investigator wishes to study the relationship between educational level and political affiliation. Educational level is broken down into four categories, (1) those who have only an elementary school education, (2) those who have completed high school, (3) those who have a bachelor's degree or equivalent, and (4) those with an advanced degree of some kind. Political affiliation is classified into the two major parties with all others going into the *other* category. Each of the 190 respondents is placed into one and only one of the 12 possible cells created by this joint classification of educational level and political affiliation. The frequencies for the 190 cases in these 12 cells are shown in Table 10–2 uner *Obtained Frequencies*.

The theoretical frequencies, $f_t$, are shown below the obtained frequencies, $f_o$, in Table 10–2. These are calculated in the same way they were calculated in Problem 35. For a given cell, the cell's row total is multiplied by its column total and the product is divided by the overall $N$. In Table 10–2, for the upper left-hand cell, the $f_t$, value, then, is given as $f_t = (35 \times 90)/190 = 16.6$; for the cell below it, $f_t = (35 \times 60)/190 = 11.0$; for the lower right-hand cell, $f_t = (40 \times 40)/190 = 8.4$. The $f_t$ values for the other cells are found in the same manner.

The calculations for chi square are shown at the bottom of Table 10–2. Each discrepancy between corresponding $f_o$ and $f_t$ cell frequencies is squared and divided by its own $f_t$ value. These figures are summed for the 12 cells to obtain the overall chi square value, $\chi^2 = 14.9$ in this case.

*Degrees of Freedom in Larger Contingency Tables.* In the four-fold table, the degrees of freedom is always 1, since only one cell frequency can be specified without fixing all the other cell frequencies, holding constant the row and column totals. In contingency tables larger than $2 \times 2$, the degrees of freedom may be calculated by the rule, $df = (r-1) \cdot (c-1)$, where $r$ is the number of rows and $c$ is the number of columns in the contingency table. Only the rows and columns of the table representing variable categories count. Thus, in Table 10–2, there are three row categories and four column categories, so $df = (3-1) \cdot (4-1) = 6$ for this contingency table. In a four-fold table, this rule would give $df = (2-1) \cdot (2-1) = 1$, which is, of course, consistent with what has already been stated about degrees of freedom in that situation.

TABLE 10–2
Computing Chi Square in a Multiple Contingency Table

| | *Obtained Frequencies* $(f_0)$ | | | | |
|---|---|---|---|---|---|
| | *Elementary School* | *High School* | *College Degree* | *Advanced Degree* | *Total* |
| Democrat......... | 16 | 36 | 24 | 15 | 90 |
| Republican....... | 15 | 19 | 13 | 9 | 60 |
| Other............ | 4 | 10 | 13 | 16 | 40 |
| Total............ | 35 | 65 | 50 | 40 | 190 |

| | *Theoretical Frequencies* $(f_t)$ | | | | |
|---|---|---|---|---|---|
| | *Elementary School* | *High School* | *College Degree* | *Advanced Degree* | *Total* |
| Democrat....... | 16.6 | 30.8 | 23.7 | 18.9 | 90 |
| Republican...... | 11.0 | 20.5 | 15.8 | 12.7† | 60 |
| Other.......... | 7.4 | 13.7 | 10.5 | 8.4 | 40 |
| Total.......... | 35.0 | 65.0 | 50.0 | 40.0 | 190 |

| *Cell* | $(f_0 - f_t)$ | $(f_0 - f_t)^2$ | $(f_0 - f_t)^2/f_t$ |
|---|---|---|---|
| 1........ | −0.6 | 0.36 | 0.00 |
| 2........ | 4.0 | 16.00 | 1.45 |
| 3........ | −3.4 | 11.56 | 1.56 |
| 4........ | 5.2 | 27.04 | 0.88 |
| 5........ | −1.5 | 2.25 | 0.11 |
| 6........ | −3.7 | 13.69 | 1.00 |
| 7........ | 0.3 | 0.09 | 0.00 |
| 8........ | −2.8 | 7.84 | 0.50 |
| 9........ | 2.5 | 6.25 | 0.60 |
| 10........ | −3.9 | 15.21 | 0.80 |
| 11........ | −3.7 | 13.69 | 1.08 |
| 12........ | 7.6 | 57.76 | 6.88 |
| | | | $\chi^2 = 14.86*$ |

$df = 3 \times 2 = 6; \chi^2_{.05} = 12.59; \chi^2_{.01} = 16.81$

† This $f_t$ had to be raised from 12.6 to 12.7 to make $\Sigma f_t = \Sigma f$. This discrepancy was due to rounding error.

That the rule $df = (r - 1) \cdot (c - 1)$ holds can be grasped intuitively by a consideration of Table 10–2. If two of the cell frequencies in the first column are fixed arbitrarily, the third one is automatically determined since the first column total must be 35. This gives two $df$ for the first column. The same holds for each of the next two columns, adding four more $df$. When the frequencies have been determined for the first three columns, with a total of six $df$ accumulated, no more degrees of freedom are available because the frequencies in the last column must be such as to make the row totals come out right. Analysis of contingency tables of various sizes will convince the reader that the $df = (r - 1) \cdot (c - 1)$ always gives the correct number of degrees of freedom.

*Interpreting the Results.* With six degrees of freedom, the values of chi square for significance at the 0.05 and 0.01 levels of significance, respectively, are (see Table D, Appendix A) $\chi^2_{.05} = 12.59$ and $\chi^2_{.01} = 16.81$. When the obtained value of $\chi^2$ for the data in Table 10-2, $\chi^2 = 14.9$, is compared with these critical values, it is found to be greater than $\chi^2_{.05}$ and less than $\chi^2_{.01}$. It is significant at the 0.05 level, therefore. This fact is indicated in Table 10-2 by one asterisk (*) beside the obtained chi square value.

It will be noted that two of the $f_t$ values in Table 10-2 are less than 10, one 7.4 and the other 8.4. This is not an ideal situation. It would have been better if the *other* category had more cases so the $f_t$ values would all have been 10 or higher. The overall $N$ is a good size, however, and there are 10 other cells with $f_t$ values greater than 10 so it probably is not unreasonable to apply chi square to this table, especially since the two low $f_t$ values are above 7.0. The obtained chi square is about half way between the 0.05 and 0.01 critical values, so to consider the result significant at the 0.05 level is probably not unreasonable even if the rule of "no $f_t$ below 10" was violated.

The significant obtained chi square suggests that there is a relationship between political preference and educational background in the population sampled, although the effect on the results of those who did not respond to the questionnaire remains in doubt. The significance of the chi square value merely indicates the presence of some type of relationship without telling what kind. It could be linear or curvilinear. Inspection of the contributions to chi square from the 12 different cells shows only one to be large, the lower right-hand cell. The number of individuals in this organization with advanced degrees who do not prefer either of the major political parties is greater than expected. Elsewhere in the table there appears to be no noteworthy discrepancy between obtained and expected frequencies. Since the one large discrepancy in this table was for one of the two cells with a low $f_t$ value, a further note of caution must be injected into any statement of conclusions drawn from this study.

## Summary of Steps in Testing the Association between Two Variables Using Frequencies in Larger Contingency Tables

1. Prepare an $r \times c$ contingency table of obtained cell frequencies where $r$ is the number of categories in the first variable and $c$ is the number of categories in the second variable.

2.  Compute the theoretical frequency for each cell of the contingency table obtained in Step 1 above by multiplying the row sum by the column sum and dividing by $N$, the sum of frequencies over the entire table.
3.  If any theoretical frequency is less than 10.0, combine categories to eliminate the small theoretical frequency if possible. Do not compute $\chi^2$ if categories cannot be combined to eliminate cells with theoretical frequencies less than 5.0.
4.  Take the difference between the obtained and theoretical frequencies for each cell in the contingency table, square the difference, and divide by the theoretical frequency.
5.  Sum the values computed in Step 4 to obtain $\chi^2$.
6.  Using $(r - 1)(c - 1)$ degrees of freedom, find the critical values of chi square, $\chi^2_{.05}$ and $\chi^2_{.01}$.
7.  Test the value of $\chi^2$ obtained in Step 5 above against the critical values of $\chi^2$ to determine whether the null hypothesis is to be rejected or not.

### Common Errors in the Use of Chi Square

Chi square is widely used, and often incorrectly so. Some of the more common errors to be avoided are the following:

1.  *Small Theoretical Frequencies.* It has already been emphasized that theoretical frequencies preferably should not be less than 10.0 in any cell and certainly under no circumstances less than 5.0. Even if one of the $f_t$ values is as low as 10.0, the others should be considerably larger. It is best to plan the study so that all $f_t$ values will be well above 10.0.

2.  *Failure to Equate Theoretical and Obtained Frequencies.* Attempts have been made to compare two sets of frequencies by means of chi square where the sums of $f_o$ and $f_t$ are different. This is unacceptable since it promotes larger discrepancies than would normally occur by chance, thereby invalidating the test.

3.  *Improper Categorization.* Small categories often must be combined with other categories to increase the $f_t$ values over the minimum. Also, variables on a more or less continuous scale are often broken down into a limited number of categories to make a chi square test. The decisions about categorization in such cases must not be made with an eye to increasing the size of the chi square value. Some uninformed individuals may even try out different categorizations to find the one that gives the highest chi square value before "deciding" how to categorize the data. Such decisions should be made in advance of data collection if possible, or at least on a rational basis that can be justified. In no case is it permissible to make such decisions in such a way as to enhance the likelihood of supporting one's favored hypothesis.

4.   *Lack of Independence.* A common and often subtle error in the application of chi square involves the situation where each subject frequency is not independent of each other data observation. Each of the $N$ trials must be free of dependence on any of the other trials for chi square to be appropriate. This requirement can be violated in many ways. In evaluating sex in relation to opinion on a given political question, for example, if subjects include husbands and wives, there is apt to be a linkage between the opinions of married individuals, invalidating the chi square test. Some investigators have applied chi square where several observations are taken from each of a number of individuals. This type of situation is very apt to introduce a lack of independence based on linkages between the trials for the same individual. For chi square to be appropriate, knowledge of the results of one trial should not enable the investigator to say what is apt to happen on some other trial because of any connection between those particular trials, e.g., they are for the same or related people. Where individuals are the source of data, it is best to have only one frequency from each subject and not have the subjects related to each other in any way that would determine the location of their frequency tallies in the contingency table.

## Problem F
### TEST THE DIFFERENCES BETWEEN OBTAINED AND NORMAL CURVE FREQUENCY DISTRIBUTION VALUES

In this problem, differences between obtained and normal curve frequencies are tested to ascertain whether or not it is reasonable to suppose that the obtained distribution could have been drawn by random sampling from a normally distributed population. If the computed chi square value proves to be larger than the selected critical value, e.g., $\chi^2_{.05}$ or $\chi^2_{.01}$, the null hypothesis is rejected and the obtained distribution cannot be considered to vary only negligibly from normality in shape. On the other hand, if the obtained chi square value is less than the selected critical value of $\chi^2$, then the null hypothesis is not rejected and the discrepancies between the obtained and the normal distribution are treated potentially or conceivably as being due to chance. Of course, a nonsignificant $\chi^2$ does not *prove* the null hypothesis any more than a nonsignificant difference between means *proves* the null hypothesis in that situation.

*An Example.* Application of the chi square test to this type of situation will be illustrated by means of a worked-out example. Methods for computing normal curve frequencies corresponding to obtained frequencies were described in Problem 15, Chapter 5. Table 5–3 gives an example of a score distribution for which normal curve frequencies have been computed both by the ordinate method and by the area method.

TABLE 10–3
Computation of Chi Square for the Difference between Obtained and Normal
Curve Frequencies

| $X$ | $f_o$ | $f_t$ | $(f_o - f_t)$ | $(f_o - f_t)^2$ | $(f_o - f_t)^2/f_t$ |
|---|---|---|---|---|---|
| 35–54.......... | 17 | 16.4 | 0.6 | 0.36 | 0.02 |
| 30–34.......... | 10 | 9.5 | 0.5 | 0.25 | 0.03 |
| 25–29.......... | 8 | 8.8 | −0.8 | 0.64 | 0.07 |
| 5–24.......... | 10 | 10.3 | −0.3 | 0.09 | 0.03 |
| Σ.............. | 45 | 45.0 | 0.0 | | $\chi^2 = 0.15$ N.S. |

With $(k - 3) = 4 - 3 = 1$ $df$, $\chi^2_{.05} = 3.84$, $\chi^2_{.01} = 6.64$

Table 10–3 reproduces the actual and theoretical frequencies for the distribution shown in Table 5–3 using the actual frequencies to obtain the $f_o$ values and the area method normal curve frequencies to obtain the $f_t$ values. Two important changes should be noted between Table 5–3 and Table 10–3. First of all, it is necessary to modify the area method frequencies so that they add up to the same number total as the obtained frequency total, i.e., $N$. In this case, the area method frequencies total is 44.9, only 0.1 off from the actual frequency total. This discrepancy is due to rounding error. It will be eliminated here by arbitrarily adding 0.1 to the top frequency of 1.1, making it 1.2. The 0.1 could have been added to any other area method frequency, but preferably not to reduce the obtained $(f_o - f_t)$ value.

The second, and more critical, difference between the values in Tables 5–3 and 10–3 is that some of the intervals at both the top and bottom of the distribution have been combined in Table 10–3. Thus, the intervals 50–54, 45–49, 40–44, and 35–39 were combined, summing their obtained frequencies of 1, 2, 5, and 9, respectively, to obtain the $f_o$ value of 17 for the interval 35–54 in Table 10–3. The corresponding area method normal curve frequencies for these intervals, 1.2, 2.3, 5.0, and 7.9, were summed to give the $f_t$ value of 16.4 for the interval 35–54 in Table 10–3. Correspondingly, the bottom intervals, 5–9, 10–14, 15–19, and 20–24 were combined to give $f_o$ and $f_t$ values of 10 and 10.3, respectively.

These intervals at the top and bottom of the distribution were combined because the $f_t$ values were too small for these end intervals.

To maintain all theoretical frequencies at 10.0 or greater requires in most cases the combining of at least some end intervals in the distribution, especially when the overall $N$ is not very large. In the case of Table 10–3, a strict adherence to the rule of 10.0 for $f_t$ values would also have required the combination of the two intervals 25–29 and 30–34. In this example, the overall $N$ is too small, making it impossible to obtain large enough $f_t$ values for an ideal use of the chi square test. The $f_t$ values of 9.5 and 8.8 for the two middle intervals in Table 10–3 are large enough, however, that the $\chi^2$ value should be usable unless it falls near the critical value.

After the intervals have been combined, as shown in Table 10–3, the next step is to take the differences between the obtained and theoretical frequencies $(f_0 - f_t)$, square them, and divide by their theoretical frequencies. These computations are shown in Table 10–3 with the sum of the last column being the chi square value. In this case, the $\chi^2$ was 0.15, a very small value since $f_0$ and $f_t$ values were very close to each other.

*Determining the Number of Degrees of Freedom.*    It is necessary to determine the number of degrees of freedom in this situation before the table of chi square can be used to find $\chi^2_{.05}$ and $\chi^2_{.01}$. These critical values of $\chi^2$, when obtained, will provide the standard of comparison against which the obtained $\chi^2 = 0.15$ will be compared. If the obtained $\chi^2$ is less than $\chi^2_{.05}$, as it is expected to be in this case, then $h_0$ cannot be rejected and there is no significant difference between the obtained and the corresponding normal curve frequencies.

If on the other hand the discrepancies between the $f_0$ and $f_t$ values had been large, such as to give an obtained chi square as large or larger than $\chi^2_{.05}$ or $\chi^2_{.01}$, then it would be appropriate to reject $h_0$. This would imply that the obtained distribution did not arise as a result of random sampling from a normal distribution with the specified $\mu$ and $\sigma$.

To determine the number of degrees of freedom in this situation it is necessary to see how many data values can be specified at liberty before the other values are of necessity fixed by the restraints created by the column total and the demand that the frequencies assigned result in the mean and standard deviation actually found in the sample. One degree of freedom is lost by the requirement that the assigned frequencies add up to $N$ since, if all but one of the frequencies had been fixed, the last one would be predetermined to make the frequency total add up to $N$. This reduces the $df$ to $k - 1$, where $k$ is the number of class intervals in this case. The requirement that the frequencies lead to a specified mean and standard deviation takes away two more degrees of freedom, leaving $k - 3$ degrees of freedom for comparing obtained and normal curve frequencies. In the problem shown in Table 10–3, this gives $4 - 3$ or 1 $df$ which leads to $\chi^2_{.05} = 3.84$ and $\chi^2_{.01} = 6.64$ (see Table D, Appendix A). Since the obtained chi square of 0.15 is well below the $\chi^2_{0.05} = 3.84$, the obtained chi square is not significant and therefore $h_0$ is not rejected in this situation.

**Exercises**

10–1.  Given the following obtained and theoretical normal curve frequencies, combine class intervals so that no $f_t$ is less than 10.0 and make a chi square test of the fit: (Problem F)

| $X$ | $f_0$ | $f_t$ |
|---|---|---|
| 110–119....... | 3 | 2.3 |
| 100–109....... | 8 | 5.9 |
| 90–99........ | 12 | 13.8 |
| 80–89........ | 20 | 24.5 |
| 70–79........ | 36 | 32.6 |
| 60–69........ | 33 | 33.0 |
| 50–59........ | 24 | 25.3 |
| 40–49........ | 16 | 14.6 |
| 30–39........ | 7 | 6.4 |
| 20–29........ | 2 | 2.6 |

10–2.  Given the following obtained and theoretical normal curve frequencies, combine class intervals so that no $f_t$ is less than 10.0 and make a chi square test of the fit: (Problem F)

| $X$ | $f_0$ | $f_t$ |
|---|---|---|
| 50–54........ | 3 | 2.6 |
| 45–49........ | 11 | 7.2 |
| 40–44........ | 22 | 18.3 |
| 35–39........ | 31 | 36.0 |
| 30–34........ | 42 | 54.3 |
| 25–29........ | 56 | 64.1 |
| 20–24........ | 70 | 55.8 |
| 15–19........ | 48 | 37.9 |
| 10–14........ | 18 | 19.8 |
| 5–9......... | 6 | 11.0 |

10–3.  In a small coeducational college a controversy has developed between student factions over whether students who do not belong to fraternities and sororities are doing their part to support student functions. The student council decides to carry out a survey. Seventy-six students are selected at random and asked whether they attended the recent college-wide dance and whether they belong to a fraternity or sorority. The results showed that of the 76 students surveyed 15 organizationals and 11 nonorganizationals went to the dance; 11 organizationals and 39 nonorganizationals did not go to the dance. Use the $\chi^2$ test to analyze the data and reach a conclusion. Would you use Yates' correction?

10–4.  A sociologist is interested in studying the relationship between age at time of marriage and durability of the marriage. Her hypothesis is that early marriages for males tend to be unstable. She takes a random sample of males from the general population in her city and determines that of 200 males married over 10 years ago, 96 were over 21 at the time of their marriage and were still married 10 years later to the same person; 24 were over 21 when married, but did not stay married for 10 years; 44 of the 200 were under 21 at time of marriage and were still married 10 years later; 36 were under 21 at the time of marriage and did not stay married 10 years. Test the hypothesis using $\chi^2$ and interpret the results of the study.

10–5.  A personality test developer wishes to exclude from his test any items on which males and females differ significantly in their responses. He obtains the following table for one of the three-choice items. Use $\chi^2$ to determine if the item should be eliminated:

|  | *No* | *?* | *Yes* | *Total* |
|---|---|---|---|---|
| Males......... | 40 | 20 | 40 | 100 |
| Females....... | 30 | 10 | 60 | 100 |
| Total......... | 70 | 30 | 100 | 200 |

10–6.  A movie producer is anxious to learn what group should be the target for his advertising relative to a new release. At a preview, he obtains the data shown below. Would the data suggest anything about his advertising plans?

|  | *under 20* | *20–29* | *30–39* | *40–49* | *over 50* | *Total* |
|---|---|---|---|---|---|---|
| Liked.......... | 33 | 31 | 20 | 10 | 11 | 105 |
| Indifferent...... | 10 | 15 | 22 | 8 | 6 | 61 |
| Disliked........ | 12 | 19 | 16 | 25 | 22 | 94 |
| Total.......... | 55 | 65 | 58 | 43 | 39 | 260 |

10–7.  A track coach suspects that the track at his school is not uniform and decides to check the records. He locates records for 180 runnings of the 100 yard dash and finds that the winners distributed themselves by lanes as follows: lane 1, 20; lane 2, 22; lane 3, 40; lane 4, 45, lane 5, 32; and lane 6, 21. Determine if there is any evidence of nonrandomness in the pattern of wins by lane. Use the 0.01 level of confidence for $\chi^2$.

10–8.  A die is suspected of being "loaded." To test this hypothesis, the die is thrown 120 times and a record kept of the number of times each face shows up. The frequencies were as follows: 1, 35; 2, 22; 3, 16; 4, 15; 5, 17; and 6, 15. Use the chi square test to determine if the obtained frequencies depart significantly from the theoretical frequencies. Use the 0.05 level of confidence.

# appendix A

# Tables

TABLE A
Table of Squares and Square Roots of the Numbers from 1 to 1000

| Number | Square | Square root | Number | Square | Square root |
|--------|--------|-------------|--------|--------|-------------|
| 1 | 1 | 1.000 | 31 | 9 61 | 5.568 |
| 2 | 4 | 1.414 | 32 | 10 24 | 5.657 |
| 3 | 9 | 1.732 | 33 | 10 89 | 5.745 |
| 4 | 16 | 2.000 | 34 | 11 56 | 5.831 |
| 5 | 25 | 2.236 | 35 | 12 25 | 5.916 |
| 6 | 36 | 2.449 | 36 | 12 96 | 6.000 |
| 7 | 49 | 2.646 | 37 | 13 69 | 6.083 |
| 8 | 64 | 2.828 | 38 | 14 44 | 6.164 |
| 9 | 81 | 3.000 | 39 | 15 21 | 6.245 |
| 10 | 1 00 | 3.162 | 40 | 16 00 | 6.325 |
| 11 | 1 21 | 3.317 | 41 | 16 81 | 6.403 |
| 12 | 1 44 | 3.464 | 42 | 17 64 | 6.481 |
| 13 | 1 69 | 3.606 | 43 | 18 49 | 6.557 |
| 14 | 1 96 | 3.742 | 44 | 19 36 | 6.633 |
| 15 | 2 25 | 3.873 | 45 | 20 25 | 6.708 |
| 16 | 2 56 | 4.000 | 46 | 21 16 | 6.782 |
| 17 | 2 89 | 4.123 | 47 | 22 09 | 6.856 |
| 18 | 3 24 | 4.243 | 48 | 23 04 | 6.928 |
| 19 | 3 61 | 4.359 | 49 | 24 01 | 7.000 |
| 20 | 4 00 | 4.472 | 50 | 25 00 | 7.071 |
| 21 | 4 41 | 4.583 | 51 | 26 01 | 7.141 |
| 22 | 4 84 | 4.690 | 52 | 27 04 | 7.211 |
| 23 | 5 29 | 4.796 | 53 | 28 09 | 7.280 |
| 24 | 5 76 | 4.899 | 54 | 29 16 | 7.348 |
| 25 | 6 25 | 5.000 | 55 | 30 25 | 7.416 |
| 26 | 6 76 | 5.099 | 56 | 31 36 | 7.483 |
| 27 | 7 29 | 5.196 | 57 | 32 49 | 7.550 |
| 28 | 7 84 | 5.292 | 58 | 33 64 | 7.616 |
| 29 | 8 41 | 5.385 | 59 | 34 81 | 7.681 |
| 30 | 9 00 | 5.477 | 60 | 36 00 | 7.746 |

TABLE A—(*continued*)

| Number | Square | Square root | Number | Square | Square root |
|--------|--------|-------------|--------|--------|-------------|
| 61 | 37 21 | 7.810 | 99 | 98 01 | 9.950 |
| 62 | 38 44 | 7.874 | 100 | 100 00 | 10.000 |
| 63 | 39 69 | 7.937 | | | |
| 64 | 40 96 | 8.000 | 101 | 1 02 01 | 10.050 |
| 65 | 42 25 | 8.062 | 102 | 1 04 04 | 10.100 |
| | | | 103 | 1 06 09 | 10.149 |
| 66 | 43 56 | 8.124 | 104 | 1 08 16 | 10.198 |
| 67 | 44 89 | 8.185 | 105 | 1 10 25 | 10.247 |
| 68 | 46 24 | 8.246 | | | |
| 69 | 47 61 | 8.307 | 106 | 1 12 36 | 10.296 |
| 70 | 49 00 | 8.367 | 107 | 1 14 49 | 10.344 |
| | | | 108 | 1 16 64 | 10.392 |
| 71 | 50 41 | 8.426 | 109 | 1 18 81 | 10.440 |
| 72 | 51 84 | 8.485 | 110 | 1 21 00 | 10.488 |
| 73 | 53 29 | 8.544 | | | |
| 74 | 54 76 | 8.602 | 111 | 1 23 21 | 10.536 |
| 75 | 56 25 | 8.660 | 112 | 1 25 44 | 10.583 |
| | | | 113 | 1 27 69 | 10.630 |
| 76 | 57 76 | 8.718 | 114 | 1 29 96 | 10.677 |
| 77 | 59 29 | 8.775 | 115 | 1 32 25 | 10.724 |
| 78 | 60 84 | 8.832 | | | |
| 79 | 62 41 | 8.888 | 116 | 1 34 56 | 10.770 |
| 80 | 64 00 | 8.944 | 117 | 1 36 89 | 10.817 |
| | | | 118 | 1 39 24 | 10.863 |
| 81 | 65 61 | 9.000 | 119 | 1 41 61 | 10.909 |
| 82 | 67 24 | 9.055 | 120 | 1 44 00 | 10.954 |
| 83 | 68 89 | 9.110 | | | |
| 84 | 70 56 | 9.165 | 121 | 1 46 41 | 11.000 |
| 85 | 72 25 | 9.220 | 122 | 1 48 84 | 11.045 |
| | | | 123 | 1 51 29 | 11.091 |
| 86 | 73 96 | 9.274 | 124 | 1 53 76 | 11.136 |
| 87 | 75 69 | 9.327 | 125 | 1 56 25 | 11.180 |
| 88 | 77 44 | 9.381 | | | |
| 89 | 79 21 | 9.434 | 126 | 1 58 76 | 11.225 |
| 90 | 81 00 | 9.487 | 127 | 1 61 29 | 11.269 |
| | | | 128 | 1 63 84 | 11.314 |
| 91 | 82 81 | 9.539 | 129 | 1 66 41 | 11.358 |
| 92 | 84 64 | 9.592 | 130 | 1 69 00 | 11.402 |
| 93 | 86 49 | 9.644 | | | |
| 94 | 88 36 | 9.695 | 131 | 1 71 61 | 11.446 |
| 95 | 90 25 | | 132 | 1 74 24 | 11.489 |
| | | | 133 | 1 76 89 | 11.533 |
| 96 | 92 16 | 9.798 | 134 | 1 79 56 | 11.576 |
| 97 | 94 09 | 9.849 | 135 | 1 82 25 | 11.619 |
| 98 | 96 04 | 9.899 | | | |

TABLE  A—(*continued*)

| Number | Square | Square root | Number | Square | Square root |
|---|---|---|---|---|---|
| 136 | 1 84 96 | 11.662 | 174 | 3 02 76 | 13.191 |
| 137 | 1 87 69 | 11.705 | 175 | 3 06 25 | 13.229 |
| 138 | 1 90 44 | 11.747 | | | |
| 139 | 1 93 21 | 11.790 | 176 | 3 09 76 | 13.266 |
| 140 | 1 96 00 | 11.832 | 177 | 3 13 29 | 13.304 |
| | | | 178 | 3 16 84 | 13.342 |
| 141 | 1 98 81 | 11.874 | 179 | 3 20 41 | 13.379 |
| 142 | 2 01 64 | 11.916 | 180 | 3 24 00 | 13.416 |
| 143 | 2 04 49 | 11.958 | | | |
| 144 | 2 07 36 | 12.000 | 181 | 3 27 61 | 13.454 |
| 145 | 2 10 25 | 12.042 | 182 | 3 31 25 | 13.491 |
| | | | 183 | 3 34 89 | 13.528 |
| 146 | 2 13 16 | 12.083 | 184 | 3 38 56 | 13.565 |
| 147 | 2 16 09 | 12.124 | 185 | 3 42 25 | 13.601 |
| 148 | 2 19 04 | 12.166 | | | |
| 149 | 2 22 01 | 12.207 | 186 | 3 45 96 | 13.638 |
| 150 | 2 25 00 | 12.247 | 187 | 3 49 69 | 13.675 |
| | | | 188 | 3 53 44 | 13.711 |
| 151 | 2 28 01 | 12.288 | 189 | 3 57 21 | 13.748 |
| 152 | 2 31 04 | 12.329 | 190 | 3 61 00 | 13.784 |
| 153 | 2 34 09 | 12.369 | | | |
| 154 | 2 37 16 | 12.410 | 191 | 3 64 81 | 13.820 |
| 155 | 2 40 25 | 12.450 | 192 | 3 68 64 | 13.856 |
| | | | 193 | 3 72 49 | 13.892 |
| 156 | 2 43 36 | 12.490 | 194 | 3 76 36 | 13.928 |
| 157 | 2 46 49 | 12.530 | 195 | 3 80 25 | 13.964 |
| 158 | 2 49 64 | 12.570 | | | |
| 159 | 2 52 81 | 12.610 | 196 | 3 84 16 | 14.000 |
| 160 | 2 56 00 | 12.649 | 197 | 3 88 09 | 14.036 |
| | | | 198 | 3 92 04 | 14.071 |
| 161 | 2 59 21 | 12.689 | 199 | 3 96 01 | 14.107 |
| 162 | 2 62 44 | 12.728 | 200 | 4 00 00 | 14.142 |
| 163 | 2 65 69 | 12.767 | | | |
| 164 | 2 68 96 | 12.806 | 201 | 4 04 01 | 14.177 |
| 165 | 2 72 25 | 12.845 | 202 | 4 08 04 | 14.213 |
| | | | 203 | 4 12 09 | 14.248 |
| 166 | 2 75 56 | 12.884 | 204 | 4 16 16 | 14.283 |
| 167 | 2 78 89 | 12.923 | 205 | 4 20 25 | 14.318 |
| 168 | 2 82 24 | 12.961 | | | |
| 169 | 2 85 61 | 13.000 | 206 | 4 24 36 | 14.353 |
| 170 | 2 89 00 | 13.038 | 207 | 4 28 49 | 14.387 |
| | | | 208 | 4 32 64 | 14.422 |
| 171 | 2 92 41 | 13.077 | 209 | 4 36 81 | 14.457 |
| 172 | 2 95 84 | 13.115 | 210 | 4 41 00 | 14.491 |
| 173 | 2 99 29 | 13.153 | | | |

TABLE A—(*continued*)

| Number | Square | Square root | Number | Square | Square root |
|--------|--------|-------------|--------|--------|-------------|
| 211 | 4 45 21 | 14.526 | 249 | 6 20 01 | 15.780 |
| 212 | 4 49 44 | 14.560 | 250 | 6 25 00 | 15.811 |
| 213 | 4 53 60 | 14.595 | | | |
| 214 | 4 57 96 | 14.629 | 251 | 6 30 01 | 15.843 |
| 215 | 4 62 25 | 14.663 | 252 | 6 35 04 | 15.875 |
| | | | 253 | 6 40 09 | 15.906 |
| 216 | 4 66 56 | 14.697 | 254 | 6 45 16 | 15.937 |
| 217 | 4 70 89 | 14.731 | 255 | 6 50 25 | 15.969 |
| 218 | 4 75 24 | 14.765 | | | |
| 219 | 4 79 61 | 14.799 | 256 | 6 55 36 | 16.000 |
| 220 | 4 84 00 | 14.832 | 257 | 6 60 49 | 16.031 |
| | | | 258 | 6 65 64 | 16.062 |
| 221 | 4 88 41 | 14.866 | 259 | 6 70 81 | 16.093 |
| 222 | 4 92 84 | 14.900 | 260 | 6 76 00 | 16.125 |
| 223 | 4 97 29 | 14.933 | | | |
| 224 | 5 01 76 | 14.967 | 261 | 6 81 21 | 16.155 |
| 225 | 5 06 25 | 15.000 | 262 | 6 86 44 | 16.186 |
| | | | 263 | 6 91 69 | 16.217 |
| 226 | 5 10 76 | 15.033 | 264 | 6 96 96 | 16.248 |
| 227 | 5 15 29 | 15.067 | 265 | 7 02 25 | 16.279 |
| 228 | 5 19 84 | 15.100 | | | |
| 229 | 5 24 41 | 15.133 | 266 | 7 07 56 | 16.310 |
| 230 | 5 29 00 | 15.166 | 267 | 7 12 89 | 16.340 |
| | | | 268 | 7 18 24 | 16.371 |
| 231 | 5 33 61 | 15.199 | 269 | 7 23 61 | 16.401 |
| 232 | 5 38 24 | 15.232 | 270 | 7 29 00 | 16.432 |
| 233 | 5 42 89 | 15.264 | | | |
| 234 | 5 47 56 | 15.297 | 271 | 7 34 41 | 16.462 |
| 235 | 5 52 25 | 15.330 | 272 | 7 39 84 | 16.492 |
| | | | 273 | 7 45 29 | 16.523 |
| 236 | 5 56 96 | 15.362 | 274 | 7 50 76 | 16.553 |
| 237 | 5 61 69 | 15.395 | 275 | 7 56 25 | 16.583 |
| 238 | 5 66 44 | 15.427 | | | |
| 239 | 5 71 21 | 15.460 | 276 | 7 61 76 | 16.613 |
| 240 | 5 76 00 | 15.492 | 277 | 7 67 29 | 16.643 |
| | | | 278 | 7 72 84 | 16.673 |
| 241 | 5 80 81 | 15.524 | 279 | 7 78 41 | 16.703 |
| 242 | 5 85 64 | 15.556 | 280 | 7 84 00 | 16.733 |
| 243 | 5 90 49 | 15.588 | | | |
| 244 | 5 95 36 | 15.620 | 281 | 7 89 61 | 16.763 |
| 245 | 6 00 25 | 15.652 | 282 | 7 95 24 | 16.793 |
| | | | 283 | 8 00 89 | 16.823 |
| 246 | 6 05 16 | 15.684 | 284 | 8 06 56 | 16.852 |
| 247 | 6 10 09 | 15.716 | 285 | 8 12 25 | 16.882 |
| 248 | 6 15 04 | 15.748 | | | |

TABLE A—(*continued*)

| Number | Square | Square root | Number | Square | Square root |
|--------|--------|-------------|--------|--------|-------------|
| 286 | 8 17 96 | 16.912 | 324 | 10 49 76 | 18.000 |
| 287 | 8 23 69 | 16.941 | 325 | 10 56 25 | 18.028 |
| 288 | 8 29 44 | 16.971 | | | |
| 289 | 8 35 21 | 17.000 | 326 | 10 62 76 | 18.055 |
| 290 | 8 41 00 | 17.029 | 327 | 10 69 29 | 18.083 |
| | | | 328 | 10 75 84 | 18.111 |
| 291 | 8 46 81 | 17.059 | 329 | 10 82 41 | 18.138 |
| 292 | 8 52 64 | 17.088 | 330 | 10 89 00 | 18.166 |
| 293 | 8 58 49 | 17.117 | | | |
| 294 | 8 64 36 | 17.146 | 331 | 10 95 61 | 18.193 |
| 295 | 8 70 25 | 17.176 | 332 | 11 01 24 | 18.221 |
| | | | 333 | 11 08 89 | 18.248 |
| 296 | 8 76 16 | 17.205 | 334 | 11 15 56 | 18.276 |
| 297 | 8 82 09 | 17.234 | 335 | 11 22 25 | 18.303 |
| 298 | 8 88 04 | 17.263 | | | |
| 299 | 8 94 01 | 17.292 | 336 | 11 28 96 | 18.330 |
| 300 | 9 00 00 | 17.321 | 337 | 11 35 69 | 18.358 |
| | | | 338 | 11 42 44 | 18.385 |
| 301 | 9 06 01 | 17.349 | 339 | 11 49 21 | 18.412 |
| 302 | 9 12 04 | 17.378 | 340 | 11 56 00 | 18.439 |
| 303 | 9 18 09 | 17.407 | | | |
| 304 | 9 24 16 | 17.436 | 341 | 11 62 81 | 18.466 |
| 305 | 9 30 25 | 17.464 | 342 | 11 69 64 | 18.493 |
| | | | 343 | 11 76 49 | 18.520 |
| 306 | 9 36 36 | 17.493 | 344 | 11 83 36 | 18.547 |
| 307 | 9 42 49 | 17.521 | 345 | 11 90 25 | 18.574 |
| 308 | 9 48 64 | 17.550 | | | |
| 309 | 9 54 81 | 17.578 | 346 | 11 97 16 | 18.601 |
| 310 | 9 61 00 | 17.607 | 347 | 12 04 09 | 18.628 |
| | | | 348 | 12 11 04 | 18.655 |
| 311 | 9 67 21 | 17.635 | 349 | 12 18 01 | 18.682 |
| 312 | 9 73 44 | 17.664 | 350 | 12 25 00 | 18.708 |
| 313 | 9 79 69 | 17.692 | | | |
| 314 | 9 85 96 | 17.720 | 351 | 12 32 01 | 18.735 |
| 315 | 9 92 25 | 17.748 | 352 | 12 39 04 | 18.762 |
| | | | 353 | 12 46 09 | 18.788 |
| 316 | 9 98 56 | 17.776 | 354 | 12 53 16 | 18.815 |
| 317 | 10 04 89 | 17.804 | 355 | 12 60 25 | 18.841 |
| 318 | 10 11 24 | 17.833 | | | |
| 319 | 10 17 61 | 17.861 | 356 | 12 67 36 | 18.868 |
| 320 | 10 24 00 | 17.889 | 357 | 12 74 49 | 18.894 |
| | | | 358 | 12 81 64 | 18.921 |
| 321 | 10 30 41 | 17.916 | 359 | 12 88 81 | 18.947 |
| 322 | 10 36 84 | 17.944 | 360 | 12 96 00 | 18.974 |
| 323 | 10 43 29 | 17.972 | | | |

TABLE A—(*continued*)

| Number | Square | Square root | Number | Square | Square root |
|--------|--------|-------------|--------|--------|-------------|
| 361 | 13 03 21 | 19.000 | 399 | 15 92 01 | 19.975 |
| 362 | 13 10 44 | 19.026 | 400 | 16 00 00 | 20.000 |
| 363 | 13 17 69 | 19.053 | | | |
| 364 | 13 24 96 | 19.079 | 401 | 16 08 01 | 20.025 |
| 365 | 13 32 25 | 19.105 | 402 | 16 16 04 | 20.050 |
| | | | 403 | 16 24 09 | 20.075 |
| 366 | 13 39 56 | 19.131 | 404 | 16 32 16 | 20.100 |
| 367 | 13 46 89 | 19.157 | 405 | 16 40 25 | 20.125 |
| 368 | 13 54 24 | 19.183 | | | |
| 369 | 13 61 61 | 19.209 | 406 | 16 48 36 | 20.149 |
| 370 | 13 69 00 | 19.235 | 407 | 16 56 49 | 20.174 |
| | | | 408 | 16 64 64 | 20.199 |
| 371 | 13 76 41 | 19.261 | 409 | 16 72 81 | 20.224 |
| 372 | 13 83 84 | 19.287 | 410 | 16 81 00 | 20.248 |
| 373 | 13 91 29 | 19.313 | | | |
| 374 | 13 98 76 | 19.339 | 411 | 16 89 21 | 20.273 |
| 375 | 14 06 25 | 19.363 | 412 | 16 97 44 | 20.298 |
| | | | 413 | 17 05 69 | 20.322 |
| 376 | 14 13 76 | 19.391 | 414 | 17 13 96 | 20.347 |
| 377 | 14 21 29 | 19.416 | 415 | 17 22 25 | 20.372 |
| 378 | 14 28 84 | 19.442 | | | |
| 379 | 14 36 41 | 19.468 | 416 | 17 30 56 | 20.396 |
| 380 | 14 44 00 | 19.494 | 417 | 17 38 89 | 20.421 |
| | | | 418 | 17 47 24 | 20.445 |
| 381 | 14 51 61 | 19.519 | 419 | 17 55 61 | 20.469 |
| 382 | 14 59 24 | 19.545 | 420 | 17 64 00 | 20.494 |
| 383 | 14 66 89 | 19.570 | | | |
| 384 | 14 74 56 | 19.596 | 421 | 17 72 41 | 20.518 |
| 385 | 14 82 25 | 19.621 | 422 | 17 80 84 | 20.543 |
| | | | 423 | 17 89 29 | 20.567 |
| 386 | 14 89 96 | 19.647 | 424 | 17 97 76 | 20.591 |
| 387 | 14 97 69 | 19.672 | 425 | 18 06 25 | 20.616 |
| 388 | 15 05 44 | 19.698 | | | |
| 389 | 15 13 21 | 19.723 | 426 | 18 14 76 | 20.640 |
| 390 | 15 21 00 | 19.748 | 427 | 18 23 29 | 20.664 |
| | | | 428 | 18 31 84 | 20.688 |
| 391 | 15 28 81 | 19.774 | 429 | 18 40 41 | 20.712 |
| 392 | 15 36 64 | 19.799 | 430 | 18 49 00 | 20.736 |
| 393 | 15 44 49 | 19.824 | | | |
| 394 | 15 52 36 | 19.849 | 431 | 18 57 61 | 20.761 |
| 395 | 15 60 25 | 19.875 | 432 | 18 66 24 | 20.785 |
| | | | 433 | 18 74 89 | 20.809 |
| 396 | 15 68 16 | 19.900 | 434 | 18 83 56 | 20.833 |
| 397 | 15 76 09 | 19.925 | 435 | 18 92 25 | 20.857 |
| 398 | 15 84 04 | 19.950 | | | |

TABLE A—(*continued*)

| Number | Square | Square root | Number | Square | Square root |
|---|---|---|---|---|---|
| 436 | 19 00 96 | 20.881 | 474 | 22 46 76 | 21.772 |
| 437 | 19 09 69 | 20.905 | 475 | 22 56 25 | 21.794 |
| 438 | 19 18 44 | 20.928 | | | |
| 439 | 19 27 21 | 20.952 | 476 | 22 65 76 | 21.817 |
| 440 | 19 36 00 | 20.976 | 477 | 22 75 29 | 21.840 |
| | | | 478 | 22 84 84 | 21.863 |
| 441 | 19 44 81 | 21.000 | 479 | 22 94 41 | 21.886 |
| 442 | 19 53 64 | 21.024 | 480 | 23 04 00 | 21.909 |
| 443 | 19 62 49 | 21.048 | | | |
| 444 | 19 71 36 | 21.071 | 481 | 23 13 61 | 21.932 |
| 445 | 19 80 25 | 21.095 | 482 | 23 23 24 | 21.954 |
| | | | 483 | 23 32 89 | 21.977 |
| 446 | 19 89 16 | 21.119 | 484 | 23 42 56 | 22.000 |
| 447 | 19 98 09 | 21.142 | 485 | 23 52 25 | 22.023 |
| 448 | 20 07 04 | 21.166 | | | |
| 449 | 20 16 01 | 21.190 | 486 | 23 61 96 | 22.045 |
| 450 | 20 25 00 | 21.213 | 487 | 23 71 69 | 22.068 |
| | | | 488 | 23 81 44 | 22.091 |
| 451 | 20 34 01 | 21.237 | 489 | 23 91 21 | 22.113 |
| 452 | 20 43 04 | 21.260 | 490 | 24 01 00 | 22.136 |
| 453 | 20 52 09 | 21.284 | | | |
| 454 | 20 61 16 | 21.307 | 491 | 24 10 81 | 22.159 |
| 455 | 20 70 25 | 21.331 | 492 | 24 20 64 | 22.181 |
| | | | 493 | 24 30 49 | 22.204 |
| 456 | 20 79 36 | 21.354 | 494 | 24 40 36 | 22.226 |
| 457 | 20 88 49 | 21.378 | 495 | 24 50 25 | 22.249 |
| 458 | 20 97 64 | 21.401 | | | |
| 459 | 21 06 81 | 21.424 | 496 | 24 60 16 | 22.271 |
| 460 | 21 16 00 | 21.448 | 497 | 24 70 09 | 22.293 |
| | | | 498 | 24 80 04 | 22.316 |
| 461 | 21 25 21 | 21.471 | 499 | 24 90 01 | 22.338 |
| 462 | 21 34 44 | 21.494 | 500 | 25 00 00 | 22.361 |
| 463 | 21 43 69 | 21.517 | | | |
| 464 | 21 52 96 | 21.541 | 501 | 25 10 01 | 22.383 |
| 465 | 21 62 25 | 21.564 | 502 | 25 20 04 | 22.405 |
| | | | 503 | 25 30 09 | 22.428 |
| 466 | 21 71 56 | 21.587 | 504 | 25 40 16 | 22.450 |
| 467 | 21 80 89 | 21.610 | 505 | 25 50 25 | 22.472 |
| 468 | 21 90 24 | 21.633 | | | |
| 469 | 21 99 61 | 21.656 | 506 | 25 60 36 | 22.494 |
| 470 | 22 09 00 | 21.679 | 507 | 25 70 49 | 22.517 |
| | | | 508 | 25 80 64 | 22.539 |
| 471 | 22 18 41 | 21.703 | 509 | 25 90 81 | 22.561 |
| 472 | 22 27 84 | 21.726 | 510 | 26 01 00 | 22.583 |
| 473 | 22 37 29 | 21.749 | | | |

TABLE A—(*continued*)

| Number | Square | Square root | Number | Square | Square root |
|--------|--------|-------------|--------|--------|-------------|
| 511 | 26 11 21 | 22.605 | 549 | 30 14 01 | 23.431 |
| 512 | 26 21 44 | 22.627 | 550 | 30 25 00 | 23.452 |
| 513 | 26 31 69 | 22.650 | | | |
| 514 | 26 41 96 | 22.672 | 551 | 30 36 01 | 23.473 |
| 515 | 26 52 25 | 22.694 | 552 | 30 47 04 | 23.495 |
| | | | 553 | 30 58 09 | 23.516 |
| 516 | 26 62 56 | 22.716 | 554 | 30 69 16 | 23.537 |
| 517 | 26 72 89 | 22.738 | 555 | 30 80 25 | 23.558 |
| 518 | 26 83 24 | 22.760 | | | |
| 519 | 26 93 61 | 22.782 | 556 | 30 91 36 | 23.580 |
| 520 | 27 04 00 | 22.804 | 557 | 31 02 49 | 23.601 |
| | | | 558 | 31 13 64 | 23.622 |
| 521 | 27 14 41 | 22.825 | 559 | 31 24 81 | 23.643 |
| 522 | 27 24 84 | 22.847 | 560 | 31 36 00 | 23.664 |
| 523 | 27 35 29 | 22.869 | | | |
| 524 | 27 45 76 | 22.891 | 561 | 31 47 21 | 23.685 |
| 525 | 27 56 25 | 22.913 | 562 | 31 58 44 | 23.707 |
| | | | 563 | 31 69 69 | 23.728 |
| 526 | 27 66 76 | 22.935 | 564 | 31 80 96 | 23.749 |
| 527 | 27 77 29 | 22.956 | 565 | 31 92 25 | 23.770 |
| 528 | 27 87 84 | 22.978 | | | |
| 529 | 27 98 41 | 23.000 | 566 | 32 03 56 | 23.791 |
| 530 | 28 09 00 | 23.022 | 567 | 32 14 89 | 23.812 |
| | | | 568 | 32 26 24 | 23.833 |
| 531 | 28 19 61 | 23.043 | 569 | 32 37 61 | 23.854 |
| 532 | 28 30 24 | 23.065 | 570 | 32 49 00 | 23.875 |
| 533 | 28 40 89 | 23.087 | | | |
| 534 | 28 51 56 | 23.108 | 571 | 32 60 41 | 23.896 |
| 535 | 28 62 25 | 23.130 | 572 | 32 71 84 | 23.917 |
| | | | 573 | 32 83 29 | 23.937 |
| 536 | 28 72 96 | 23.152 | 574 | 32 94 76 | 23.958 |
| 537 | 28 83 69 | 23.173 | 575 | 33 06 25 | 23.979 |
| 538 | 28 94 44 | 23.195 | | | |
| 539 | 29 05 21 | 23.216 | 576 | 33 17 76 | 24.000 |
| 540 | 29 16 00 | 23.238 | 577 | 33 29 29 | 24.021 |
| | | | 578 | 33 40 84 | 24.042 |
| 541 | 29 26 81 | 23.259 | 579 | 33 52 41 | 24.062 |
| 542 | 29 37 64 | 23.281 | 580 | 33 64 00 | 24.083 |
| 543 | 29 48 49 | 23.302 | | | |
| 544 | 29 59 36 | 23.324 | 581 | 33 75 61 | 24.104 |
| 545 | 29 70 25 | 23.345 | 582 | 33 87 24 | 24.125 |
| | | | 583 | 33 98 89 | 24.145 |
| 546 | 29 81 16 | 23.367 | 584 | 34 10 56 | 24.166 |
| 547 | 29 92 09 | 23.388 | 585 | 34 22 25 | 24.187 |
| 548 | 30 03 04 | 23.409 | | | |

TABLE A—(*continued*)

| Number | Square | Square root | Number | Square | Square root |
|--------|--------|-------------|--------|--------|-------------|
| 586 | 34 33 96 | 24.207 | 624 | 38 93 76 | 24.980 |
| 587 | 34 45 69 | 24.228 | 625 | 39 06 25 | 25.000 |
| 588 | 34 57 44 | 24.249 | | | |
| 589 | 34 69 21 | 24.269 | 626 | 39 18 76 | 25.020 |
| 590 | 34 81 00 | 24.290 | 627 | 39 31 29 | 25.040 |
| | | | 628 | 39 43 84 | 25.060 |
| 591 | 34 92 81 | 24.310 | 629 | 39 56 41 | 25.080 |
| 592 | 35 04 64 | 24.331 | 630 | 39 69 00 | 25.100 |
| 593 | 35 16 49 | 24.352 | | | |
| 594 | 35 28 36 | 24.372 | 631 | 39 81 61 | 25.120 |
| 595 | 35 40 25 | 24.393 | 632 | 39 94 24 | 25.140 |
| | | | 633 | 40 06 89 | 25.159 |
| 596 | 35 52 16 | 24.413 | 634 | 40 19 56 | 25.179 |
| 597 | 35 64 09 | 24.434 | 635 | 40 32 25 | 25.199 |
| 598 | 35 76 04 | 24.454 | | | |
| 599 | 35 88 01 | 24.474 | 636 | 40 44 96 | 25.219 |
| 600 | 36 00 00 | 24.495 | 637 | 40 57 69 | 25.239 |
| | | | 638 | 40 70 44 | 25.259 |
| 601 | 36 12 01 | 24.515 | 639 | 40 83 21 | 25.278 |
| 602 | 36 24 04 | 24.536 | 640 | 40 96 00 | 25.298 |
| 603 | 36 36 09 | 24.556 | | | |
| 604 | 36 48 16 | 24.576 | 641 | 41 08 81 | 25.318 |
| 605 | 36 60 25 | 24.597 | 642 | 41 21 64 | 25.338 |
| | | | 643 | 41 34 49 | 25.357 |
| 606 | 36 72 36 | 24.617 | 644 | 41 47 36 | 25.377 |
| 607 | 36 84 49 | 24.637 | 645 | 41 60 25 | 25.397 |
| 608 | 36 96 64 | 24.658 | | | |
| 609 | 37 08 81 | 24.678 | 646 | 41 73 16 | 25.417 |
| 610 | 37 21 00 | 24.698 | 647 | 41 86 09 | 25.436 |
| | | | 648 | 41 99 04 | 25.456 |
| 611 | 37 33 21 | 24.718 | 649 | 42 12 01 | 25.475 |
| 612 | 37 45 44 | 24.739 | 650 | 42 25 00 | 25.495 |
| 613 | 37 57 69 | 24.759 | | | |
| 614 | 37 69 96 | 24.779 | 651 | 42 38 01 | 25.515 |
| 615 | 37 82 25 | 24.799 | 652 | 42 51 04 | 25.534 |
| | | | 653 | 42 64 09 | 25.554 |
| 616 | 37 94 56 | 24.819 | 654 | 42 77 16 | 25.573 |
| 617 | 38 06 89 | 24.839 | 655 | 42 90 25 | 25.593 |
| 618 | 38 19 24 | 24.860 | | | |
| 619 | 38 31 61 | 24.880 | 656 | 43 03 36 | 25.612 |
| 620 | 38 44 00 | 24.900 | 657 | 43 16 49 | 25.632 |
| | | | 658 | 43 29 64 | 25.652 |
| 621 | 38 56 41 | 24.920 | 659 | 43 42 81 | 25.671 |
| 622 | 38 68 84 | 24.940 | 660 | 43 56 00 | 25.690 |
| 623 | 38 81 29 | 24.960 | | | |

TABLE A—(*continued*)

| Number | Square | Square root | Number | Square | Square root |
|--------|--------|-------------|--------|--------|-------------|
| 661 | 43 69 21 | 25.710 | 699 | 48 86 01 | 26.439 |
| 662 | 43 82 44 | 25.729 | 700 | 49 00 00 | 26.458 |
| 663 | 43 95 69 | 25.749 | | | |
| 664 | 44 08 96 | 25.768 | 701 | 49 14 01 | 26.476 |
| 665 | 44 22 25 | 25.788 | 702 | 49 28 04 | 26.495 |
| | | | 703 | 49 42 09 | 26.514 |
| 666 | 44 35 56 | 25.807 | 704 | 49 56 16 | 26.533 |
| 667 | 44 48 89 | 25.826 | 705 | 49 70 25 | 26.552 |
| 668 | 44 62 24 | 25.846 | | | |
| 669 | 44 75 61 | 25.865 | 706 | 49 84 36 | 26.571 |
| 670 | 44 89 00 | 25.884 | 707 | 49 98 49 | 26.589 |
| | | | 708 | 50 12 64 | 26.608 |
| 671 | 45 02 41 | 25.904 | 709 | 50 26 81 | 26.627 |
| 672 | 45 15 84 | 25.923 | 710 | 50 41 00 | 26.646 |
| 673 | 45 29 29 | 25.942 | | | |
| 674 | 45 42 76 | 25.962 | 711 | 50 55 21 | 26.665 |
| 675 | 45 56 25 | 25.981 | 712 | 50 69 44 | 26.683 |
| | | | 713 | 50 83 69 | 26.702 |
| 676 | 45 69 76 | 26.000 | 714 | 50 97 96 | 26.721 |
| 677 | 45 83 29 | 26.019 | 715 | 51 12 25 | 26.739 |
| 678 | 45 96 84 | 26.038 | | | |
| 679 | 46 10 41 | 26.058 | 716 | 51 26 56 | 26.758 |
| 680 | 46 24 00 | 26.077 | 717 | 51 40 89 | 26.777 |
| | | | 718 | 51 55 24 | 26.796 |
| 681 | 46 37 61 | 26.096 | 719 | 51 69 61 | 26.814 |
| 682 | 46 51 24 | 26.115 | 720 | 51 84 00 | 26.833 |
| 683 | 46 64 89 | 26.134 | | | |
| 684 | 46 78 56 | 26.153 | 721 | 51 98 41 | 26.851 |
| 685 | 46 92 25 | 26.173 | 722 | 52 12 84 | 26.870 |
| | | | 723 | 52 27 29 | 26.889 |
| 686 | 47 05 96 | 26.192 | 724 | 52 41 76 | 26.907 |
| 687 | 47 19 69 | 26.211 | 725 | 52 56 25 | 26.926 |
| 688 | 47 33 44 | 26.230 | | | |
| 689 | 47 47 21 | 26.249 | 726 | 52 70 76 | 26.944 |
| 690 | 47 61 00 | 26.268 | 727 | 52 85 29 | 26.963 |
| | | | 728 | 52 99 84 | 26.981 |
| 691 | 47 74 81 | 26.287 | 729 | 53 14 41 | 27.000 |
| 692 | 47 88 64 | 26.306 | 730 | 53 29 00 | 27.019 |
| 693 | 48 02 49 | 26.325 | | | |
| 694 | 48 16 36 | 26.344 | 731 | 53 43 61 | 27.037 |
| 695 | 48 30 25 | 26.363 | 732 | 53 58 24 | 27.055 |
| | | | 733 | 53 72 89 | 27.074 |
| 696 | 48 44 16 | 26.382 | 734 | 53 87 56 | 27.092 |
| 697 | 48 58 09 | 26.401 | 735 | 54 02 25 | 27.111 |
| 698 | 48 72 04 | 26.420 | | | |

TABLE A—(*continued*)

| Number | Square | Square root | Number | Square | Square root |
|--------|--------|-------------|--------|--------|-------------|
| 736 | 54 16 96 | 27.129 | 774 | 59 90 76 | 27.821 |
| 737 | 54 31 69 | 27.148 | 775 | 60 06 25 | 27.839 |
| 738 | 54 46 44 | 27.166 | | | |
| 739 | 54 61 21 | 27.185 | 776 | 60 21 76 | 27.857 |
| 740 | 54 76 00 | 27.203 | 777 | 60 37 29 | 27.875 |
| | | | 778 | 60 52 84 | 27.893 |
| 741 | 54 90 81 | 27.221 | 779 | 60 68 41 | 27.911 |
| 742 | 55 05 64 | 27.240 | 780 | 60 84 00 | 27.928 |
| 743 | 55 20 49 | 27.258 | | | |
| 744 | 55 35 36 | 27.276 | 781 | 60 99 61 | 27.946 |
| 745 | 55 50 25 | 27.295 | 782 | 61 15 24 | 27.964 |
| | | | 783 | 61 30 89 | 27.982 |
| 746 | 55 65 16 | 27.313 | 784 | 61 46 56 | 28.000 |
| 747 | 55 80 09 | 27.331 | 785 | 61 62 25 | 28.018 |
| 748 | 55 95 04 | 27.350 | | | |
| 749 | 56 10 01 | 27.368 | 786 | 61 77 96 | 28.036 |
| 750 | 56 25 00 | 27.386 | 787 | 61 93 69 | 28.054 |
| | | | 788 | 62 09 44 | 28.071 |
| 751 | 56 40 01 | 27.404 | 789 | 62 25 21 | 28.089 |
| 752 | 56 55 04 | 27.423 | 790 | 62 41 00 | 28.107 |
| 753 | 56 70 09 | 27.441 | | | |
| 754 | 56 85 16 | 27.459 | 791 | 62 56 81 | 28.125 |
| 755 | 57 00 25 | 27.477 | 792 | 62 72 64 | 28.142 |
| | | | 793 | 62 88 49 | 28.160 |
| 756 | 57 15 36 | 27.495 | 794 | 63 04 36 | 28.178 |
| 757 | 57 30 49 | 27.514 | 795 | 63 20 25 | 28.196 |
| 758 | 57 45 64 | 27.532 | | | |
| 759 | 57 60 81 | 27.550 | 796 | 63 36 16 | 28.213 |
| 760 | 57 76 00 | 27.568 | 797 | 63 52 09 | 28.231 |
| | | | 798 | 63 68 04 | 28.249 |
| 761 | 57 91 21 | 27.586 | 799 | 63 84 01 | 28.267 |
| 762 | 58 06 44 | 27.604 | 800 | 64 00 00 | 28.284 |
| 763 | 58 21 69 | 27.622 | | | |
| 764 | 58 36 96 | 27.641 | 801 | 64 16 01 | 28.302 |
| 765 | 58 52 25 | 27.659 | 802 | 64 32 04 | 28.320 |
| | | | 803 | 64 48 09 | 28.337 |
| 766 | 58 67 56 | 27.677 | 804 | 64 64 16 | 28.355 |
| 767 | 58 82 89 | 27.695 | 805 | 64 80 25 | 28.373 |
| 768 | 58 98 24 | 27.713 | | | |
| 769 | 59 13 61 | 27.731 | 806 | 64 96 36 | 28.390 |
| 770 | 59 29 00 | 27.749 | 807 | 65 12 49 | 28.408 |
| | | | 808 | 65 28 64 | 28.425 |
| 771 | 59 44 41 | 27.767 | 809 | 65 44 81 | 28.443 |
| 772 | 59 59 84 | 27.785 | 810 | 65 61 00 | 28.460 |
| 773 | 59 75 29 | 27.803 | | | |

TABLE A—(*continued*)

| Number | Square | Square root | Number | Square | Square root |
|--------|--------|-------------|--------|--------|-------------|
| 811 | 65 77 21 | 28.478 | 849 | 72 08 01 | 29.138 |
| 812 | 65 93 44 | 28.496 | 850 | 72 25 00 | 29.155 |
| 813 | 66 09 69 | 28.513 | | | |
| 814 | 66 25 96 | 28.531 | 851 | 72 42 01 | 29.172 |
| 815 | 66 42 25 | 28.548 | 852 | 72 59 04 | 29.189 |
| | | | 853 | 72 76 09 | 29.206 |
| 816 | 66 58 56 | 28.566 | 854 | 72 93 16 | 29.223 |
| 817 | 66 74 89 | 28.583 | 855 | 73 10 25 | 29.240 |
| 818 | 66 91 24 | 28.601 | | | |
| 819 | 67 07 61 | 28.618 | 856 | 73 27 36 | 29.257 |
| 820 | 67 24 00 | 28.636 | 857 | 73 44 49 | 29.275 |
| | | | 858 | 73 61 64 | 29.292 |
| 821 | 67 40 41 | 28.653 | 859 | 73 78 81 | 29.309 |
| 822 | 67 56 84 | 28.671 | 860 | 73 96 00 | 29.326 |
| 823 | 67 73 29 | 28.688 | | | |
| 824 | 67 89 76 | 28.705 | 861 | 74 13 21 | 29.343 |
| 825 | 68 06 25 | 28.723 | 862 | 74 30 44 | 29.360 |
| | | | 863 | 74 47 69 | 29.377 |
| 826 | 68 22 76 | 28.740 | 864 | 74 64 96 | 29.394 |
| 827 | 68 39 29 | 28.758 | 865 | 74 82 25 | 29.411 |
| 828 | 68 55 84 | 28.775 | | | |
| 829 | 68 72 41 | 28.792 | 866 | 74 99 56 | 29.428 |
| 830 | 68 89 00 | 28.810 | 867 | 75 16 89 | 29.445 |
| | | | 868 | 75 34 24 | 29.462 |
| 831 | 69 05 61 | 28.827 | 869 | 75 51 61 | 29.479 |
| 832 | 69 22 24 | 28.844 | 870 | 75 69 00 | 29.496 |
| 833 | 69 38 89 | 28.862 | | | |
| 834 | 69 55 56 | 28.879 | 871 | 75 86 41 | 29.513 |
| 835 | 69 72 25 | 28.896 | 872 | 76 03 84 | 29.530 |
| | | | 873 | 76 21 29 | 29.547 |
| 836 | 69 88 96 | 28.914 | 874 | 76 38 76 | 29.563 |
| 837 | 70 05 69 | 28.931 | 875 | 76 56 25 | 29.580 |
| 838 | 70 22 44 | 28.948 | | | |
| 839 | 70 39 21 | 28.965 | 876 | 76 73 76 | 29.597 |
| 840 | 70 56 00 | 28.983 | 877 | 76 91 29 | 29.614 |
| | | | 878 | 77 08 84 | 29.631 |
| 841 | 70 72 81 | 29.000 | 879 | 77 26 41 | 29.648 |
| 842 | 70 89 64 | 29.017 | 880 | 77 44 00 | 29.665 |
| 843 | 71 06 49 | 29.034 | | | |
| 844 | 71 23 36 | 29.052 | 881 | 77 61 61 | 29.682 |
| 845 | 71 40 25 | 29.069 | 882 | 77 79 24 | 29.698 |
| | | | 883 | 77 96 89 | 29.715 |
| 846 | 71 57 16 | 29.086 | 884 | 78 14 56 | 29.732 |
| 847 | 71 74 09 | 29.103 | 885 | 78 32 25 | 29.749 |
| 848 | 71 91 04 | 29.120 | | | |

TABLE A—(*continued*)

| Number | Square | Square root | Number | Square | Square root |
|---|---|---|---|---|---|
| 886 | 78 49 96 | 29.766 | 924 | 85 37 76 | 30.397 |
| 887 | 78 67 69 | 29.783 | 925 | 85 56 25 | 30.414 |
| 888 | 78 85 44 | 29.799 | | | |
| 889 | 79 03 21 | 29.816 | 926 | 85 74 76 | 30.430 |
| 890 | 79 21 00 | 29.833 | 927 | 85 93 29 | 30.447 |
| | | | 928 | 86 11 84 | 30.463 |
| 891 | 79 38 81 | 29.850 | 929 | 86 30 41 | 30.480 |
| 892 | 79 56 64 | 29.866 | 930 | 86 49 00 | 30.496 |
| 893 | 79 74 49 | 29.833 | | | |
| 894 | 79 92 36 | 29.900 | 931 | 86 67 61 | 30.512 |
| 895 | 80 10 25 | 29.916 | 932 | 86 86 24 | 30.529 |
| | | | 933 | 87 04 89 | 30.545 |
| 896 | 80 28 16 | 29.933 | 934 | 87 23 56 | 30.561 |
| 897 | 80 46 09 | 29.950 | 935 | 87 42 25 | 30.578 |
| 898 | 80 64 04 | 29.967 | | | |
| 899 | 80 82 01 | 29.983 | 936 | 87 60 96 | 30.594 |
| 900 | 81 00 00 | 30.000 | 937 | 87 79 69 | 30.610 |
| | | | 938 | 87 98 44 | 30.627 |
| 901 | 81 18 01 | 30.017 | 939 | 88 17 21 | 30.643 |
| 902 | 81 36 04 | 30.033 | 940 | 88 36 00 | 30.659 |
| 903 | 81 54 09 | 30.050 | | | |
| 904 | 81 72 16 | 30.067 | 941 | 88 54 81 | 30.676 |
| 905 | 81 90 25 | 30.083 | 942 | 88 73 64 | 30.692 |
| | | | 943 | 88 92 49 | 30.708 |
| 906 | 82 08 36 | 30.100 | 944 | 89 11 36 | 30.725 |
| 907 | 82 26 49 | 30.116 | 945 | 89 30 25 | 30.741 |
| 908 | 82 44 64 | 30.133 | | | |
| 909 | 82 62 81 | 30.150 | 946 | 89 49 16 | 30.757 |
| 910 | 82 81 00 | 30.166 | 947 | 89 68 09 | 30.773 |
| | | | 948 | 89 87 04 | 30.790 |
| 911 | 82 99 21 | 30.183 | 949 | 90 06 01 | 30.806 |
| 912 | 83 17 44 | 30.199 | 950 | 90 25 00 | 30.822 |
| 913 | 83 35 69 | 30.216 | | | |
| 914 | 83 53 96 | 30.232 | 951 | 90 44 01 | 30.838 |
| 915 | 83 72 25 | 30.249 | 952 | 90 63 04 | 30.854 |
| | | | 953 | 90 82 09 | 30.871 |
| 916 | 83 90 56 | 30.265 | 954 | 91 01 16 | 30.887 |
| 917 | 84 08 89 | 30.282 | 955 | 91 20 25 | 30.903 |
| 918 | 84 27 24 | 30.299 | | | |
| 919 | 84 45 61 | 30.315 | 956 | 91 39 36 | 30.919 |
| 920 | 84 64 00 | 30.332 | 957 | 91 58 49 | 30.935 |
| | | | 958 | 91 77 64 | 30.952 |
| 921 | 84 82 41 | 30.348 | 959 | 91 96 81 | 30.968 |
| 922 | 85 00 84 | 30.364 | 960 | 92 16 00 | 30.984 |
| 923 | 85 19 29 | 30.381 | | | |

TABLE A—(*concluded*)

| Number | Square | Square root | Number | Square | Square root |
|--------|--------|-------------|--------|--------|-------------|
| 961 | 92 35 21 | 31.000 | 981 | 96 23 61 | 31.321 |
| 962 | 92 54 44 | 31.016 | 982 | 96 43 24 | 31.337 |
| 963 | 92 73 69 | 31.032 | 983 | 96 62 89 | 31.353 |
| 964 | 92 92 96 | 31.048 | 984 | 96 82 56 | 31.369 |
| 965 | 93 12 25 | 31.064 | 985 | 97 02 25 | 31.385 |
| 966 | 93 31 56 | 31.081 | 986 | 97 21 96 | 31.401 |
| 967 | 93 50 89 | 31.097 | 987 | 97 41 69 | 31.417 |
| 968 | 93 70 24 | 31.113 | 988 | 97 61 44 | 31.432 |
| 969 | 93 89 61 | 31.129 | 989 | 97 81 21 | 31.448 |
| 970 | 94 09 00 | 31.145 | 990 | 98 01 00 | 31.464 |
| 971 | 94 28 41 | 31.161 | 991 | 98 20 81 | 31.480 |
| 972 | 94 47 84 | 31.177 | 992 | 98 40 64 | 31.496 |
| 973 | 94 67 29 | 31.193 | 993 | 98 60 49 | 31.512 |
| 974 | 94 86 76 | 31.209 | 994 | 98 80 36 | 31.528 |
| 975 | 95 06 25 | 31.225 | 995 | 99 00 25 | 31.544 |
| 976 | 95 25 76 | 31.241 | 996 | 99 20 16 | 31.559 |
| 977 | 95 45 29 | 31.257 | 997 | 99 40 09 | 31.575 |
| 978 | 95 64 84 | 31.273 | 998 | 99 60 04 | 31.591 |
| 979 | 95 84 41 | 31.289 | 999 | 99 80 01 | 31.607 |
| 980 | 96 04 00 | 31.305 | 1000 | 100 00 00 | 31.623 |

TABLE B
Normal Curve Table

| Z | Area $-\infty$ to Z | y | Z | Area $-\infty$ to Z | y | Z | Area $-\infty$ to Z | y | Z | Area $-\infty$ to Z | y |
|---|---|---|---|---|---|---|---|---|---|---|---|
| 0.00 | .5000 | .3989 | 0.50 | .6915 | .3521 | 1.00 | .8413 | .2420 | 1.50 | .9332 | .1295 |
| 0.01 | .5040 | .3989 | 0.51 | .6950 | .3503 | 1.01 | .8438 | .2396 | 1.51 | .9345 | .1276 |
| 0.02 | .5080 | .3989 | 0.52 | .6985 | .3485 | 1.02 | .8461 | .2371 | 1.52 | .9357 | .1257 |
| 0.03 | .5120 | .3988 | 0.53 | .7019 | .3467 | 1.03 | .8485 | .2347 | 1.53 | .9370 | .1238 |
| 0.04 | .5160 | .3986 | 0.54 | .7054 | .3448 | 1.04 | .8508 | .2323 | 1.54 | .9382 | .1219 |
| 0.05 | .5199 | .3984 | 0.55 | .7088 | .3429 | 1.05 | .8531 | .2299 | 1.55 | .9394 | .1200 |
| 0.06 | .5239 | .3982 | 0.56 | .7123 | .3410 | 1.06 | .8554 | .2275 | 1.56 | .9406 | .1182 |
| 0.07 | .5279 | .3980 | 0.57 | .7157 | .3391 | 1.07 | .8577 | .2251 | 1.57 | .9418 | .1163 |
| 0.08 | .5319 | .3977 | 0.58 | .7190 | .3372 | 1.08 | .8599 | .2227 | 1.58 | .9429 | .1145 |
| 0.09 | .5359 | .3973 | 0.59 | .7224 | .3352 | 1.09 | .8621 | .2203 | 1.59 | .9441 | .1127 |
| 0.10 | .5398 | .3970 | 0.60 | .7257 | .3332 | 1.10 | .8643 | .2179 | 1.60 | .9452 | .1109 |
| 0.11 | .5438 | .3965 | 0.61 | .7291 | .3312 | 1.11 | .8665 | .2155 | 1.61 | .9463 | .1092 |
| 0.12 | .5478 | .3961 | 0.62 | .7324 | .3292 | 1.12 | .8686 | .2131 | 1.62 | .9474 | .1074 |
| 0.13 | .5517 | .3956 | 0.63 | .7357 | .3271 | 1.13 | .8708 | .2107 | 1.63 | .9484 | .1057 |
| 0.14 | .5557 | .3951 | 0.64 | .7389 | .3251 | 1.14 | .8729 | .2083 | 1.64 | .9495 | .1040 |
| 0.15 | .5596 | .3945 | 0.65 | .7422 | .3230 | 1.15 | .8749 | .2059 | 1.65 | .9505 | .1023 |
| 0.16 | .5636 | .3939 | 0.66 | .7454 | .3209 | 1.16 | .8770 | .2036 | 1.66 | .9515 | .1006 |
| 0.17 | .5675 | .3932 | 0.67 | .7486 | .3187 | 1.17 | .8790 | .2012 | 1.67 | .9525 | .0989 |
| 0.18 | .5714 | .3925 | 0.68 | .7517 | .3166 | 1.18 | .8810 | .1989 | 1.68 | .9535 | .0973 |
| 0.19 | .5753 | .3918 | 0.69 | .7549 | .3144 | 1.19 | .8830 | .1965 | 1.69 | .9545 | .0957 |
| 0.20 | .5793 | .3910 | 0.70 | .7580 | .3123 | 1.20 | .8849 | .1942 | 1.70 | .9554 | .0940 |
| 0.21 | .5832 | .3902 | 0.71 | .7611 | .3101 | 1.21 | .8869 | .1919 | 1.71 | .9564 | .0925 |
| 0.22 | .5871 | .3894 | 0.72 | .7642 | .3079 | 1.22 | .8888 | .1895 | 1.72 | .9573 | .0909 |
| 0.23 | .5910 | .3885 | 0.73 | .7673 | .3056 | 1.23 | .8907 | .1872 | 1.73 | .9582 | .0893 |
| 0.24 | .5948 | .3876 | 0.74 | .7704 | .3034 | 1.24 | .8925 | .1849 | 1.74 | .9591 | .0878 |
| 0.25 | .5987 | .3867 | 0.75 | .7734 | .3011 | 1.25 | .8944 | .1826 | 1.75 | .9599 | .0863 |
| 0.26 | .6026 | .3857 | 0.76 | .7764 | .2989 | 1.26 | .8962 | .1804 | 1.76 | .9608 | .0848 |
| 0.27 | .6064 | .3847 | 0.77 | .7794 | .2966 | 1.27 | .8980 | .1781 | 1.77 | .9616 | .0833 |
| 0.28 | .6103 | .3836 | 0.78 | .7823 | .2943 | 1.28 | .8997 | .1758 | 1.78 | .9625 | .0818 |
| 0.29 | .6141 | .3825 | 0.79 | .7852 | .2920 | 1.29 | .9015 | .1736 | 1.79 | .9633 | .0804 |
| 0.30 | .6179 | .3814 | 0.80 | .7881 | .2897 | 1.30 | .9032 | .1714 | 1.80 | .9641 | .0790 |
| 0.31 | .6217 | .3802 | 0.81 | .7910 | .2874 | 1.31 | .9049 | .1691 | 1.81 | .9649 | .0775 |
| 0.32 | .6255 | .3790 | 0.82 | .7939 | .2850 | 1.32 | .9066 | .1669 | 1.82 | .9656 | .0761 |
| 0.33 | .6293 | .3778 | 0.83 | .7967 | .2827 | 1.33 | .9082 | .1647 | 1.83 | .9664 | .0748 |
| 0.34 | .6331 | .3765 | 0.84 | .7995 | .2803 | 1.34 | .9099 | .1626 | 1.84 | .9671 | .0734 |
| 0.35 | .6368 | .3752 | 0.85 | .8023 | .2780 | 1.35 | .9115 | .1604 | 1.85 | .9678 | .0721 |
| 0.36 | .6406 | .3739 | 0.86 | .8051 | .2756 | 1.36 | .9131 | .1582 | 1.86 | .9686 | .0707 |
| 0.37 | .6443 | .3726 | 0.87 | .8078 | .2732 | 1.37 | .9147 | .1561 | 1.87 | .9693 | .0694 |
| 0.38 | .6480 | .3712 | 0.88 | .8106 | .2709 | 1.38 | .9162 | .1539 | 1.88 | .9699 | .0681 |
| 0.39 | .6517 | .3697 | 0.89 | .8133 | .2685 | 1.39 | .9177 | .1518 | 1.89 | .9706 | .0669 |
| 0.40 | .6554 | .3683 | 0.90 | .8159 | .2661 | 1.40 | .9192 | .1497 | 1.90 | .9713 | .0656 |
| 0.41 | .6591 | .3668 | 0.91 | .8186 | .2637 | 1.41 | .9207 | .1476 | 1.91 | .9719 | .0644 |
| 0.42 | .6628 | .3653 | 0.92 | .8212 | .2613 | 1.42 | .9222 | .1456 | 1.92 | .9726 | .0632 |
| 0.43 | .6664 | .3637 | 0.93 | .8238 | .2589 | 1.43 | .9236 | .1435 | 1.93 | .9732 | .0620 |
| 0.44 | .6700 | .3621 | 0.94 | .8264 | .2565 | 1.44 | .9251 | .1415 | 1.94 | .9738 | .0608 |
| 0.45 | .6736 | .3605 | 0.95 | .8289 | .2541 | 1.45 | .9265 | .1394 | 1.95 | .9744 | .0596 |
| 0.46 | .6772 | .3589 | 0.96 | .8315 | .2516 | 1.46 | .9279 | .1374 | 1.96 | .9750 | .0584 |
| 0.47 | .6808 | .3572 | 0.97 | .8340 | .2492 | 1.47 | .9292 | .1354 | 1.97 | .9756 | .0573 |
| 0.48 | .6844 | .3555 | 0.98 | .8365 | .2468 | 1.48 | .9306 | .1334 | 1.98 | .9761 | .0562 |
| 0.49 | .6879 | .3538 | 0.99 | .8389 | .2444 | 1.49 | .9319 | .1315 | 1.99 | .9767 | .0551 |

TABLE B—*(continued)*

| Z | Area −∞ to Z | y | Z | Area −∞ to Z | y | Z | Area −∞ to Z | y | Z | Area −∞ to Z | y |
|---|---|---|---|---|---|---|---|---|---|---|---|
| 2.00 | .9772 | .0540 | 2.50 | .9938 | .0175 | 3.00 | .9987 | .0044 | 3.50 | .9998 | .0009 |
| 2.01 | .9778 | .0529 | 2.51 | .9940 | .0171 | 3.01 | .9987 | .0043 | 3.51 | .9998 | .0008 |
| 2.02 | .9783 | .0519 | 2.52 | .9941 | .0167 | 3.02 | .9987 | .0042 | 3.52 | .9998 | .0008 |
| 2.03 | .9788 | .0508 | 2.53 | .9943 | .0163 | 3.03 | .9988 | .0040 | 3.53 | .9998 | .0008 |
| 2.04 | .9793 | .0498 | 2.54 | .9945 | .0158 | 3.04 | .9988 | .0039 | 3.54 | .9998 | .0008 |
| 2.05 | .9798 | .0488 | 2.55 | .9946 | .0154 | 3.05 | .9989 | .0038 | 3.55 | .9998 | .0007 |
| 2.06 | .9803 | .0478 | 2.56 | .9948 | .0151 | 3.06 | .9989 | .0037 | 3.56 | .9998 | .0007 |
| 2.07 | .9808 | .0468 | 2.57 | .9949 | .0147 | 3.07 | .9989 | .0036 | 3.57 | .9998 | .0007 |
| 2.08 | .9812 | .0459 | 2.58 | .9951 | .0143 | 3.08 | .9990 | .0035 | 3.58 | .9998 | .0007 |
| 2.09 | .9817 | .0449 | 2.59 | .9952 | .0139 | 3.09 | .9990 | .0034 | 3.59 | .9998 | .0006 |
| 2.10 | .9821 | .0440 | 2.60 | .9953 | .0136 | 3.10 | .9990 | .0033 | 3.60 | .9998 | .0006 |
| 2.11 | .9826 | .0431 | 2.61 | .9955 | .0132 | 3.11 | .9991 | .0032 | 3.61 | .9999 | .0006 |
| 2.12 | .9830 | .0422 | 2.62 | .9956 | .0129 | 3.12 | .9991 | .0031 | 3.62 | .9999 | .0006 |
| 2.13 | .9834 | .0413 | 2.63 | .9957 | .0126 | 3.13 | .9991 | .0030 | 3.63 | .9999 | .0006 |
| 2.14 | .9838 | .0404 | 2.64 | .9959 | .0122 | 3.14 | .9992 | .0029 | 3.64 | .9999 | .0005 |
| 2.15 | .9842 | .0395 | 2.65 | .9960 | .0119 | 3.15 | .9992 | .0028 | 3.65 | .9999 | .0005 |
| 2.16 | .9846 | .0387 | 2.66 | .9961 | .0116 | 3.16 | .9992 | .0027 | 3.66 | .9999 | .0005 |
| 2.17 | .9850 | .0379 | 2.67 | .9962 | .0113 | 3.17 | .9992 | .0026 | 3.67 | .9999 | .0005 |
| 2.18 | .9854 | .0371 | 2.68 | .9963 | .0110 | 3.18 | .9993 | .0025 | 3.68 | .9999 | .0005 |
| 2.19 | .9857 | .0363 | 2.69 | .9964 | .0107 | 3.19 | .9993 | .0025 | 3.69 | .9999 | .0004 |
| 2.20 | .9861 | .0355 | 2.70 | .9965 | .0104 | 3.20 | .9993 | .0024 | 3.70 | .9999 | .0004 |
| 2.21 | .9864 | .0347 | 2.71 | .9966 | .0101 | 3.21 | .9993 | .0023 | 3.71 | .9999 | .0004 |
| 2.22 | .9868 | .0339 | 2.72 | .9967 | .0099 | 3.22 | .9994 | .0022 | 3.72 | .9999 | .0004 |
| 2.23 | .9871 | .0332 | 2.73 | .9968 | .0096 | 3.23 | .9994 | .0022 | 3.73 | .9999 | .0004 |
| 2.24 | .9875 | .0325 | 2.74 | .9969 | .0093 | 3.24 | .9994 | .0021 | 3.74 | .9999 | .0004 |
| 2.25 | .9878 | .0317 | 2.75 | .9970 | .0091 | 3.25 | .9994 | .0020 | 3.75 | .9999 | .0004 |
| 2.26 | .9881 | .0310 | 2.76 | .9971 | .0088 | 3.26 | .9994 | .0020 | 3.76 | .9999 | .0003 |
| 2.27 | .9884 | .0303 | 2.77 | .9972 | .0086 | 3.27 | .9995 | .0019 | 3.77 | .9999 | .0003 |
| 2.28 | .9887 | .0297 | 2.78 | .9973 | .0084 | 3.28 | .9995 | .0018 | 3.78 | .9999 | .0003 |
| 2.29 | .9890 | .0290 | 2.79 | .9974 | .0081 | 3.29 | .9995 | .0018 | 3.79 | .9999 | .0003 |
| 2.30 | .9893 | .0283 | 2.80 | .9974 | .0079 | 3.30 | .9995 | .0017 | 3.80 | .9999 | .0003 |
| 2.31 | .9896 | .0277 | 2.81 | .9975 | .0077 | 3.31 | .9995 | .0017 | 3.81 | .9999 | .0003 |
| 2.32 | .9898 | .0270 | 2.82 | .9976 | .0075 | 3.32 | .9996 | .0016 | 3.82 | .9999 | .0003 |
| 2.33 | .9901 | .0264 | 2.83 | .9977 | .0073 | 3.33 | .9996 | .0016 | 3.83 | .9999 | .0003 |
| 2.34 | .9904 | .0258 | 2.84 | .9977 | .0071 | 3.34 | .9996 | .0015 | 3.84 | .9999 | .0003 |
| 2.35 | .9906 | .0252 | 2.85 | .9978 | .0069 | 3.35 | .9996 | .0015 | 3.85 | .9999 | .0002 |
| 2.36 | .9909 | .0246 | 2.86 | .9979 | .0067 | 3.36 | .9996 | .0014 | 3.86 | .9999 | .0002 |
| 2.37 | .9911 | .0241 | 2.87 | .9979 | .0065 | 3.37 | .9996 | .0014 | 3.87 | 1.0000 | .0002 |
| 2.38 | .9913 | .0235 | 2.88 | .9980 | .0063 | 3.38 | .9996 | .0013 | 3.88 | 1.0000 | .0002 |
| 2.39 | .9916 | .0229 | 2.89 | .9981 | .0061 | 3.39 | .9997 | .0013 | 3.89 | 1.0000 | .0002 |
| 2.40 | .9918 | .0224 | 2.90 | .9981 | .0060 | 3.40 | .9997 | .0012 | 3.90 | 1.0000 | .0002 |
| 2.41 | .9920 | .0219 | 2.91 | .9982 | .0058 | 3.41 | .9997 | .0012 | 3.91 | 1.0000 | .0002 |
| 2.42 | .9922 | .0213 | 2.92 | .9982 | .0056 | 3.42 | .9997 | .0012 | 3.92 | 1.0000 | .0002 |
| 2.43 | .9925 | .0208 | 2.93 | .9983 | .0055 | 3.43 | .9997 | .0011 | 3.93 | 1.0000 | .0002 |
| 2.44 | .9927 | .0203 | 2.94 | .9984 | .0053 | 3.44 | .9997 | .0011 | 3.94 | 1.0000 | .0002 |
| 2.45 | .9929 | .0198 | 2.95 | .9984 | .0051 | 3.45 | .9997 | .0010 | 3.95 | 1.0000 | .0002 |
| 2.46 | .9931 | .0194 | 2.96 | .9985 | .0050 | 3.46 | .9997 | .0010 | 3.96 | 1.0000 | .0002 |
| 2.47 | .9932 | .0189 | 2.97 | .9985 | .0048 | 3.47 | .9997 | .0010 | 3.97 | 1.0000 | .0002 |
| 2.48 | .9934 | .0184 | 2.98 | .9986 | .0047 | 3.48 | .9998 | .0009 | 3.98 | 1.0000 | .0001 |
| 2.49 | .9936 | .0180 | 2.99 | .9986 | .0046 | 3.49 | .9998 | .0009 | 3.99 | 1.0000 | .0001 |

TABLE C
Table of the t-Distribution

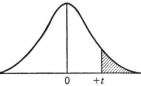

The quantity tabled is the value of $t$ which for a given number of degrees of freedom is exceeded with the given probability. The area is shown in the diagram.

| Degrees of freedom | Probability | | | | |
|---|---|---|---|---|---|
| | .10 | .05 | .025 | .01 | .005 |
| 1 | 3.078 | 6.314 | 12.706 | 31.821 | 63.657 |
| 2 | 1.886 | 2.920 | 4.303 | 6.965 | 9.925 |
| 3 | 1.638 | 2.353 | 3.182 | 4.541 | 5.841 |
| 4 | 1.533 | 2.132 | 2.776 | 3.747 | 4.604 |
| 5 | 1.476 | 2.015 | 2.571 | 3.365 | 4.032 |
| 6 | 1.440 | 1.943 | 2.447 | 3.143 | 3.707 |
| 7 | 1.415 | 1.895 | 2.365 | 2.998 | 3.499 |
| 8 | 1.397 | 1.860 | 2.306 | 2.896 | 3.355 |
| 9 | 1.383 | 1.833 | 2.262 | 2.821 | 3.250 |
| 10 | 1.372 | 1.812 | 2.228 | 2.764 | 3.169 |
| 11 | 1.363 | 1.796 | 2.201 | 2.718 | 3.106 |
| 12 | 1.356 | 1.782 | 2.179 | 2.681 | 3.055 |
| 13 | 1.350 | 1.771 | 2.160 | 2.650 | 3.012 |
| 14 | 1.345 | 1.761 | 2.145 | 2.624 | 2.997 |
| 15 | 1.341 | 1.753 | 2.131 | 2.602 | 2.947 |
| 16 | 1.337 | 1.746 | 2.120 | 2.583 | 2.921 |
| 17 | 1.333 | 1.740 | 2.110 | 2.567 | 2.898 |
| 18 | 1.330 | 1.734 | 2.101 | 2.552 | 2.878 |
| 19 | 1.328 | 1.729 | 2.093 | 2.539 | 2.861 |
| 20 | 1.325 | 1.725 | 2.086 | 2.528 | 2.845 |
| 21 | 1.323 | 1.721 | 2.080 | 2.518 | 2.831 |
| 22 | 1.321 | 1.717 | 2.074 | 2.508 | 2.819 |
| 23 | 1.319 | 1.714 | 2.069 | 2.500 | 2.807 |
| 24 | 1.318 | 1.711 | 2.064 | 2.492 | 2.797 |
| 25 | 1.316 | 1.708 | 2.060 | 2.485 | 2.787 |
| 26 | 1.315 | 1.706 | 2.056 | 2.479 | 2.779 |
| 27 | 1.314 | 1.703 | 2.052 | 2.473 | 2.771 |
| 28 | 1.313 | 1.701 | 2.048 | 2.467 | 2.763 |
| 29 | 1.311 | 1.699 | 2.045 | 2.462 | 2.756 |
| 30 | 1.310 | 1.697 | 2.042 | 2.457 | 2.750 |
| 40 | 1.303 | 1.684 | 2.021 | 2.423 | 2.704 |
| 60 | 1.296 | 1.671 | 2.000 | 2.390 | 2.660 |
| 120 | 1.289 | 1.658 | 1.980 | 2.358 | 2.617 |
| ∞ | 1.282 | 1.645 | 1.960 | 2.326 | 2.576 |

Source: Fisher and Yates, *Statistical Tables for Biological, Agricultural and Medical Research*, 6th ed., 1974c. Abridged with permission from Longman Group Limited, Essex, England.

TABLE D
Table of the Chi Square Distribution* (The tabled value is the value $\chi^2$ which, for a given number of degrees of freedom, is exceeded with the given probability.)

| d.f. | Probability | | | | |
|---|---|---|---|---|---|
| | .10 | .05 | .025 | .01 | .005 |
| 1 | 2.71 | 3.84 | 5.02 | 6.63 | 7.88 |
| 2 | 4.61 | 5.99 | 7.38 | 9.21 | 10.60 |
| 3 | 6.25 | 7.81 | 9.35 | 11.34 | 12.84 |
| 4 | 7.78 | 9.49 | 11.14 | 13.28 | 14.86 |
| 5 | 9.24 | 11.07 | 12.83 | 15.09 | 16.75 |
| 6 | 10.64 | 12.59 | 14.45 | 16.81 | 18.55 |
| 7 | 12.02 | 14.07 | 16.01 | 18.48 | 20.28 |
| 8 | 13.36 | 15.51 | 17.53 | 20.09 | 21.96 |
| 9 | 14.68 | 16.92 | 19.02 | 21.67 | 23.59 |
| 10 | 15.99 | 18.31 | 20.48 | 23.21 | 25.19 |
| 11 | 17.28 | 19.68 | 21.92 | 24.73 | 26.76 |
| 12 | 18.55 | 21.03 | 23.34 | 26.22 | 28.30 |
| 13 | 19.81 | 22.36 | 24.74 | 27.69 | 29.82 |
| 14 | 21.06 | 23.68 | 26.12 | 29.14 | 31.32 |
| 15 | 22.31 | 25.00 | 27.49 | 30.58 | 32.80 |
| 16 | 23.54 | 26.30 | 28.85 | 32.00 | 34.27 |
| 18 | 25.99 | 28.87 | 31.53 | 34.81 | 37.16 |
| 20 | 28.41 | 31.41 | 34.17 | 37.57 | 40.00 |
| 24 | 33.20 | 36.42 | 39.36 | 42.98 | 45.56 |
| 30 | 40.26 | 43.77 | 46.98 | 50.89 | 53.67 |
| 40 | 51.81 | 55.76 | 59.34 | 63.69 | 66.77 |
| 60 | 74.40 | 79.08 | 83.30 | 88.38 | 91.95 |
| 120 | 140.23 | 146.57 | 152.21 | 158.95 | 163.64 |

* Table D is abridged from Table IV of Fisher and Yates, "Statistical Tables for Biological, Agricultural, and Medical Research," published by Oliver & Boyd, Ltd., Edinburgh, by permission of the authors and publishers.

For larger values of $n$, the expression $\sqrt{2\chi^2} - \sqrt{2n-1}$ may be used as a normal deviate with standard deviation equal to 1, remembering that the probability for $\chi^2$ corresponds with that of a single tail of the normal curve.

# TABLE E
## Critical Values of F* (0.05 level in roman type, 0.01 level in bold face)

Degrees of freedom for greater mean square [numerator]

| denom | 1 | 2 | 3 | 4 | 5 | 6 | 7 | 8 | 9 | 10 | 11 | 12 | 14 | 16 | 20 | 24 | 30 | 40 | 50 | 75 | 100 | 200 | 500 | ∞ |
|---|---|---|---|---|---|---|---|---|---|---|---|---|---|---|---|---|---|---|---|---|---|---|---|---|
| 1 | 161 **4,052** | 200 **4,999** | 216 **5,403** | 225 **5,625** | 230 **5,764** | 234 **5,859** | 237 **5,928** | 239 **5,981** | 241 **6,022** | 242 **6,056** | 243 **6,082** | 244 **6,106** | 245 **6,142** | 246 **6,169** | 248 **6,208** | 249 **6,234** | 250 **6,261** | 251 **6,286** | 252 **6,302** | 253 **6,323** | 253 **6,334** | 254 **6,352** | 254 **6,361** | 254 **6,366** |
| 2 | 18.51 **98.49** | 19.00 **99.00** | 19.16 **99.17** | 19.25 **99.25** | 19.30 **99.30** | 19.33 **99.33** | 19.36 **99.36** | 19.37 **99.37** | 19.38 **99.39** | 19.39 **99.40** | 19.40 **99.41** | 19.41 **99.42** | 19.42 **99.43** | 19.43 **99.44** | 19.44 **99.45** | 19.45 **99.46** | 19.46 **99.47** | 19.47 **99.48** | 19.47 **99.48** | 19.48 **99.49** | 19.49 **99.49** | 19.49 **99.49** | 19.50 **99.50** | 19.50 **99.50** |
| 3 | 10.13 **34.12** | 9.55 **30.82** | 9.28 **29.46** | 9.12 **28.71** | 9.01 **28.24** | 8.94 **27.91** | 8.88 **27.67** | 8.84 **27.49** | 8.81 **27.34** | 8.78 **27.23** | 8.76 **27.13** | 8.74 **27.05** | 8.71 **26.92** | 8.69 **26.83** | 8.66 **26.69** | 8.64 **26.60** | 8.62 **26.50** | 8.60 **26.41** | 8.58 **26.35** | 8.57 **26.27** | 8.56 **26.23** | 8.54 **26.18** | 8.54 **26.14** | 8.53 **26.12** |
| 4 | 7.71 **21.20** | 6.94 **18.00** | 6.59 **16.69** | 6.39 **15.98** | 6.26 **15.52** | 6.16 **15.21** | 6.09 **14.98** | 6.04 **14.80** | 6.00 **14.66** | 5.96 **14.54** | 5.93 **14.45** | 5.91 **14.37** | 5.87 **14.24** | 5.84 **14.15** | 5.80 **14.02** | 5.77 **13.93** | 5.74 **13.83** | 5.71 **13.74** | 5.70 **13.69** | 5.68 **13.61** | 5.66 **13.57** | 5.65 **13.52** | 5.64 **13.48** | 5.63 **13.46** |
| 5 | 6.61 **16.26** | 5.79 **13.27** | 5.41 **12.06** | 5.19 **11.39** | 5.05 **10.97** | 4.95 **10.67** | 4.88 **10.45** | 4.82 **10.29** | 4.78 **10.15** | 4.74 **10.05** | 4.70 **9.96** | 4.68 **9.89** | 4.64 **9.77** | 4.60 **9.68** | 4.56 **9.55** | 4.53 **9.47** | 4.50 **9.38** | 4.46 **9.29** | 4.44 **9.24** | 4.42 **9.17** | 4.40 **9.13** | 4.38 **9.07** | 4.37 **9.04** | 4.36 **9.02** |
| 6 | 5.99 **13.74** | 5.14 **10.92** | 4.76 **9.78** | 4.53 **9.15** | 4.39 **8.75** | 4.28 **8.47** | 4.21 **8.26** | 4.15 **8.10** | 4.10 **7.98** | 4.06 **7.87** | 4.03 **7.79** | 4.00 **7.72** | 3.96 **7.60** | 3.92 **7.52** | 3.87 **7.39** | 3.84 **7.31** | 3.81 **7.23** | 3.77 **7.14** | 3.75 **7.09** | 3.72 **7.02** | 3.71 **6.99** | 3.69 **6.94** | 3.68 **6.90** | 3.67 **6.88** |
| 7 | 5.59 **12.25** | 4.74 **9.55** | 4.35 **8.45** | 4.12 **7.85** | 3.97 **7.46** | 3.87 **7.19** | 3.79 **7.00** | 3.73 **6.84** | 3.68 **6.71** | 3.63 **6.62** | 3.60 **6.54** | 3.57 **6.47** | 3.52 **6.35** | 3.49 **6.27** | 3.44 **6.15** | 3.41 **6.07** | 3.38 **5.98** | 3.34 **5.90** | 3.32 **5.85** | 3.29 **5.78** | 3.28 **5.75** | 3.25 **5.70** | 3.24 **5.67** | 3.23 **5.65** |
| 8 | 5.32 **11.26** | 4.46 **8.65** | 4.07 **7.59** | 3.84 **7.01** | 3.69 **6.63** | 3.58 **6.37** | 3.50 **6.19** | 3.44 **6.03** | 3.39 **5.91** | 3.34 **5.82** | 3.31 **5.74** | 3.28 **5.67** | 3.23 **5.56** | 3.20 **5.48** | 3.15 **5.36** | 3.12 **5.28** | 3.08 **5.20** | 3.05 **5.11** | 3.03 **5.06** | 3.00 **5.00** | 2.98 **4.96** | 2.96 **4.91** | 2.94 **4.88** | 2.93 **4.86** |
| 9 | 5.12 **10.56** | 4.26 **8.02** | 3.86 **6.99** | 3.63 **6.42** | 3.48 **6.06** | 3.37 **5.80** | 3.29 **5.62** | 3.23 **5.47** | 3.18 **5.35** | 3.13 **5.26** | 3.10 **5.18** | 3.07 **5.11** | 3.02 **5.00** | 2.98 **4.92** | 2.93 **4.80** | 2.90 **4.73** | 2.86 **4.64** | 2.82 **4.56** | 2.80 **4.51** | 2.77 **4.45** | 2.76 **4.41** | 2.73 **4.36** | 2.72 **4.33** | 2.71 **4.31** |
| 10 | 4.96 **10.04** | 4.10 **7.56** | 3.71 **6.55** | 3.48 **5.99** | 3.33 **5.64** | 3.22 **5.39** | 3.14 **5.21** | 3.07 **5.06** | 3.02 **4.95** | 2.97 **4.85** | 2.94 **4.78** | 2.91 **4.71** | 2.86 **4.60** | 2.82 **4.52** | 2.77 **4.41** | 2.74 **4.33** | 2.70 **4.25** | 2.67 **4.17** | 2.64 **4.12** | 2.61 **4.05** | 2.59 **4.01** | 2.56 **3.96** | 2.55 **3.93** | 2.54 **3.91** |
| 11 | 4.84 **9.65** | 3.98 **7.20** | 3.59 **6.22** | 3.36 **5.67** | 3.20 **5.32** | 3.09 **5.07** | 3.01 **4.88** | 2.95 **4.74** | 2.90 **4.63** | 2.86 **4.54** | 2.82 **4.46** | 2.79 **4.40** | 2.74 **4.29** | 2.70 **4.21** | 2.65 **4.10** | 2.61 **4.02** | 2.57 **3.94** | 2.53 **3.86** | 2.50 **3.80** | 2.47 **3.74** | 2.45 **3.70** | 2.42 **3.66** | 2.41 **3.62** | 2.40 **3.60** |
| 12 | 4.75 **9.33** | 3.88 **6.93** | 3.49 **5.95** | 3.26 **5.41** | 3.11 **5.06** | 3.00 **4.82** | 2.92 **4.65** | 2.85 **4.50** | 2.80 **4.39** | 2.76 **4.30** | 2.72 **4.22** | 2.69 **4.16** | 2.64 **4.05** | 2.60 **3.98** | 2.54 **3.86** | 2.50 **3.78** | 2.46 **3.70** | 2.42 **3.61** | 2.40 **3.56** | 2.36 **3.49** | 2.35 **3.46** | 2.32 **3.41** | 2.31 **3.38** | 2.30 **3.36** |
| 13 | 4.67 **9.07** | 3.80 **6.70** | 3.41 **5.74** | 3.18 **5.20** | 3.02 **4.86** | 2.92 **4.62** | 2.84 **4.44** | 2.77 **4.30** | 2.72 **4.19** | 2.67 **4.10** | 2.63 **4.02** | 2.60 **3.96** | 2.55 **3.85** | 2.51 **3.78** | 2.46 **3.67** | 2.42 **3.59** | 2.38 **3.51** | 2.34 **3.42** | 2.32 **3.37** | 2.28 **3.30** | 2.26 **3.27** | 2.24 **3.21** | 2.22 **3.18** | 2.21 **3.16** |

[Degrees of freedom for lesser mean square (denominator)]

*Source: Reprinted by permission from *Statistical Methods* by George W. Snedecor and William G. Cochran, sixth edition © 1967 by Iowa State University Press, Ames, Iowa. The values in the table are the critical values of F for the degrees of freedom listed over the columns (the degrees of freedom for the greater mean square or numerator of the F ratio) and the degrees of freedom listed for the rows (the degrees of freedom for the lesser mean square for the denominator of the F ratio). The critical value for the .05 level of significance is presented first (roman type) followed by the critical value at the .01 level (bold face). If the observed value is *greater than or equal to* the tabled value, reject $H_0$.

TABLE E—(continued)

Degrees of freedom for lesser mean square [denominator]

Degrees of freedom for greater mean square [numerator]

| | 1 | 2 | 3 | 4 | 5 | 6 | 7 | 8 | 9 | 10 | 11 | 12 | 14 | 16 | 20 | 24 | 30 | 40 | 50 | 75 | 100 | 200 | 500 | ∞ | |
|---|---|---|---|---|---|---|---|---|---|---|---|---|---|---|---|---|---|---|---|---|---|---|---|---|---|
| 14 | 4.60 8.86 | 3.74 6.51 | 3.34 5.56 | 3.11 5.03 | 2.96 4.69 | 2.85 4.46 | 2.77 4.28 | 2.70 4.14 | 2.65 4.03 | 2.60 3.94 | 2.56 3.86 | 2.53 3.80 | 2.48 3.70 | 2.44 3.62 | 2.39 3.51 | 2.35 3.43 | 2.31 3.34 | 2.27 3.26 | 2.24 3.21 | 2.21 3.14 | 2.19 3.11 | 2.16 3.06 | 2.14 3.02 | 2.13 3.00 | 14 |
| 15 | 4.54 8.68 | 3.68 6.36 | 3.29 5.42 | 3.06 4.89 | 2.90 4.56 | 2.79 4.32 | 2.70 4.14 | 2.64 4.00 | 2.59 3.89 | 2.55 3.80 | 2.51 3.73 | 2.48 3.67 | 2.43 3.56 | 2.39 3.48 | 2.33 3.36 | 2.29 3.29 | 2.25 3.20 | 2.21 3.12 | 2.18 3.07 | 2.15 3.00 | 2.12 2.97 | 2.10 2.92 | 2.08 2.89 | 2.07 2.87 | 15 |
| 16 | 4.49 8.53 | 3.63 6.23 | 3.24 5.29 | 3.01 4.77 | 2.85 4.44 | 2.74 4.20 | 2.66 4.03 | 2.59 3.89 | 2.54 3.78 | 2.49 3.69 | 2.45 3.61 | 2.42 3.55 | 2.37 3.45 | 2.33 3.37 | 2.28 3.25 | 2.24 3.18 | 2.20 3.10 | 2.16 3.01 | 2.13 2.96 | 2.09 2.89 | 2.07 2.86 | 2.04 2.80 | 2.02 2.77 | 2.01 2.75 | 16 |
| 17 | 4.45 8.40 | 3.59 6.11 | 3.20 5.18 | 2.96 4.67 | 2.81 4.34 | 2.70 4.10 | 2.62 3.93 | 2.55 3.79 | 2.50 3.68 | 2.45 3.59 | 2.41 3.52 | 2.38 3.45 | 2.33 3.35 | 2.29 3.27 | 2.23 3.16 | 2.19 3.08 | 2.15 3.00 | 2.11 2.92 | 2.08 2.86 | 2.04 2.79 | 2.02 2.76 | 1.99 2.70 | 1.97 2.67 | 1.96 2.65 | 17 |
| 18 | 4.41 8.28 | 3.55 6.01 | 3.16 5.09 | 2.93 4.58 | 2.77 4.25 | 2.66 4.01 | 2.58 3.85 | 2.51 3.71 | 2.46 3.60 | 2.41 3.51 | 2.37 3.44 | 2.34 3.37 | 2.29 3.27 | 2.25 3.19 | 2.19 3.07 | 2.15 3.00 | 2.11 2.91 | 2.07 2.83 | 2.04 2.78 | 2.00 2.71 | 1.98 2.68 | 1.95 2.62 | 1.93 2.59 | 1.92 2.57 | 18 |
| 19 | 4.38 8.18 | 3.52 5.93 | 3.13 5.01 | 2.90 4.50 | 2.74 4.17 | 2.63 3.94 | 2.55 3.77 | 2.48 3.63 | 2.43 3.52 | 2.38 3.43 | 2.34 3.36 | 2.31 3.30 | 2.26 3.19 | 2.21 3.12 | 2.15 3.00 | 2.11 2.92 | 2.07 2.84 | 2.02 2.76 | 2.00 2.70 | 1.96 2.63 | 1.94 2.60 | 1.91 2.54 | 1.90 2.51 | 1.88 2.49 | 19 |
| 20 | 4.35 8.10 | 3.49 5.85 | 3.10 4.94 | 2.87 4.43 | 2.71 4.10 | 2.60 3.87 | 2.52 3.71 | 2.45 3.56 | 2.40 3.45 | 2.35 3.37 | 2.31 3.30 | 2.28 3.23 | 2.23 3.13 | 2.18 3.05 | 2.12 2.94 | 2.08 2.86 | 2.04 2.77 | 1.99 2.69 | 1.96 2.63 | 1.92 2.56 | 1.90 2.53 | 1.87 2.47 | 1.85 2.44 | 1.84 2.42 | 20 |
| 21 | 4.32 8.02 | 3.47 5.78 | 3.07 4.87 | 2.84 4.37 | 2.68 4.04 | 2.57 3.81 | 2.49 3.65 | 2.42 3.51 | 2.37 3.40 | 2.32 3.31 | 2.28 3.24 | 2.25 3.17 | 2.20 3.07 | 2.15 2.99 | 2.09 2.88 | 2.05 2.80 | 2.00 2.72 | 1.96 2.63 | 1.93 2.58 | 1.89 2.51 | 1.87 2.47 | 1.84 2.42 | 1.82 2.38 | 1.81 2.36 | 21 |
| 22 | 4.30 7.94 | 3.44 5.72 | 3.05 4.82 | 2.82 4.31 | 2.66 3.99 | 2.55 3.76 | 2.47 3.59 | 2.40 3.45 | 2.35 3.35 | 2.30 3.26 | 2.26 3.18 | 2.23 3.12 | 2.18 3.02 | 2.13 2.94 | 2.07 2.83 | 2.03 2.75 | 1.98 2.67 | 1.93 2.58 | 1.91 2.53 | 1.87 2.46 | 1.84 2.42 | 1.81 2.37 | 1.80 2.33 | 1.78 2.31 | 22 |
| 23 | 4.28 7.88 | 3.42 5.66 | 3.03 4.76 | 2.80 4.26 | 2.64 3.94 | 2.53 3.71 | 2.45 3.54 | 2.38 3.41 | 2.32 3.30 | 2.28 3.21 | 2.24 3.14 | 2.20 3.07 | 2.14 2.97 | 2.10 2.89 | 2.04 2.78 | 2.00 2.70 | 1.96 2.62 | 1.91 2.53 | 1.88 2.48 | 1.84 2.41 | 1.82 2.37 | 1.79 2.32 | 1.77 2.28 | 1.76 2.26 | 23 |
| 24 | 4.26 7.82 | 3.40 5.61 | 3.01 4.72 | 2.78 4.22 | 2.62 3.90 | 2.51 3.67 | 2.43 3.50 | 2.36 3.36 | 2.30 3.25 | 2.26 3.17 | 2.22 3.09 | 2.18 3.03 | 2.13 2.93 | 2.09 2.85 | 2.02 2.74 | 1.98 2.66 | 1.94 2.58 | 1.89 2.49 | 1.86 2.44 | 1.82 2.36 | 1.80 2.33 | 1.76 2.27 | 1.74 2.23 | 1.73 2.21 | 24 |
| 25 | 4.24 7.77 | 3.38 5.57 | 2.99 4.68 | 2.76 4.18 | 2.60 3.86 | 2.49 3.63 | 2.41 3.46 | 2.34 3.32 | 2.28 3.21 | 2.24 3.13 | 2.20 3.05 | 2.16 2.99 | 2.11 2.89 | 2.06 2.81 | 2.00 2.70 | 1.96 2.62 | 1.92 2.54 | 1.87 2.45 | 1.84 2.40 | 1.80 2.32 | 1.77 2.29 | 1.74 2.23 | 1.72 2.19 | 1.71 2.17 | 25 |
| 26 | 4.22 7.72 | 3.37 5.53 | 2.98 4.64 | 2.74 4.14 | 2.59 3.82 | 2.47 3.59 | 2.39 3.42 | 2.32 3.29 | 2.27 3.17 | 2.22 3.09 | 2.18 3.02 | 2.15 2.96 | 2.10 2.86 | 2.05 2.77 | 1.99 2.66 | 1.95 2.58 | 1.90 2.50 | 1.85 2.41 | 1.82 2.36 | 1.78 2.28 | 1.76 2.25 | 1.72 2.19 | 1.70 2.15 | 1.69 2.13 | 26 |

TABLE E—(continued)

Degrees of freedom for greater mean square [numerator]

| (denom) | 1 | 2 | 3 | 4 | 5 | 6 | 7 | 8 | 9 | 10 | 11 | 12 | 14 | 16 | 20 | 24 | 30 | 40 | 50 | 75 | 100 | 200 | 500 | ∞ |
|---|---|---|---|---|---|---|---|---|---|---|---|---|---|---|---|---|---|---|---|---|---|---|---|---|
| 27 | 4.21 / 7.68 | 3.35 / 5.49 | 2.96 / 4.60 | 2.73 / 4.11 | 2.57 / 3.79 | 2.46 / 3.56 | 2.37 / 3.39 | 2.30 / 3.26 | 2.25 / 3.14 | 2.20 / 3.06 | 2.16 / 2.98 | 2.13 / 2.93 | 2.08 / 2.83 | 2.03 / 2.74 | 1.97 / 2.63 | 1.93 / 2.55 | 1.88 / 2.47 | 1.84 / 2.38 | 1.80 / 2.33 | 1.76 / 2.25 | 1.74 / 2.21 | 1.71 / 2.16 | 1.68 / 2.12 | 1.67 / 2.10 |
| 28 | 4.20 / 7.64 | 3.34 / 5.45 | 2.95 / 4.57 | 2.71 / 4.07 | 2.56 / 3.76 | 2.44 / 3.53 | 2.36 / 3.36 | 2.29 / 3.23 | 2.24 / 3.11 | 2.19 / 3.03 | 2.15 / 2.95 | 2.12 / 2.90 | 2.06 / 2.80 | 2.02 / 2.71 | 1.96 / 2.60 | 1.91 / 2.52 | 1.87 / 2.44 | 1.81 / 2.35 | 1.78 / 2.30 | 1.75 / 2.22 | 1.72 / 2.18 | 1.69 / 2.13 | 1.67 / 2.09 | 1.65 / 2.06 |
| 29 | 4.18 / 7.60 | 3.33 / 5.42 | 2.93 / 4.54 | 2.70 / 4.04 | 2.54 / 3.73 | 2.43 / 3.50 | 2.35 / 3.33 | 2.28 / 3.20 | 2.22 / 3.08 | 2.18 / 3.00 | 2.14 / 2.92 | 2.10 / 2.87 | 2.05 / 2.77 | 2.00 / 2.68 | 1.94 / 2.57 | 1.90 / 2.49 | 1.85 / 2.41 | 1.80 / 2.32 | 1.77 / 2.27 | 1.73 / 2.19 | 1.71 / 2.15 | 1.68 / 2.10 | 1.65 / 2.06 | 1.64 / 2.03 |
| 30 | 4.17 / 7.56 | 3.32 / 5.39 | 2.92 / 4.51 | 2.69 / 4.02 | 2.53 / 3.70 | 2.42 / 3.47 | 2.34 / 3.30 | 2.27 / 3.17 | 2.21 / 3.06 | 2.16 / 2.98 | 2.12 / 2.90 | 2.09 / 2.84 | 2.04 / 2.74 | 1.99 / 2.66 | 1.93 / 2.55 | 1.89 / 2.47 | 1.84 / 2.38 | 1.79 / 2.29 | 1.76 / 2.24 | 1.72 / 2.16 | 1.69 / 2.13 | 1.66 / 2.07 | 1.64 / 2.03 | 1.62 / 2.01 |
| 32 | 4.15 / 7.50 | 3.30 / 5.34 | 2.90 / 4.46 | 2.67 / 3.97 | 2.51 / 3.66 | 2.40 / 3.42 | 2.32 / 3.25 | 2.25 / 3.12 | 2.19 / 3.01 | 2.14 / 2.94 | 2.10 / 2.86 | 2.07 / 2.80 | 2.02 / 2.70 | 1.97 / 2.62 | 1.91 / 2.51 | 1.86 / 2.42 | 1.82 / 2.34 | 1.76 / 2.25 | 1.74 / 2.20 | 1.69 / 2.12 | 1.67 / 2.08 | 1.64 / 2.02 | 1.61 / 1.98 | 1.59 / 1.96 |
| 34 | 4.13 / 7.44 | 3.28 / 5.29 | 2.88 / 4.42 | 2.65 / 3.93 | 2.49 / 3.61 | 2.38 / 3.38 | 2.30 / 3.21 | 2.23 / 3.08 | 2.17 / 2.97 | 2.12 / 2.89 | 2.08 / 2.82 | 2.05 / 2.76 | 2.00 / 2.66 | 1.95 / 2.58 | 1.89 / 2.47 | 1.84 / 2.38 | 1.80 / 2.30 | 1.74 / 2.21 | 1.71 / 2.15 | 1.67 / 2.08 | 1.64 / 2.04 | 1.61 / 1.98 | 1.59 / 1.94 | 1.57 / 1.91 |
| 36 | 4.11 / 7.39 | 3.26 / 5.25 | 2.86 / 4.38 | 2.63 / 3.89 | 2.48 / 3.58 | 2.36 / 3.35 | 2.28 / 3.18 | 2.21 / 3.04 | 2.15 / 2.94 | 2.10 / 2.86 | 2.06 / 2.78 | 2.03 / 2.72 | 1.98 / 2.62 | 1.93 / 2.54 | 1.87 / 2.43 | 1.82 / 2.35 | 1.78 / 2.26 | 1.72 / 2.17 | 1.69 / 2.12 | 1.65 / 2.04 | 1.62 / 2.00 | 1.59 / 1.94 | 1.56 / 1.90 | 1.55 / 1.87 |
| 38 | 4.10 / 7.35 | 3.25 / 5.21 | 2.85 / 4.34 | 2.62 / 3.86 | 2.46 / 3.54 | 2.35 / 3.32 | 2.26 / 3.15 | 2.19 / 3.02 | 2.14 / 2.91 | 2.09 / 2.82 | 2.05 / 2.75 | 2.02 / 2.69 | 1.96 / 2.59 | 1.92 / 2.51 | 1.85 / 2.40 | 1.80 / 2.32 | 1.76 / 2.22 | 1.71 / 2.14 | 1.67 / 2.08 | 1.63 / 2.00 | 1.60 / 1.97 | 1.57 / 1.90 | 1.54 / 1.86 | 1.53 / 1.84 |
| 40 | 4.08 / 7.31 | 3.23 / 5.18 | 2.84 / 4.31 | 2.61 / 3.83 | 2.45 / 3.51 | 2.34 / 3.29 | 2.25 / 3.12 | 2.18 / 2.99 | 2.12 / 2.88 | 2.07 / 2.80 | 2.04 / 2.73 | 2.00 / 2.66 | 1.95 / 2.56 | 1.90 / 2.49 | 1.84 / 2.37 | 1.79 / 2.29 | 1.74 / 2.20 | 1.69 / 2.11 | 1.66 / 2.05 | 1.61 / 1.97 | 1.59 / 1.94 | 1.55 / 1.88 | 1.53 / 1.84 | 1.51 / 1.81 |
| 42 | 4.07 / 7.27 | 3.22 / 5.15 | 2.83 / 4.29 | 2.59 / 3.80 | 2.44 / 3.49 | 2.32 / 3.26 | 2.24 / 3.10 | 2.17 / 2.96 | 2.11 / 2.86 | 2.06 / 2.77 | 2.02 / 2.70 | 1.99 / 2.64 | 1.94 / 2.54 | 1.89 / 2.46 | 1.82 / 2.35 | 1.78 / 2.26 | 1.73 / 2.17 | 1.68 / 2.08 | 1.64 / 2.02 | 1.60 / 1.94 | 1.57 / 1.91 | 1.54 / 1.85 | 1.51 / 1.80 | 1.49 / 1.78 |
| 44 | 4.06 / 7.24 | 3.21 / 5.12 | 2.82 / 4.26 | 2.58 / 3.78 | 2.43 / 3.46 | 2.31 / 3.24 | 2.23 / 3.07 | 2.16 / 2.94 | 2.10 / 2.84 | 2.05 / 2.75 | 2.01 / 2.68 | 1.98 / 2.62 | 1.92 / 2.52 | 1.88 / 2.44 | 1.81 / 2.32 | 1.76 / 2.24 | 1.72 / 2.15 | 1.66 / 2.06 | 1.63 / 2.00 | 1.58 / 1.92 | 1.56 / 1.88 | 1.52 / 1.82 | 1.50 / 1.78 | 1.48 / 1.75 |
| 46 | 4.05 / 7.21 | 3.20 / 5.10 | 2.81 / 4.24 | 2.57 / 3.76 | 2.42 / 3.44 | 2.30 / 3.22 | 2.22 / 3.05 | 2.14 / 2.92 | 2.09 / 2.82 | 2.04 / 2.73 | 2.00 / 2.66 | 1.97 / 2.60 | 1.91 / 2.50 | 1.87 / 2.42 | 1.80 / 2.30 | 1.75 / 2.22 | 1.71 / 2.13 | 1.65 / 2.04 | 1.62 / 1.98 | 1.57 / 1.90 | 1.54 / 1.86 | 1.51 / 1.80 | 1.48 / 1.76 | 1.46 / 1.72 |
| 48 | 4.04 / 7.19 | 3.19 / 5.08 | 2.80 / 4.22 | 2.56 / 3.74 | 2.41 / 3.42 | 2.30 / 3.20 | 2.21 / 3.04 | 2.14 / 2.90 | 2.08 / 2.80 | 2.03 / 2.71 | 1.99 / 2.64 | 1.96 / 2.58 | 1.90 / 2.48 | 1.86 / 2.40 | 1.79 / 2.28 | 1.74 / 2.20 | 1.70 / 2.11 | 1.64 / 2.02 | 1.61 / 1.96 | 1.56 / 1.88 | 1.53 / 1.84 | 1.50 / 1.78 | 1.47 / 1.73 | 1.45 / 1.70 |

Degrees of freedom for lesser mean square [denominator]

TABLE E—(concluded)

Degrees of freedom for greater mean square [numerator]

Degrees of freedom for lesser mean square [denominator]

| df | 1 | 2 | 3 | 4 | 5 | 6 | 7 | 8 | 9 | 10 | 11 | 12 | 14 | 16 | 20 | 24 | 30 | 40 | 50 | 75 | 100 | 200 | 500 | ∞ |
|---|---|---|---|---|---|---|---|---|---|---|---|---|---|---|---|---|---|---|---|---|---|---|---|---|
| 50 | 4.03/7.17 | 3.18/5.06 | 2.79/4.20 | 2.56/3.72 | 2.40/3.41 | 2.29/3.18 | 2.20/3.02 | 2.13/2.88 | 2.07/2.78 | 2.02/2.70 | 1.98/2.62 | 1.95/2.56 | 1.90/2.46 | 1.85/2.39 | 1.78/2.26 | 1.74/2.18 | 1.69/2.10 | 1.63/2.00 | 1.60/1.94 | 1.55/1.86 | 1.52/1.82 | 1.48/1.76 | 1.46/1.71 | 1.44/1.68 |
| 55 | 4.02/7.12 | 3.17/5.01 | 2.78/4.16 | 2.54/3.68 | 2.38/3.37 | 2.27/3.15 | 2.18/2.98 | 2.11/2.85 | 2.05/2.75 | 2.00/2.66 | 1.97/2.59 | 1.93/2.53 | 1.88/2.43 | 1.83/2.35 | 1.76/2.23 | 1.72/2.15 | 1.67/2.06 | 1.61/1.96 | 1.58/1.90 | 1.52/1.82 | 1.50/1.78 | 1.46/1.71 | 1.43/1.66 | 1.41/1.64 |
| 60 | 4.00/7.08 | 3.15/4.98 | 2.76/4.13 | 2.52/3.65 | 2.37/3.34 | 2.25/3.12 | 2.17/2.95 | 2.10/2.82 | 2.04/2.72 | 1.99/2.63 | 1.95/2.56 | 1.92/2.50 | 1.86/2.40 | 1.81/2.32 | 1.75/2.20 | 1.70/2.12 | 1.65/2.03 | 1.59/1.93 | 1.56/1.87 | 1.50/1.79 | 1.48/1.74 | 1.44/1.68 | 1.41/1.63 | 1.39/1.60 |
| 65 | 3.99/7.04 | 3.14/4.95 | 2.75/4.10 | 2.51/3.62 | 2.36/3.31 | 2.24/3.09 | 2.15/2.93 | 2.08/2.79 | 2.02/2.70 | 1.98/2.61 | 1.94/2.54 | 1.90/2.47 | 1.85/2.37 | 1.80/2.30 | 1.73/2.18 | 1.68/2.09 | 1.63/2.00 | 1.57/1.90 | 1.54/1.84 | 1.49/1.76 | 1.46/1.71 | 1.42/1.64 | 1.39/1.60 | 1.37/1.56 |
| 70 | 3.98/7.01 | 3.13/4.92 | 2.74/4.08 | 2.50/3.60 | 2.35/3.29 | 2.23/3.07 | 2.14/2.91 | 2.07/2.77 | 2.01/2.67 | 1.97/2.59 | 1.93/2.51 | 1.89/2.45 | 1.84/2.35 | 1.79/2.28 | 1.72/2.15 | 1.67/2.07 | 1.62/1.98 | 1.56/1.88 | 1.53/1.82 | 1.47/1.74 | 1.45/1.69 | 1.40/1.62 | 1.37/1.56 | 1.35/1.53 |
| 80 | 3.96/6.96 | 3.11/4.88 | 2.72/4.04 | 2.48/3.56 | 2.33/3.25 | 2.21/3.04 | 2.12/2.87 | 2.05/2.74 | 1.99/2.64 | 1.95/2.55 | 1.91/2.48 | 1.88/2.41 | 1.82/2.32 | 1.77/2.24 | 1.70/2.11 | 1.65/2.03 | 1.60/1.94 | 1.54/1.84 | 1.51/1.78 | 1.45/1.70 | 1.42/1.65 | 1.38/1.57 | 1.35/1.52 | 1.32/1.49 |
| 100 | 3.94/6.90 | 3.09/4.82 | 2.70/3.98 | 2.46/3.51 | 2.30/3.20 | 2.19/2.99 | 2.10/2.82 | 2.03/2.69 | 1.97/2.59 | 1.92/2.51 | 1.88/2.43 | 1.85/2.36 | 1.79/2.26 | 1.75/2.19 | 1.68/2.06 | 1.63/1.98 | 1.57/1.89 | 1.51/1.79 | 1.48/1.73 | 1.42/1.64 | 1.39/1.59 | 1.34/1.51 | 1.30/1.46 | 1.28/1.43 |
| 125 | 3.92/6.84 | 3.07/4.78 | 2.68/3.94 | 2.44/3.47 | 2.29/3.17 | 2.17/2.95 | 2.08/2.79 | 2.01/2.65 | 1.95/2.56 | 1.90/2.47 | 1.86/2.40 | 1.83/2.33 | 1.77/2.23 | 1.72/2.15 | 1.65/2.03 | 1.60/1.94 | 1.55/1.85 | 1.49/1.75 | 1.45/1.68 | 1.39/1.59 | 1.36/1.54 | 1.31/1.46 | 1.27/1.40 | 1.25/1.37 |
| 150 | 3.91/6.81 | 3.06/4.75 | 2.67/3.91 | 2.43/3.44 | 2.27/3.14 | 2.16/2.92 | 2.07/2.76 | 2.00/2.62 | 1.94/2.53 | 1.89/2.44 | 1.85/2.37 | 1.82/2.30 | 1.76/2.20 | 1.71/2.12 | 1.64/2.00 | 1.59/1.91 | 1.54/1.83 | 1.47/1.72 | 1.44/1.66 | 1.37/1.56 | 1.34/1.51 | 1.29/1.43 | 1.25/1.37 | 1.22/1.33 |
| 200 | 3.89/6.76 | 3.04/4.71 | 2.65/3.88 | 2.41/3.41 | 2.26/3.11 | 2.14/2.90 | 2.05/2.73 | 1.98/2.60 | 1.92/2.50 | 1.87/2.41 | 1.83/2.34 | 1.80/2.28 | 1.74/2.17 | 1.69/2.09 | 1.62/1.97 | 1.57/1.88 | 1.52/1.79 | 1.45/1.69 | 1.42/1.62 | 1.35/1.53 | 1.32/1.48 | 1.26/1.39 | 1.22/1.33 | 1.19/1.28 |
| 400 | 3.86/6.70 | 3.02/4.66 | 2.62/3.83 | 2.39/3.36 | 2.23/3.06 | 2.12/2.85 | 2.03/2.69 | 1.96/2.55 | 1.90/2.46 | 1.85/2.37 | 1.81/2.29 | 1.78/2.23 | 1.72/2.12 | 1.67/2.04 | 1.60/1.92 | 1.54/1.84 | 1.49/1.74 | 1.42/1.64 | 1.38/1.57 | 1.32/1.47 | 1.28/1.42 | 1.22/1.32 | 1.16/1.24 | 1.13/1.19 |
| 1000 | 3.85/6.66 | 3.00/4.62 | 2.61/3.80 | 2.38/3.34 | 2.22/3.04 | 2.10/2.82 | 2.02/2.66 | 1.95/2.53 | 1.89/2.43 | 1.84/2.34 | 1.80/2.26 | 1.76/2.20 | 1.70/2.09 | 1.65/2.01 | 1.58/1.89 | 1.53/1.81 | 1.47/1.71 | 1.41/1.61 | 1.36/1.54 | 1.30/1.44 | 1.26/1.38 | 1.19/1.28 | 1.13/1.19 | 1.08/1.11 |
| ∞ | 3.84/6.64 | 2.99/4.60 | 2.60/3.78 | 2.37/3.32 | 2.21/3.02 | 2.09/2.80 | 2.01/2.64 | 1.94/2.51 | 1.88/2.41 | 1.83/2.32 | 1.79/2.24 | 1.75/2.18 | 1.69/2.07 | 1.64/1.99 | 1.57/1.87 | 1.52/1.79 | 1.46/1.69 | 1.40/1.59 | 1.35/1.52 | 1.28/1.41 | 1.24/1.36 | 1.17/1.25 | 1.11/1.15 | 1.00/1.00 |

# appendix B

# Answers to Odd-Numbered Exercises

*Exercise 2–1, Frequency Distribution, Data B*

Range $= 145 - 58 + 1 = 88$

With $i = 10$, begin lowest class interval at 50. This will give 10 class intervals, the minimum. With $i = 5$, begin lowest class interval at 55. This will give 19 class intervals. The frequency distribution for $i = 10$ is given below:

Frequency Distribution

| X | Tally | f |
|---|---|---|
| 140–149 | / | 1 |
| 130–139 | /// | 3 |
| 120–129 | //// | 4 |
| 110–119 | ////////// | 14 |
| 100–109 | ///////////////// | 21 |
| 90–99 | /////////////// | 19 |
| 80–89 | /////// | 11 |
| 70–79 | //// | 4 |
| 60–69 | // | 2 |
| 50–59 | / | 1 |
| N | | 80 |

276

Frequency Polygon and Histogram, Data B

Frequency

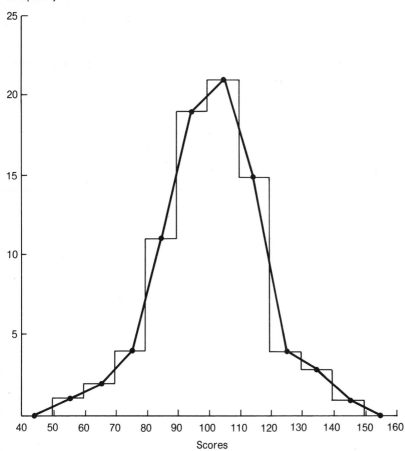

Scores

*Exercise 2–3, Data D*

Range $= 33 - 2 + 1 = \cancel{34}$ $32$

With $i = 3$, begin lowest class interval at 0. This will give 12 class intervals. With $i = 2$, begin lowest class interval at 2. This will give 17 class intervals. The frequency distribution for $i = 3$ is given below:

Frequency Distribution

| $X$ | Tally | $f$ |
|---|---|---|
| 33–35 | / | 1 |
| 30–32 | / | 1 |
| 27–29 | / | 1 |
| 24–26 | | 0 |
| 21–23 | | 0 |
| 18–20 | // | 2 |
| 15–17 | // | 2 |
| 12–14 | //// | 4 |
| 9–11 | ////// | 8 |
| 6–8 | ///// | 7 |
| 3–5 | /// | 3 |
| 0–2 | / | 1 |
| $N$ | | 30 |

Frequency Polygon and Histogram, Data D

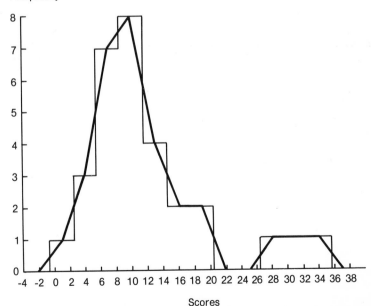

*Exercise 2–5, Data F*

Range $= 25 - 3 + 1 = 23$

With $i = 2$, begin lowest class interval at 2. No other class interval is acceptable. The frequency distribution for $i = 2$ is given below:

Frequency Distribution

| $X$ | Tally | $f$ |
|-----|-------|-----|
| 24–25 | / | 1 |
| 22–23 | // | 2 |
| 20–21 | / | 1 |
| 18–19 | / | 1 |
| 16–17 | // | 2 |
| 14–15 | // | 2 |
| 12–13 | //// | 4 |
| 10–11 | //// //// | 9 |
| 8–9 | //// //// | 9 |
| 6–7 | //// //// / | 11 |
| 4–5 | //// / | 6 |
| 2–3 | // | 2 |
| $N$ | | 50 |

Frequency Polygon and Histogram, Data F

Frequency

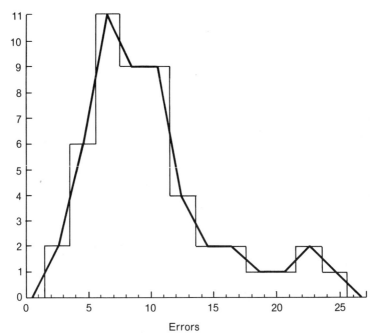

Errors

*Exercise 2–7, Data H*

Range $= 3.7 - 1.9 + .1 = 1.9$

Either $i = 0.1$ or $i = 0.2$ would be acceptable as a class interval size. The frequency distribution for $i = 0.2$ is given below:

Frequency Distribution

| X | Tally | f |
|---|---|---|
| 3.6–3.7 | / | 1 |
| 3.4–3.5 | / | 1 |
| 3.2–3.3 | // | 2 |
| 3.0–3.1 | //// | 4 |
| 2.8–2.9 | //// / | 6 |
| 2.6–2.7 | //// | 4 |
| 2.4–2.5 | //// //// | 9 |
| 2.2–2.3 | //// | 4 |
| 2.0–2.1 | /// | 3 |
| 1.8–1.9 | / | 1 |
| N | | 35 |

Frequency Polygon and Histogram, Data H

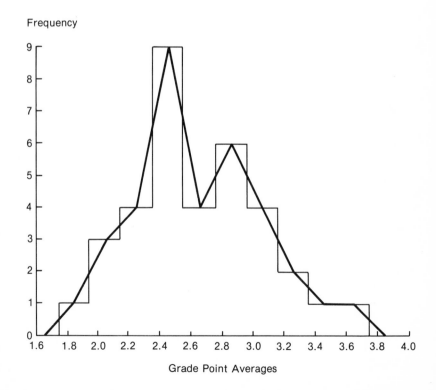

*Chapter 3*

3–1.   74.5

3–3.   $\text{Mdn} = 59.5 + \left[\dfrac{(80.5 - 49)}{33}\right] \times 10 = 59.5 + \dfrac{315}{33}$

$= 59.5 + 9.545 = 69.045 = 69.0$

3–5.   $\text{Mdn} = 24.5 + \left[\dfrac{(5 - 3)}{3}\right] \times 1 = 24.5 + \dfrac{2}{3}$

$= 24.5 + .67 = 25.17 = 25.2$

3–7.   $\text{Mdn} = 34.5 + \left[\dfrac{(6 - 4)}{4}\right] \times 1 = 34.5 + \dfrac{2}{4}$

$= 34.5 + .5 = 35.0$

3–9.   From the bottom up: $\text{Mdn} = 99.5 + \left[\dfrac{(40 - 37)}{21}\right] \times 10$

$= 99.5 + \dfrac{30}{21} = 99.5 + 1.43 = 100.93 = 100.9$

From the top down: $\text{Mdn} = 109.5 - \left[\dfrac{(40 - 22)}{21}\right] \times 10$

$= 109.5 - \dfrac{180}{21} = 109.5 - 8.57 = 100.93 = 100.9$

3–11.   $M = 7.4$

3–13.   $M = M_G + \left[\dfrac{\Sigma fx'}{N}\right] \cdot i$

$= 37.0 + \left[\dfrac{(-122)}{160}\right] \times 5 = 37.0 - \dfrac{610}{160}$

$= 37.0 - 3.81 = 33.19 = 33.2$

3–15.   $M = 24.5 + \left[\dfrac{718}{161}\right] \times 10 = 24.5 + \dfrac{7180}{161}$

$= 24.5 + 44.596 = 69.096 = 69.1$

3–17.   $M = 104.5 + \left[\dfrac{(-31)}{80}\right] \times 10 = 104.5 - \dfrac{310}{80}$

$= 104.5 - 3.875 = 100.625 = 100.6$

*Chapter 4*

4–1.   $Q_1 = P_{25} = 15.00$; any number between 10.5 and 19.5 satis-
fies the definition, so 15.00, the middle point in this range is
taken as $Q_1$; $Q_3 = P_{75} = 30.5 + \left(\dfrac{2}{3}\right) \cdot 1 = 30.5 + 0.67 =$
$31.17$; $Q = \dfrac{(31.17 - 15.00)}{2} = \dfrac{16.17}{2} = 8.08 = 8.1$.

4-3.   $Q_1 = 3.5 + \left[\dfrac{2.5}{4}\right] \cdot 1 = 3.5 + 0.625 = 4.125;$

$Q_3 = 11.5 + \left[\dfrac{1.5}{3}\right] \cdot 1 = 11.5 + 0.5 = 12.00;$

$Q = \dfrac{(12.000 - 4.125)}{2} = \dfrac{7.875}{2} = 3.9375 = 3.9.$

4-5.   $Q_1 = 89.5 + \left[\dfrac{(20 - 18)}{19}\right] \cdot 10 = 89.5 + \dfrac{20}{19}$

$= 89.5 + 1.05 = 90.55;$

$Q_3 = 119.5 - \left[\dfrac{(20 - 8)}{14}\right] \cdot 10 = 119.5 - \dfrac{120}{14}$

$= 119.5 - 8.57 = 110.93;$

$Q = \dfrac{(110.93 - 90.55)}{2} = \dfrac{20.38}{2} = 10.19 = 10.2.$

The distribution is very, very slightly negatively skewed, because $Q_3 - Q_2 = 10.0$ is less than $Q_2 - Q_1 = 10.4$.

4-7.   $Q_1 = 83.5 + \left[\dfrac{(52.25 - 38)}{16}\right] \cdot 1 = 83.5 + \dfrac{14.25}{16}$

$= 83.5 + 0.89 = 84.39;$

$Q_3 = 89.5 + \left[\dfrac{(156.75 - 156)}{15}\right] \cdot 1 = 89.5 + 0.05 = 89.55;$

$Q = \dfrac{(89.55 - 84.39)}{2} = \dfrac{5.16}{2} = 2.58 = 2.6.$

4-9.   Deviation score method: S.D. $= \sqrt{\dfrac{106}{10}} = 3.25576 = 3.3.$

Raw score method: S.D. $= \dfrac{1}{10}\sqrt{10(746) - (80)^2}$

$= \left(\dfrac{1}{10}\right)\sqrt{1060} = 3.25576 = 3.3.$

$M = 8.0$; when 2 points are added to each score, the new mean is 10.0 and the new S.D. is $3.25576 = 3.3$; when every score is multiplied by 3, the new mean is $3(8) = 24$; the new S.D. is $3(3.25576) = 9.76728 = 9.8$. Adding a constant to the scores adds that same constant to the mean while leaving the S.D. unchanged; multiplying each score by a constant multiplies both $M$ and S.D. by that same constant.

4–11.

```
          3  9 · 5  2  7  5 +
        |15 62 · 43 00 00 00
          9
(69)      6 62
          6 21
(785)       41    43
            39    25
(7902)       2  18 .00
             1  58 04
(79047)        59 96 00
               55 33 29
(790545)        4 62 71 00
                3 95 27 25
                  67 43 75
```

Answer: 39.528

4–13.  S.D. $= 10 \sqrt{\dfrac{623}{161} - \left(\dfrac{-87}{161}\right)^2} = 10 \sqrt{3.8696 - 0.2920}$

$\qquad = 10 \sqrt{3.5776} = 10(1.89145) = 18.9$

4–15.

| $X$ | $f$ | $x'$ | $fx'$ | $fx'^2$ |
|---|---|---|---|---|
| 55–59.... | 3 | 4 | 12 | 48 |
| 50–54.... | 8 | 3 | 24 | 72 |
| 45–49.... | 14 | 2 | 28 | 56 |
| 40–44.... | 19 | 1 | 19 | 19 |
| 35–39.... | 25 | 0 | 0 | 0 |
| 30–34.... | 17 | −1 | −17 | 17 |
| 25–29.... | 15 | −2 | −30 | 60 |
| 20–24.... | 10 | −3 | −30 | 90 |
| 15–19.... | 6 | −4 | −24 | 96 |
| 10–14.... | 2 | −5 | −10 | 50 |
| Σ.... | 119 | | −28 | 508 |

$M = 37.0 + \left(\dfrac{-28}{119}\right) \cdot 5$

$\quad = 37.0 - \dfrac{140}{119}$

$\quad = 37.00 - 1.18 = 35.82$

$M = 35.8$

S.D. $= 5 \sqrt{\dfrac{508}{119} - \left(\dfrac{-28}{119}\right)^2}$

$\quad = 5 \sqrt{4.2689 - .0553}$

$\quad = 5 \sqrt{4.2136}$

$\quad = 10.265$

S.D. $= 10.3$

*Chapter 5*

5–1.  $100 \dfrac{\left(140 + \dfrac{16}{2}\right)}{209} = \dfrac{14800}{209} = 70.8 = 71$

5–3.  $100 \dfrac{\left[9 + \left(\dfrac{42.0 - 39.5}{10}\right)16\right]}{161} = \dfrac{100(9 + 4)}{161} = \dfrac{1300}{161}$

$\qquad\qquad\qquad\qquad\qquad\qquad = 8.1 = 8$

5-5.    $(161)(0.68) = 109.48$ scores below $P_{68}$

$$P_{68} = 69.5 + \left(\frac{109.48 - 82}{36}\right) \cdot 10$$

$$= 69.5 + \frac{264.8}{36} \quad \overset{274.8}{} = 69.5 + 7.33 \quad 7.63$$

$$= 76.83 = 76.8 \quad 77.13 = 77.1$$

5-7.    $Z = \dfrac{(21 - 26.1)}{5.7} = \dfrac{-5.1}{5.7} = -0.89$

5-9.    $Z = \dfrac{52 - 69.1}{18.9} = \dfrac{-17.1}{18.9} = -0.90$

5-11.    $Z = \dfrac{73 - 65}{20} = 0.40;$

Area $= 0.6554;$
$f = 131.1$

5-13.    $Z = \dfrac{98 - 80}{15} = 1.2;$

Area $= 0.1151;$
$f = 17.3$

5-15.    $T = 10(-1.2) + 50 = -12 + 50 = 38$

5-17.    $S = 100(-3.12) + 500 = -312 + 500 = 188$

5-19.    Expected frequencies in the following calculations have been adjusted slightly for rounding error to make the sum of $f_t$ values equal exactly to $N = 161$.

| $X$ | $f_0$ | $X_{11}$ | $Z$ | Area $-\infty$ to $Z$ | Interval Area | Computed $f_t$ | Adjusted $f_t$ |
|---|---|---|---|---|---|---|---|
| 110–119... | 3 | 109.5 | 2.14 | 0.9838 | 0.0162 | 2.61 | 2.6 |
| 100–109... | 8 | 99.5 | 1.61 | 0.9463 | 0.0375 | 6.04 | 6.0 |
| 90–99.... | 12 | 89.5 | 1.08 | 0.8599 | 0.0864 | 13.91 | 13.9 |
| 80–89.... | 20 | 79.5 | 0.55 | 0.7088 | 0.1511 | 24.32 | 24.3 |
| 70–79.... | 36 | 69.5 | 0.02 | 0.5080 | 0.2008 | 32.33 | 32.3 |
| 60–69.... | 33 | 59.5 | −0.51 | 0.3050 | 0.2030 | 32.68 | 32.7 |
| 50–59.... | 24 | 49.5 | −1.04 | 0.1492 | 0.1558 | 25.08 | 25.1 |
| 40–49.... | 16 | 39.5 | −1.57 | 0.0582 | 0.0910 | 14.65 | 14.7 |
| 30–39.... | 7 | 29.5 | −2.10 | 0.0179 | 0.0403 | 6.49 | 6.5 |
| 20–29.... | 2 | | | | 0.0179 | 2.88 | 2.9 |
| $N$...... | 161 | | | | | 160.99 | 161.0 |

5–21.  The proportion of scores below 52.0 is

$$\frac{\left[25 + \left(\dfrac{52.0 - 49.5}{10}\right) \cdot 24\right]}{161} = \frac{[25 + (2.5)(2.4)]}{161}$$

$$= \frac{31}{161} = 0.193;$$

the $Z$ score with area 0.193 below it (see normal curve table) is $-0.87$.

5–23.  Proportion of scores below

$$86.0 = \frac{\left[118 + \left(\dfrac{86.0 - 79.5}{10}\right) \cdot 20\right]}{161} = \frac{(118 + 13)}{161}$$

$$= \frac{131}{161} = 0.813;$$

the $Z$ score with area 0.813 below it is $+0.89$.

*Chapter 6*

6–1.

| X | Y | $X^2$ | $Y^2$ | $XY$ |
|---|---|---|---|---|
| 6 | 7 | 36 | 49 | 42 |
| 5 | 4 | 25 | 16 | 20 |
| 4 | 2 | 16 | 4 | 8 |
| 5 | 5 | 25 | 25 | 25 |
| 6 | 5 | 36 | 25 | 30 |
| 2 | 4 | 4 | 16 | 8 |
| 9 | 8 | 81 | 64 | 72 |
| 7 | 9 | 49 | 81 | 63 |
| 1 | 4 | 1 | 16 | 4 |
| 8 | 7 | 64 | 49 | 56 |
| 6 | 5 | 36 | 25 | 30 |
| 5 | 6 | 25 | 36 | 30 |
| Σ... 64 | 66 | 398 | 406 | 388 |

$$r = \frac{12(388) - (64)(66)}{\sqrt{12(398) - (64)^2}\ \sqrt{12(406) - (66)^2}}$$

$$= \frac{4656 - 4224}{\sqrt{4776 - 4096}\ \sqrt{4872 - 4356}}$$

$$r = \frac{432}{\sqrt{680}\ \sqrt{516}} = \frac{432}{\sqrt{350880}}$$

$$r = \frac{432}{592.35} = 0.729 = 0.73$$

6–3.  $r = \dfrac{N\Sigma XY - \Sigma X\Sigma Y}{\sqrt{N\Sigma X^2 - (\Sigma X)^2}\ \sqrt{N\Sigma Y^2 - (\Sigma Y)^2}}$

$r = \dfrac{(20)(119458) - (1638)(1415)}{\sqrt{20(137942) - (1638)^2}\ \sqrt{20(108317) - (1415)^2}}$

$r = \dfrac{2389160 - 2317770}{\sqrt{2758840 - 2683044}\ \sqrt{2166340 - 2002225}}$

$r = \dfrac{71390}{\sqrt{75796}\ \sqrt{164115}}$

$r = \dfrac{71390}{(275.3)(405.1)} = \dfrac{71390}{111524.0}$

$r = 0.64$

6–5.  $r = \dfrac{60(365803) - (4915)(4378)}{\sqrt{60(415411) - (4915)^2}\ \sqrt{60(338730) - (4378)^2}}$

$r = \dfrac{21948180 - 21517870}{\sqrt{24924660 - 24157225}\ \sqrt{20323800 - 19166884}}$

$r = \dfrac{430310}{\sqrt{767435}\ \sqrt{1156916}}$

$r = \dfrac{430310}{(876.0)(1075.6)} = \dfrac{430310}{942226}$

$r = 0.46$

6–7.  $\Sigma X = 69$;  $\Sigma Y = 63$;  $\Sigma X^2 = 605$;  $\Sigma Y^2 = 585$;  $\Sigma XY = 561$;
$r = 0.81$

6–9.  $y' = 0.33x$;  $-6.7$

6–11.  $Z'_y = 0.5Z_x$;  $Z'_y$ for $X = 60$ is $-0.67$;
$Y' = 0.33X + 23.3$;  $Y'$ for $X = 60$ is $43.3$

6–13.  (See accompanying scatter plot)

$r = \dfrac{\dfrac{\Sigma x'y'}{N} - \dfrac{\Sigma fx'}{N}\dfrac{\Sigma fy'}{N}}{\sqrt{\dfrac{\Sigma fx'^2}{N} - \left(\dfrac{\Sigma fx'}{N}\right)^2}\ \sqrt{\dfrac{\Sigma fy'^2}{N} - \left(\dfrac{\Sigma fy'}{N}\right)^2}}$

$r = \dfrac{\dfrac{275}{60} - \left(\dfrac{-1}{60}\right)\left(\dfrac{-108}{60}\right)}{\sqrt{\dfrac{501}{60} - \left(\dfrac{-1}{60}\right)^2}\ \sqrt{\dfrac{984}{60} - \left(\dfrac{-108}{60}\right)^2}}$

$r = \dfrac{4.583 - (-.017)(-1.800)}{\sqrt{8.350 - .000}\ \sqrt{16.400 - 3.240}}$

## Scatter Plot of O vs. C

| O | 50-54 | 55-59 | 60-64 | 65-69 | 70-74 | 75-79 | 80-84 | 85-89 | 90-94 | 95-99 | 100-104 | 105-109 | 110-114 | $f_y$ | $y'$ | $f_y'$ | $f_y'^2$ | $x'y'$ |
|---|---|---|---|---|---|---|---|---|---|---|---|---|---|---|---|---|---|---|
| 105-109 | | | | | | −5 / 1 | | | | | | | | 1 | 5 | 5 | 25 | −5 |
| 100-104 | | | | | | | | | | 12 / 1 | | | | 3 | 4 | 12 | 48 | 0 |
| 95-99 | −18 / 1 | | | | | | | 3 / 1 | 12 / 1 | 12 / 1 | | | | 3 | 3 | 9 | 27 | −3 |
| 90-94 | | | | | | −4 / 1 | | 2 / 1 | 4 / 1 | 6 / 1 | 8 / 1 | | | 6 | 2 | 12 | 24 | 20 |
| 85-89 | | | | | | 1 | 1 | 1 / 1 | 6 / ||| | | | | | 6 | 1 | 6 | 6 | 12 |
| 80-84 | | | | | | 1 | || | 1 | | 1 | ||| | | 1 | 8 | 0 | 0 | 0 | 0 |
| 75-79 | | | | | | 1 / 1 | | | −4 / ||| −12 | | −8 / 1 | | | 2 | −1 | −2 | 2 | 7 |
| 70-74 | | | | | 1 | 2 | | | −6 / 1 | | | | | 5 | −2 | −10 | 20 | −18 |
| 65-69 | | | | 2 | 6 / 1 | 3 | | | −8 / 1 | | | | | 5 | −3 | −15 | 45 | 3 |
| 60-64 | | | | | 6 / 1 | 4 / 1 | | | −10 / 1 | | | | | 3 | −4 | −12 | 48 | −4 |
| 55-59 | | 25 / 1 | | 15 / || 30 | 10 / || 20 | | | | | | | | | 6 | −5 | −30 | 150 | 65 |
| 50-54 | | 30 / 1 | | 18 / 1 | 12 / ||| 36 | 6 / 1 | | | | | | | | 6 | −6 | −36 | 216 | 90 |
| 45-49 | 42 / 1 | | | | 14 / 1 | | 1 | | | | | | | 3 | −7 | −21 | 147 | 56 |
| 40-44 | | | | | 16 / 1 | | | | | | | | | 1 | −8 | −8 | 64 | 16 |
| 35-39 | 54 / 1 | | | | | | | | −18 / 1 | | | | | 2 | −9 | −18 | 162 | 36 |
| $f_x$ | 4 | 2 | 0 | 3 | 10 | 8 | 7 | 4 | 11 | 3 | 6 | 1 | 1 | 60 | | −108 | 984 | 275 |
| $x'$ | −6 | −5 | −4 | −3 | −2 | −1 | 0 | 1 | 2 | 3 | 4 | 5 | 6 | | | | | |
| $f_x'$ | −24 | −10 | 0 | −9 | −20 | −8 | 0 | 4 | 22 | 9 | 24 | 5 | 6 | −1 | | | | |
| $f_x'^2$ | 144 | 50 | 0 | 27 | 40 | 8 | 0 | 4 | 44 | 27 | 96 | 25 | 36 | 501 | | | | |
| $x'y'$ | 84 | 55 | 0 | 48 | 84 | 7 | 0 | 6 | −44 | 18 | 12 | 5 | 0 | 275 | | | | |

$$r = \frac{4.583 - .031}{\sqrt{8.350} \, \sqrt{13.160}}$$

$$r = \frac{4.552}{(2.89)(3.63)}$$

$$r = \frac{4.552}{10.49}$$

$$r = 0.434 = 0.43$$

*Chapter 7*

7-1.  Make a one-tailed test at the five percent level; $h_0$ states that the probability is $\frac{1}{2}$ that the circle with a dot will be judged larger on any given trial; the probability of getting a 13–4 split under $h_0$ is 0.0245 which is significant at the 0.05 level of confidence, causing $h_0$ to be rejected. It is concluded that the results are consistent with Professor X's theory.

7-3.  For the binomial $p = 0.0182$; for normal curve approximation, $p = 0.0187$.

7-5.  $Z = 2.184$; interpolated ordinate $= 0.0368$; base $= 1.0/2.06$; probability $= 0.0179$.

7-7.  The first three terms of the binomial expansion for $(\frac{9}{10} + \frac{1}{10})^{10}$ give the probability of 8 or more bags passing the test. These terms are 0.3487, 0.3874, and 0.1937, summing to 0.9298. This gives 0.0702 as the probability of 3 or more failures under $h_0$: $p = 0.9$ for the bag being good. Since 0.07 is greater than 0.05, $h_0$ cannot be rejected at the 0.05 level. The process cannot be termed "out of control," but bears watching.

7-9.  Apply the normal curve approximation, two-tailed test, to the 25 cases that changed opinion. This gives $n = 25$ with a 16–9 split; $Z = 1.20$, which is not significant. There is no evidence of significant change in opinion due to the interpolated film.

*Chapter 8*

8-1.  $M = -4.33$; $\Sigma x^2 = 290.67$; $s = 8.56$; $s_M = 2.47$  $t = -1.75$; Not significant since $t_{.05}$ with 11 $df$ is 1.80 even for a one-tailed test. Cannot conclude that friend is correct, although he may be since the trend is in the predicted direction and approaching significance at the 0.05 level with a one-tailed test.

8–3. The 95 percent C. I. is $M \pm t_{.05} \, s_M = 117 \pm (2.01)(1.14) = 117 \pm 2.3 = 114.7$ to $119.3$.

8–5. $t = 3.11$ and with 98 degrees of freedom, $t_{.01} = 2.63$; hence $r$ is significant at the 0.01 level of confidence.

8–7. $F = s_x^2/s_y^2 = 228.9/228.7 = 1.00$; $F$ is not significant; proceed with $t$ test.

8–9. $h_0$: $\mu_1 = \mu_2$;

$$s_{D_M} = \sqrt{\frac{\Sigma x_1^2 + \Sigma x_2^2}{N_1 + N_2 - 2}\left(\frac{N_1 + N_2}{N_1 N_2}\right)}$$

$$= \sqrt{\frac{4574.2 + 4119.6}{20 + 18 - 2}\left(\frac{20 + 18}{20 \cdot 18}\right)}$$

$$= 5.05$$

$$t = \frac{(15.28 - 11.30)}{5.05} = 0.79 \text{ N.S.}$$

8–11. $M_d = 3.0$, S.D. $= 1.58$; $s_{d_M} = 0.60$, $t = 5.0**$; $t_{.05}$ with 7 $df = 3.50$; significant at 0.01 level, two-tailed test; playing tape during sleep with poem on it seems to have reduced the number of trials needed to learn poem later in waking state.

8–13. $\Sigma D_E = 82$, $\Sigma D_E^2 = 850$, $\Sigma x_E^2 = 289.67$; $\Sigma D_C = 18$, $\Sigma D_C^2 = 262$, $\Sigma x_C^2 = 232.55$; $M_{D_E} = 6.83$, $M_{D_C} = 1.64$, $s_{D_M} = 2.33$ $t = 2.23*$; $t_{.05}$ with 21 $df = 2.08$.

8–15. $h_0 = M_d = 0$; $M_d = 7.36$, $s_d = 6.53$, $s_{d_M} = 1.97$, $t = 3.74**$, significant at 0.01 level, since $t_{.01}$ for 10 $df = 3.17$, two-tailed test.

*Chapter 9*

9–1. One-Way Anova Summary:

| Source of Variation | Sum of Squares | Degrees of Freedom | Mean Square (variance estimate) | F |
|---|---|---|---|---|
| Between Groups........ | 2452.5 | 2 | 1226.25 | 110.9** |
| Within Groups........ | 298.5 | 27 | 11.06 | |
| Total............. | 2751.0 | 29 | | |

$F_{.05} = 3.35$; $F_{.01} = 5.49$

9–3. $s_1^2 = 16.5$, $s_2^2 = 9.7$, $s_3^2 = 6.9$; largest $F$ is 2.4; $F_{.05} = 3.18$; assumption is tenable.

9–5. $t_{12} = 8.3$, $t_{13} = 14.8$, $t_{23} = 6.6$; Scheffé $t'_{.05} = 2.59$; $t'_{.01} = 3.31$; all $t$ values significant at the 0.01 level.

9–7.   One-Way Anova Summary:

| Source of Variation | Sum of Squares | Degrees of Freedom | Mean Square (variance estimate) | F |
|---|---|---|---|---|
| Between Groups..... | 155.2 | 2 | 77.58 | 21.1** |
| Within Groups...... | 110.1 | 30 | 3.67 | |
| Total........... | 265.3 | 32 | | |

$F_{.05} = 3.32$; $F_{.01} = 5.39$

9–9.   $t_{12} = 1.52$, $t_{13} = 6.31$, $t_{23} = 4.70$; Scheffé $t'_{.05} = 2.58$, $t'_{.01} = 3.28$; $t_{13}$ and $t_{14}$ are significant at the 0.01 level.

9–11.   Two-Way Anova Summary:

| Source of Variation | Sum of Squares | Degrees of Freedom | Mean Square (variance estimate) | F |
|---|---|---|---|---|
| Between Rows......... | 39.9 | 1 | 39.90 | 3.81 |
| Between Columns....... | 385.8 | 2 | 192.90 | 18.41** |
| Interaction............ | 161.4 | 2 | 80.70 | 7.70** |
| Within Groups......... | 565.9 | 54 | 10.48 | |
| Total.............. | 1153.0 | 59 | | |

Rows, $F_{.05} = 4.02$, $F_{.01} = 7.12$; Columns, $F_{.05} = 3.17$, $F_{.01} = 5.01$; Interaction, $F_{.05} = 3.17$, $F_{.01} = 5.01$

9–13.   The column effect was significant at the one-percent level so $t$ tests of column means are in order. $t_{12} = 5.39$, $t_{13} = 0.25$, $t_{23} = 5.15$; Scheffé $t'_{.05} = 2.52$, $t'_{.01} = 3.17$; $t_{12}$ and $t_{23}$ significant at 0.01 level showing diet and diet plus exercise are superior to exercise alone. The difference between the sexes was not significant. There was a significant interaction effect. Inspection of the subgroup means suggests that women do more poorly with exercise alone than men do.

*Chapter 10*

10–1.

| X | $f_0$ | $f_t$ | $(f_0 - f_t)$ | $(f_0 - f_t)^2$ | $(f_0 - f_t)^2/f_t$ |
|---|---|---|---|---|---|
| 90–119...... | 23 | 22.0 | 1.0 | 1.00 | 0.05 |
| 80–89....... | 20 | 24.5 | −4.5 | 20.25 | 0.83 |
| 70–79....... | 36 | 32.6 | 3.4 | 11.56 | 0.35 |
| 60–69....... | 33 | 33.0 | 0.0 | 0.00 | 0.00 |
| 50–59....... | 24 | 25.3 | −1.3 | 1.69 | 0.07 |
| 20–49....... | 25 | 23.6 | 1.4 | 1.96 | 0.08 |
| Total..... | 161 | 161.0 | 0.0 | | $\chi^2 = 1.38$ N.S. |

$df = 6 - 3 = 3$; $\chi^2_{.05} = 7.82$; $\chi^2_{.01} = 11.34$

10–3.

| | Obtained Frequencies | | | Theoretical Frequencies | | |
|---|---|---|---|---|---|---|
| | Org. | Non-Org. | Total | Org. | Non-Org. | Total |
| Go......... | 15 | 11 | 26 | 8.9 | 17.1 | 26.0 |
| No Go..... | 11 | 39 | 50 | 17.1 | 32.9 | 50.0 |
| Total..... | 26 | 50 | 76 | 26.0 | 50.0 | 76.0 |

$$\chi^2 = \frac{(6.1 - 0.5)^2}{8.9} + \frac{(6.1 - 0.5)^2}{17.1} + \frac{(6.1 - 0.5)^2}{17.1} + \frac{(6.1 - 0.5)^2}{32.9} = 8.14**$$

Yates' Correction must be applied; One $f_t$ below 10.0; The charge against non orgs appears to be supported.

10–5.   $\chi^2 = 8.76*$, with $\chi^2_{.05} = 5.99$ and $\chi^2_{.01} = 9.21$ for 2 *df*. The item should be eliminated to meet the test developer's requirements. One $f_t$ is below 10.0 and the overall $N$ is not large; also, $\chi^2$ is significant only at the 0.05 level, so the outcome is not overwhelmingly conclusive.

10–7.   $\chi^2 = 19.1**$, with $\chi^2_{.01} = 15.086$ for 5 *df*; reject $h_0$. The middle lanes appear to be better than the outer lanes.

# Index

*This book has been set in 11 and 10 point Baskerville, leaded 1 point. Chapter numbers are 14 and 30 point Scotch Roman; chapter titles are 18 point Scotch Roman. The size of the type page is 27 × 45½ picas.*